The Behavioural Biology of Chickens

FSC
www.fsc.org
MIX
Paper from
responsible sources
FSC® C013604

The Behavioural Biology of Chickens

Christine J. Nicol

School of Veterinary Science, University of Bristol, UK

www.cabi.org

CABI is a trading name of CAB International

CABI	CABI
Nosworthy Way	38 Chauncy Street
Wallingford	Suite 1002
Oxfordshire OX10 8DE	Boston, MA 02111
UK	USA

Tel: +44 (0)1491 832111	Tel: +1 800 552 3083 (toll free)
Fax: +44 (0)1491 833508	Tel: +1 (0)617 395 4051
E-mail: info@cabi.org	E-mail: cabi-nao@cabi.org
Website: www.cabi.org	

A catalogue record for this book is available from the British Library, London, UK.

Library of Congress Cataloging-in-Publication Data

Nicol, Christine, author.
  The behavioural biology of chickens / Christine Nicol.
      p. ; cm.
  Includes bibliographical references and index.
    ISBN 978-1-78064-249-9 (hardback : alk. paper) -- ISBN 978-1-78064-250-5 (pbk. : alk. paper) 1. Chickens--Behavior. 2. Chickens--Physiology. I. C.A.B. International, issuing body. II. Title.
    [DNLM: 1. Chickens--physiology. 2. Behavior, Animal--physiology. 3. Ethology--methods. SF 768.2.P6]

    SF493.5.N53 2015
    636.5--dc23
                          2015000780

ISBN-13: 978 1 78064 249 9 (hbk)
          978 1 78064 250 5 (pbk)

Commissioning editor: Caroline Makepeace
Assistant editor: Alexandra Lainsbury
Production editor: Lauren Povey

Typeset by SPi, Pondicherry, India.
Printed and bound in the UK by CPI Group (UK) Ltd, Croydon, CR0 4YY.

Contents

Acknowledgements

I would like to thank from the bottom of my heart:

Paula Baker, Gina Caplen, Jo Edgar, Tim Guilford, Mike Mendl, Isabelle Pettersson, Jon Walton, Claire Weeks, Lorna Wilson and Rory Wilson for reading through chapter drafts.

Siobhan Abeyesinghe, Steve Brown, William Browne, Gina Caplen, Jonathan Cooper, Anna Davies, Marian Dawkins, Greg Dixon, Jo Edgar, Mary Foster, Raf Freire, Anne-Marie Gilani, Laura Green, Tim Guilford, Sue Haslam, Becky Hothersall, Frances Kim-Madslien, Toby Knowles, Cecilia Lindberg, Justin McKinstry, Ralph Merrill, Jo Murrell, Mohammed Nasr, Liz Paul, Stuart Pope, Gemma Richards, Bas Rodenburg, Chris Sherwin, Poppy Statham, Mike Toscano, Christopher Wathes, Claire Weeks, Lindsay Wilkins and Patrick Zimmerman for making chicken behaviour research *such fun* and for being great colleagues and friends.

Andrew Joret and Richard Kempsey for their longstanding willingness to put theory into practice.

The Tubney Trust, the RSPCA, Defra and the BBSRC for funding work referred to in this book.

Ingrid de Jong, Per Jensen and Lucia Regolin for sending me their valuable photos.

Alexandra Lainsbury and Caroline Makepeace at CABI for their support and help throughout a longer-than-anticipated writing period.

Peter Kloesges who built me a place to write.

Jane Gotto who helped me regain the perspective I needed.

Lady Hestercombe and Miss Petherton, stand-out individuals.

My family – Doug, Ursula, Lorna, Rory and Ruby - for their love and encouragement, which I will always return.

1

Genetics and Domestication

Chicken Numbers

It is hard to start a book on chickens without first commenting on just how many of them there are in the world. The figures lend themselves to the style of writing used in astronomy. Just as we might struggle to appreciate that the nearest sun to our own, Proxima Centauri, is a mere 24,807,799,074,834 miles (or 4.22 light years) away, it is difficult to visualize the scale of the recent explosion in chicken numbers. In 1961 when the United Nations Food and Agriculture Organization first compiled the data, the number of chickens in the world at any one time was less than 4 billion (http://faostat.fao.org/site/573/default.aspx#ancor), a figure that had increased to nearly 22 billion by 2013. The fivefold increase in chicken numbers is overshadowed by a 12-fold increase in meat yield over the same period, rising from 7.7 million tonnes in 1961 to more than 92 million tonnes in 2012. It is predicted to exceed 100 million tonnes within a few more years. The consequences for the individual chickens have been profound. Each one reaches a much heavier weight in a far shorter time than would have been conceivable to previous generations, and the average total lifespan of a commercial meat chicken is now just 6 weeks. Such a short lifespan means that most farmers can raise and replace several flocks in a given year, so the total number of chickens killed for meat in 2012 was nearly 60 billion (59,794,239,000). A further 4 or 5 billion egg-laying hens produced over 1 trillion (1,197,137,680,000) eggs.

Chicken Origins

The ancestral origins of these billions of chickens fascinated Darwin. He wrote:

> ... when we compare the gamecock, so pertinacious in battle, with other breeds so little quarrelsome, with "everlasting layers" which never desire to sit, and with the bantam so small and elegant...we cannot suppose that all the breeds were suddenly produced as perfect and as useful as we now see them; indeed, in several cases, we know that this has not been their history. The key is man's power of accumulative selection: nature gives successive variations; man adds them up in certain directions useful to him. In this sense he may be said to make for himself useful breeds.
> (Darwin, 1858)

Later, he concluded that domestic chickens originated solely from the red junglefowl (*Gallus gallus gallus*) based on their shared physical characteristics (Darwin, 1868).

An early review of archaeological evidence (West and Zhou, 1988) provided tantalizing glimpses of the process of breed creation envisaged by Darwin, as well as broad support for his view that chickens were first domesticated in South-east Asia. Evidence from remains found in Neolithic sites suggests that chickens originated as a domestic species at least 8000 years ago. Certainly, the archaeological data strongly support the idea that domestic chickens were taken from South-east Asia to become well established in China by 6000 BC. From China, they seem to have accompanied human migrations to Korea and Japan, and separately to Europe via Russia during the Neolithic and Bronze Ages.

Because various species of junglefowl still live in the wild throughout South-east Asia, it has become possible to examine the genetic similarities between these species and compare these results with the archaeological evidence for the origin of domestic chickens. One widely used method of establishing relationships between species (their phylogenies) is to compare microsatellite regions in DNA sequences. Microsatellites have core motifs of two to six base pairs that can be repeated up to 100 times, they have a high mutation rate and they are relatively evenly distributed in the genome, all of these features making them a useful marker of genetic relationships. A second common way of estimating phylogeny is to compare regions of non-coding mitochondrial DNA (mtDNA). mtDNA is an appropriate marker because it is transmitted without recombination via the maternal line and has a rapid rate of (supposedly) neutral mutations. In both cases, the essential idea is that larger differences in DNA sequence between samples taken from different populations or species signify more distant relationships. Unfortunately for our purposes, in reality the rate of microsatellite or mtDNA mutation may not be as regular as assumed. Mutation rates can vary between species due to factors such as metabolic rate or individual lifespan. The rate of mtDNA mutation for example is not, as once conceived, a steady ticking clock with a 2% substitution rate per million years (Nabholz et al., 2009). Thus, we must be careful to regard the degree of divergence of microsatellite regions or mtDNA between species not as absolute proof of phylogeny but as a useful indicator.

An early study examining the mtDNA of a handful of birds concluded, as Darwin had once envisaged, that there *was* a monophyletic origin for all domestic chickens, almost certainly from a red junglefowl population (Fumihito et al., 1996). However, a more recent and extensive study of the mtDNA of over 800 chickens and 66 red junglefowl suggested multiple origins from different Asian regions (Liu et al., 2006). The Eurasian chickens

in their sample appeared to derive from three separate monophyletic groups, while other distinct branches led to groups of chickens now found only in certain regions of South-east Asia, Japan or China. Of course, studies of genetic origins and relationships depend on the material included in the first place. The wider the net is cast, the more complex the relationships that are revealed. For example, it is now thought that Indian populations of domestic chicken derived from a variety of South-east Asian red junglefowl subspecies, including *G. g. spadiceus* and *G. g. gallus*, but that the distinct Indian red junglefowl, *G. g. murghi*, also made an important genetic contribution (Kanginakudru et al., 2008). When the entire mtDNA sequence is examined alongside certain regions from the cell nucleus, there is evidence for an even more complex multiple origin, with the Indian grey junglefowl *Gallus sonneratii* seemingly contributing to the chicken's yellow legs (Eriksson et al., 2008).

In a further study, Sawai et al. (2010) confirmed that a high proportion of DNA sequences at adjacent locations were shared between red junglefowl and domestic chickens, but they also found that a small proportion of these sequences were shared between green junglefowl and chickens, although green junglefowl are not generally thought to have contributed greatly to the modern chicken. One possible explanation for these sometimes confusing results is that modern domestic chickens may occasionally have bred with the original parental species, a process called introgression. The extent to which this might have occurred is unknown. Modern chickens prefer to mate with chickens of similar appearance to themselves (Tiemann and Rehkämper, 2012), but they are not averse to mating with more distant chicken and junglefowl relatives. Such a process would introduce genes that would not originally have been present in that parental genotype and greatly confuse attempts to derive phylogenies.

Perhaps the biggest achievement of these molecular approaches is their remarkable ability to estimate when chickens first became domesticated. Given that the oldest archaeological remains of chickens have been dated to around 10,000 years ago (8000 BC), it is remarkable that the molecular evidence hints at an origin perhaps as long ago as 58,000 years (Sawai et al., 2010), although the authors do admit to an uncertainty factor in this estimate of around 16,000 years either side. As humans are thought to have migrated from Africa to South-east Asia around 70,000 years ago, it seems it was not long after their arrival that they started domesticating the local wild bird population. But what was the initial purpose of this domestication?

Almost certainly, chickens were not first kept or bred for their eggs. Since junglefowl produce only five to six eggs in any given breeding season, a traditional cooked breakfast would have been a rare luxury. Perhaps they were first sheltered and bred as a scavenging animal that could be eaten when other food sources became scarce. Some archaeologists believe that chickens were first domesticated because of the fighting ability of the male birds, or for ceremonial purposes. The sound of the cockerel's crow may even have been valued as a navigational guide for seafaring islanders in the Pacific Islands (Sykes, 2012). Recent genetic data show that the distribution of one branch of the chicken evolutionary tree is closely related to the distribution of the practice of cockfighting (Liu *et al.*, 2006), although it is not known when this practice first emerged. By 5000 BC, chickens feature in Greek legends, and their fighting abilities were depicted visually on a variety of artefacts. Archaeological evidence shows a much higher prevalence of male than female chickens in many European Roman sites, which may be because the males were kept for fighting. Sykes (2012) noted a correspondence between the rise in cockfighting and a decline in interpersonal violence during the transition from the Iron Age to the Roman period, and speculated that this may have been more than coincidence, with some aspects of human conflict displaced on to animals. Chickens were also used as portents in fortune-telling by many civilizations in Africa, Europe and America. The appearance of domestic chickens differs from that of the junglefowl (Fig. 1.1), and an interest in breeding chickens for their appearance may also have been a component of the chickens' worldwide spread.

While their role in magic rituals may have declined, the human fascination with the colour and plumage of the domestic chicken remains as strong as ever. Today, there are more than 500 fancy breeds with appearances that range from the sublime to the ridiculous.

Commercial Chicken Breeding

In the past few centuries, chickens have become a food staple in nearly all countries of the world. Although backyard and fancy breeds of chickens

Fig. 1.1. The red junglefowl chick at the front appears strikingly different from the domesticated white leghorn chick of the same age. However, the behavioural differences between the wild-type and domesticated fowl are largely a matter of degree. Domestic chickens and red junglefowl have identical behavioural repertoires but different response thresholds. (Photo courtesy of Professor Per Jensen, Linköping University, Sweden.)

remain around the world, the vast majority of domestic chickens are now bred commercially for the purposes of food production. There was a dramatic divergence in the appearance of meat (broiler) and specialized egg-layers in the late 19th and early 20th centuries. The different genetic backgrounds of modern broiler and layer breeds reflect inherent trade-offs in the way bodily resources can be allocated to either reproduction or growth. Over the past 70 years, there have been truly incredible changes in the manner and extent to which chickens act as a human food resource, made possible largely because the genetic contribution to variation (heritability) of many of the characteristics that affect chicken production is reasonably high. The moderate to high heritability of traits such as weight gain, feed-conversion efficiency and egg size has been exploited to produce the commercial breeds we see today.

The selection methods used to produce commercial chicken breeds have also changed dramatically over the past century. Modern broiler and layer breeds were initially established by crossing dual-purpose Asian and European breeds. In the 1930s, birds were selected largely on their observable individual characteristics (phenotype) without knowledge of their underlying genotype. The strategy for producing fast-growing meat chickens was simply to select the heaviest male and female birds and breed from them, deriving a pure line of selected birds. However, it was soon discovered that crossing pure lines to obtain hybrid meat or laying birds had many advantages. It was also soon noted that selecting for a single trait often led to problems – chickens selected to grow fast and reach heavy weights, for example, had fewer resources to allocate to reproduction and could be behaviourally disinterested, or incapable of breeding. Gradually, index or matrix selection methods were adopted, such that a number of important traits were selected simultaneously, with more important traits weighted more strongly. Another dramatic change was to use selection methods that took account of genotype, not just phenotype, using methods of family selection and progeny testing. The principle was to rear several families under identical conditions, compare the properties of the offspring with their known pedigrees, determine which of these offspring had the most optimal 'breeding values' and retain them to produce the next generation.

There is an element of art as well as science in the practice of progeny testing, particularly for traits with relatively low heritability, where the environment plays an important role in the variation that is observed among animals. Many behaviour and welfare traits fall into this category. For this reason, commercial breeders generally evaluate performance not only in specialized research breeding facilities but also under a variety of different field conditions, the choice of exactly what conditions to include representing the art side of the process.

Specialized parental and grandparental lines now exist for both meat and egg types of chicken. These elite breeding populations are called nucleus flocks and are extremely valuable. For this reason, biosecurity is an absolute priority, and the flocks are kept in separate and often remote regions of the world. Large effective populations are kept to avoid problems with inbreeding. Almost all the world's branded strains of meat chicken are supplied by just three breeding companies: Cobb-Vantress, Erich Wesjohann (acquired Aviagen and Ross brand) and Groupe Grimaud (Hubbard brand), and a similarly small handful of companies supply layer-strain chicks: Hendrix (brands: Isa, Babcock, Shaver, Hisex, Bovans and Dekalb), Erich Wesjohann (brands: Lohmann, Hyline and H&N Brown Nick) and Groupe Grimaud (Novogen).

Effects of Domestication on the Chicken Genome

The advent of new genomic technologies means that we can retrospectively examine how the domestication process has had its effects on chickens. The genetic variation of the chicken genome is as diverse as that of the ancestral species, suggesting that past artificial selection during domestication involved only a small number of genes. Following the publication of the initial chicken genome (International Chicken Genome Sequencing Consortium, 2004) and further resequencing of other chicken genomes (an increasingly cost-effective enterprise that has corrected previous errors, with second and third versions now available), an insight is emerging as to the specific genes that have been altered during domestication. One large-scale project obtained samples from birds representing eight populations of domestic chickens as well as populations of red junglefowl (Rubin *et al.*, 2010). By pooling whole DNA from these different samples, the authors identified sequence variations that characterized each population. They found that the locus encoding the thyroid stimulating hormone receptor (TSHR) on chromosome 5 distinguished all the domestic chickens from the red junglefowl, with most domestic chickens possessing a glycine-to-arginine substitution in a key region. Because TSHR is closely involved in the photoperiodic control of reproduction, these findings suggest that (nearly) all domestic chickens carry a mutant allele that may enable them to reproduce throughout the year. Decoupling reproduction from photoperiod would be disastrous for a red junglefowl, but the additional food and protection provided by human caretakers would allow this trait to spread in domesticated birds. The same study also found that populations of commercial broilers were distinguished by regions containing genes involved in growth, appetite regulation and glucose metabolism.

Although extensive in its scope, the study of Rubin *et al.* (2010) examined only a small number of domestic chicken breeds. Elferink *et al.* (2012) therefore analysed 67 different chicken breeds that had been selected for a variety of different purposes, as well as non-commercial traditional Dutch and Chinese breeds. These researchers identified 26 regions showing strong evidence of selection, some of which contained genes such as an insulin-like growth factor (IGF1), described previously by Rubin *et al.* (2010). A highlight was the discovery of regions not previously known to be involved in domestication, which contained, for example, genes coding for skeletal function.

Effects of Domestication on Chicken Behaviour

The primary purpose of domestication was to breed chickens that were more useful to humans for food or entertainment, but, deliberately or inadvertently, selection may also have altered the behaviour of chickens. Working out which aspects of behaviour have been affected, and to what degree, is tricky because the environment plays such a large part in the control of behaviour. Thus, variations in early life experience and learning can explain many of the differences in behavioural response that we see among adults, but there is also an undeniable element of genetic control over both simple and complex behavioural responses.

When we compare the behaviour of modern commercial strains of chickens with the ancestral junglefowl, we have to consider both the genetic and environmental influences. One way of doing so is to compare the populations of interest in the same environment. Although no one has tried to house junglefowl in flocks of many thousands in commercial laying hen or broiler accommodation (it would probably be a disaster, and I certainly do not suggest it), there have been studies examining how modern chickens behave and fare when kept under semi-natural conditions. In the mid-1970s, a small population of domestic chickens was released on an uninhabited island off the west coast of Scotland. An area of oats was sown to provide a source of food during the winter, and the behaviour and reproductive success of the birds was monitored throughout 1976. The nest site choice of the birds was variable: some selected damp or unsuitable sites resulting in nest abandonment or poor hatching rates; but some birds selected good sites, made only short forays from the nest each day, and showed tactical running and stopping movements to distract predators during their circuitous return to the nest. Within a clutch, all eggs were laid in the same nest, but between clutches nest sites were always altered. Although mortality was not negligible due to the presence of mink on the island, many hens successfully raised their broods (Duncan *et al.*, 1978).

This study suggested that few behavioural traits have been lost and that at least some strains of domestic chicken have a tendency to perform (nearly) all the activities that are observed in junglefowl. One notable exception, apparent in many modern strains, is the loss of broodiness, which seems to have occurred as an inadvertent effect of selection for high egg production. However, the loss of broodiness is not a recent phenomenon. Darwin wrote:

> Natural instincts are lost under domestication: a remarkable instance of this is seen in those breeds of fowls which very rarely or never become "broody," that is, never wish to sit on

their eggs. Familiarity alone prevents our seeing how largely and how permanently the minds of our domestic animals have been modified...

(Darwin, 1868)

And yet, in the case of the chicken, Darwin's conclusion is exaggerated. As the work by Duncan *et al.* (1978) showed, many less-commercial strains of chickens still go broody and sit on their eggs. In our work at Bristol University, UK, we particularly appreciate the broody tendencies of the little bantam hens that rear chicks for our studies of parental behaviour (Nicol and Pope, 1996). These strains also form more cohesive social groups (Keeling and Duncan, 1991).

With the exception of brooding, the capacity to do most other behaviours does not appear to have been lost completely, even in the most highly selected commercial strains, and there is certainly no good evidence that the minds of domestic chickens have been drastically altered. Indeed, Appleby *et al.* (2004) emphasized that the repertoires of domestic chickens and junglefowl are identical, with no behaviours totally eliminated by domestication and no entirely new behaviours produced. Generally, the differences between domestic and wild chickens can be attributed to

modified stimulus thresholds whereby animals respond at lower or higher levels of stimulation than would the wild ancestor. For example, cockerels crow more than junglefowl (Wood-Gush, 1959) and domestic laying hens perch during the day but not as frequently as junglefowl (Eklund and Jensen, 2011). More generally, domesticated animals (of all species) show reduced fear and anti-predator responses, alongside increased sociability. Direct comparisons of the behaviour of domestic laying hens and red junglefowl partly confirm this view. Hens are less fearful and responsive to predatory stimuli than junglefowl (Schütz *et al.*, 2001; Campler *et al.*, 2009). One of the main traits selected during the domestication process, deliberately or by happenstance, was almost certainly a reduced fear of humans. When red junglefowl are artificially selected for reduced fear towards humans, this is correlated with a number of other changes in phenotype including a reduced tendency to explore and peck for mealworms hidden in pots of shavings (Agnvall *et al.*, 2012), and increased bodyweight (Agnvall *et al.*, 2014). Domestic hens are also generally less active (Schütz *et al.*, 2001) and can be less synchronized in their activity (Eklund and Jensen, 2011) (Fig. 1.2), characteristics that may help them conserve

Fig. 1.2. Red junglefowl perch together on structures that many domestic hens would struggle to utilize. Red junglefowl tend to be more active and more synchronized in their behaviour than domestic chickens. (Photo courtesy of Dr Becky Hothersall, University of Bristol, UK.)

energy when their need to compete for food is much lower than for their wild cousins.

In contrast to the general pattern seen in domesticated mammals, however, chickens do not appear to show increased sociability. Indeed, they are more aggressive than junglefowl under certain conditions (Väisänen et al., 2005). One reason for this may lie in the domestication history of the chicken, where selection for extreme aggression in fighting males may have been favoured. Another unintended consequence of selection has been an increased tendency in many laying hen strains to peck at and damage the plumage of their companions. This is not an aggressive behaviour (although it can have devastating consequences). Rather, it is a redirection of foraging or exploratory pecking under conditions where foraging opportunities are limited. The tendency to perform feather pecking thus seems to have been selected inadvertently as a trait that hitchhiked on the back of selection for early onset of egg laying and the ability to use calcium to produce good-quality eggshells, as egg production and feather-pecking tendencies show striking correlations (Väisänen et al., 2005; Su et al., 2006).

Future Trends in Chicken Breeding

It has been suggested that commercial breeding programmes will continue to add traits to the selection index, and that more account will be taken of bird welfare and sustainability, alongside maintaining or further increasing bird productivity. To achieve these goals, behavioural traits will inevitably need to be added to the selection index, but how exactly this will be achieved remains to be seen. In addition to the continued selection for egg production traits for birds in all housing systems, Lohmann report that selection priorities for hens destined for non-cage systems include strong plumage, docile behaviour and nest acceptance (Preisinger and Icken, 2013). As we have seen, selection for behavioural traits has taken place historically without any knowledge of or contribution from genetic technologies, but future selection programmes are likely to make increasing use of genetic information.

One step in identifying the (many) genes responsible for a complex behavioural trait is to establish where on the chromosomes these genes are located. Experimental crosses can identify regions called quantitative trait loci (QTLs) that jointly control such multi-factorial traits. QTLs have been identified in laying hens for traits involved with fearfulness (Buitenhuis et al., 2004; Schütz et al., 2004) and giving and receiving feather pecks (Buitenhuis et al., 2003a,b). However, so far, the studies are too few to make general conclusions about the genetic control of these traits. The regions of genetic involvement identified for these traits appear to relate rather specifically to the line and age of bird studied and cannot be generalized too widely at this stage. QTLs for behaviour in broilers have not yet been examined. QTL mapping at least highlights potentially relevant regions of the genome, but using this information in breeding programmes requires much further knowledge and development. Genomic selection can work only if good markers are available to establish the regions of the genome that code for the traits of interest. When a certain trait is correlated with the occurrence of a marker, we infer that a relevant gene for that trait is located close by.

It is safe to predict that breeding programmes will become increasingly dependent on the use of genetic markers, but these will need to be identified using quicker and more comprehensive methods than experimental crossing of lines. Large numbers of QTLs are necessary to explain genetic variation in most complex traits of interest in chicken breeding. The identification of just a few QTLs via crossing programmes or using marker-assisted selection therefore has limited value. Only when huge numbers of QTLs can be identified will genome selection for behavioural traits become more feasible. This goal is creeping ever nearer now that microarray technology is available. Chips or panels that contain information about vast numbers of single- nucleotide polymorphisms (SNPs) on a solid surface are now available for the chicken, although some of the whole-genome panels remain commercially restricted in their use. Whole-genome SNP panels will allow the rapid detection of mutations involved in simple and complex traits and will further enable the process of gene discovery.

The process of genomic selection does not depend solely on phenotypic information or even on progeny testing, so direct selection can occur for traits that are sex-linked (e.g. selection for egg production in females can be pursued in males) and for traits that are difficult or time-consuming to measure (e.g. disease resistance, longevity). It can also be conducted very early in life, increasing the speed of the entire selection process. Breeding values can be estimated even without knowing precisely where specific genes are located. With vast numbers of SNPs available, there is likely to be one in close proximity to any gene of interest. The genomes of (large groups of) animals that possess general traits of interest (e.g. elite populations of highly productive chickens) can therefore be used as a reference population and statistical models built to estimate the effect of each SNP on the trait(s) of interest. The genomes of other groups of animals can then be compared and examined for linkage disequilibrium (non-random variation) in their SNPs. The animals that have the polymorphisms positively associated with the required traits can be used in future breeding programmes.

Epigenetic mechanisms, whereby environmental effects can modify DNA and affect gene expression, are only just starting to be investigated in chickens. In principle, some of these epigenetic effects could be inherited and may need to be considered in future breeding programmes. Jensen (2014) reviewed work on chickens that suggested that exposure to stress can cause modified gene expression. In white leghorn chickens, differences in gene expression in the hypothalamus were observed between stressed and control parents, and these differences were also observed in their offspring (hatched and reared separately from the parents). The mechanisms for these effects are not yet fully understood, although glucocorticoid receptor sites are present throughout the genome, and changes in these receptors could alter the methylation status of promoter regions (i.e. sequences of DNA that turn nearby genes on or off).

The process of genomic selection is rather blind, associating variation in genetic sequences with phenotypic traits of interest, without much concern about how the sequence variation produces its phenotypic effect. This process therefore needs to be more integrated with studies of gene localization (QTL analysis) and gene function before we really understand what we are selecting for. In mammals, the role of many genes has been elucidated using transgenic technologies. It seems that we are now on the brink of a new wave of avian transgenic studies in which the chicken will feature heavily. Programmable gene editing technologies have been developed for chickens and will allow genetically modified strains to be developed rapidly both for commercial food production and for the production of valuable biomaterials in their eggs (Park *et al.*, 2014).

Genetic Aspects of Chicken Disease and Welfare

It would be comforting to think that the welfare of any chicken could be assured if it was kept in a clean, well-managed system, with sufficient opportunities to perform its essential behavioural needs. The reality, however, is more complex than this, and there are some welfare concerns that can be laid at the door marked 'genetic selection'. Highly selected strains seem to have compromised immune function, with the most productive strains showing the greatest decline in humoral immune capacity (Bridle *et al.*, 2006). This strongly suggests that selective breeding has reduced resistance to infectious disease (Zekarias *et al.*, 2002), although an unequivocal conclusion is difficult to reach, as highly productive strains kept under intensive conditions are also subjected to dramatic vaccination programmes that could certainly upset general immune function. Recently, some groups have begun to investigate the feasibility of selecting broilers with enhanced immune function (e.g. an increased pro-inflammatory cytokine response) or a direct resistance to specific pathogen challenge (Cavero *et al.*, 2009), not with the primary aim of improving bird welfare but to reduce the risk of human food poisoning and/or the need to use antibiotics routinely. However, the jury is still out on these endeavours. There is a constant arms race taking place between host and microorganism populations and, to some extent, the pathogens have the upper hand with their rapid generational turnover and higher mutation rates. Disease resistance is such a complex trait that it may not be easy to improve it without knock-on consequences on other traits, including productivity and stress reactivity. Another scenario sees resistance against some pathogens acquired but at the expense of resistance to others (Jie and Liu, 2011).

Other welfare concerns that result partially from intensive selection for production include skeletal problems in laying hens and leg disorders in broilers. These negative traits may result from hitchhiking of undesirable alleles with the alleles under selection, but, in theory, it may be possible to resolve these problems more easily than selecting for disease resistance. The intensive demands of egg production have profound effects on bone composition. Hens kept in conventional cages suffer from disuse osteoporosis with a decrease in the amount of structural bone and a high likelihood of bone fracture at the point where birds are removed from cages at the end of the laying period. Some work points to a genetic component to bone strength (Schreiweis *et al.*, 2005; Dunn *et al.*, 2007), but selection for increased bone strength by itself will not solve the welfare problem of fractures in laying hens. Indeed, bone strength plays only a minor part in the fractures sustained by hens kept in larger furnished cages or in non-cage systems. Recent work shows that, contrary to popular scientific opinion, bone strength and bone mineral density do not decrease with bird age; under some circumstances, bones become stronger and more dense as birds age. The occurrence of bone fractures then arises because of a sharp decline in bone elasticity combined with a reduction in collagen content and an increase in mature collagen crosslinking with age (Tarlton *et al.*, 2013). Essentially, bones become stronger but stiffer, more brittle and less compliant. This is reflected in the high prevalence of fractures, particularly of the keel bone, observed in all commercial systems, which can in some flocks affect as many as 90% of hens by the end of lay (Wilkins *et al.*, 2011). There is also likely to be a genetic component to bone suppleness, but, at present, knowledge in this area is in its infancy and no serious steps have yet been taken to adjust breeding goals to improve the skeletal integrity of laying hens.

In contrast, and in response to concerns about the levels of lameness and leg disorders in broiler chickens, there has been a welcome shift in the breeding practices of some of the biggest broiler producers. Dawkins and Layton (2012) pointed out that positive population correlations between growth rate and leg problems do not apply to all individuals – there are birds that seem to grow quickly without becoming lame. This has been confirmed by statistical analysis of the correlations between body weight and leg health traits in 146,000 Cobb-Vantress birds (Rekaya *et al.*, 2013). There are also birds that grow quickly but that seem to have a greater resistance to skin conditions such as footpad dermatitis (Kjaer *et al.*, 2006). The existence of such birds indicates that traits for good production and improved welfare do not have to be in opposition when selection takes place. There is good evidence that adopting multi-trait selection techniques and incorporating welfare outcomes into overall breeding goals can make a real difference. Reviewing the outcomes from the Aviagen UK breeding strategy over the past 25 years, Kapell *et al.* (2012) noted a decline in crooked toes, deformities of the long bones and tibial dyschondroplasia (a form of growth-plate abnormality). The prevalence of tibial dyschondroplasia, for example, has decreased by about –0.4 to –1.2% per year, although the problem has not yet been eliminated.

The results of the selection programme as summarized by Kapell *et al.* (2012) provide a highly encouraging demonstration that the leg health of broiler chickens can be improved, but questions remain. First, correlations between different leg health traits are not strong (Rekaya *et al.*, 2013) and yet all must be considered in a breeding programme. Secondly, the birds in the selection programme of Kapell *et al.* (2012) were housed on farms with exceptionally stringent biosecurity procedures and the prevalence of many of these leg disorders on 'real' farms is likely to be much higher. A comprehensive survey of 176 UK broiler farms assessed the gait score (an index of walking ability) of 50,000 birds and found that approximately 27% had difficulty walking, as indicated by a score of 3 or more (Knowles *et al.*, 2008). Another substantial survey of 89 commercial broiler flocks across France, the UK, the Netherlands and Italy detected 15.6% of birds with similarly high gait scores (Bassler *et al.*, 2013). Of course, tibial dyschondroplasia is not the only cause of walking difficulty, and both genetic and environmental factors have an influence. None the less, the scale of the broiler welfare problem revealed by these comprehensive on-farm assessments is surprising. Genetic selection undoubtedly has an important part to play in finding a solution and will be retained as a breeding goal so long as consumers demonstrate a continued interest and commitment to chicken welfare.

Genetic Aspects of Chicken Behaviour and Welfare

Laying hens have a tendency to peck and injure each other to such a degree that the welfare of pecked birds can be severely affected and great economic loss incurred (Nicol et al., 2013; Rodenburg et al., 2013). This important topic is reviewed in full in Chapter 9, but it serves here as an example of an unwanted behaviour with a clear genetic basis. Heritability estimates for feather-pecking behaviour vary from 0.12 to 0.38 (Kjaer and Sorensen, 1997; Rodenburg et al., 2003), and strain differences are regularly associated with different rates of feather pecking and different degrees of plumage loss and damage (Kjaer and Sorensen, 2002; Oden et al., 2002; Keeling et al., 2004; Bright, 2007; Uitdehaag et al., 2008). In this context, genetic selection would seem to have a potentially critical role in improving the welfare of laying hens.

Various selection strategies have been attempted, but success has been mixed. Bessei et al. (1999) selected birds based on the number of pecks they made to a static bundle of feathers attached to a strain gauge, but the results did not correlate well with pecks directed to live birds (reported by Buitenhuis and Kjaer, 2008). Kjaer and Sorensen (1997) estimated the heritability of giving and receiving feather pecks in commercial white leghorn birds and then instigated a selection experiment for the trait 'number of bouts of feather pecking per bird per hour'. Individuals showing the highest and lowest values for this trait were selected. After three generations, the two lines differed substantially in this trait (Kjaer et al., 2001), with further divergence between the lines apparent by the fifth generation, after a backcross (Su et al., 2005). After eight generations, progress in reducing feather pecking in the low-pecking line had slowed, while pecking rates in the high-pecking line had increased further due to ongoing selection of polygenic variables linked with feather-pecking activity but also to rapid selection of a single allele linked to extreme pecking and hyperactivity (Labouriau et al., 2009).

Although this work shows clearly that genetic selection can influence feather-pecking levels, application to the commercial situation has been limited by the laborious nature of the phenotyping process, where the pecking behaviour of large numbers of individual birds must somehow be monitored. In addition, commercial companies are aware of, and cautious about, the many correlations between feather pecking and other traits. Characteristics associated with feather pecking include the stress response to physical restraint (measured by heart rate variability and corticosterone increase) (Kjaer and Guemene, 2009; Kjaer and Jorgensen, 2011), fearfulness (Rodenburg et al., 2010a; Nordqvist et al., 2011), locomotor activity (Rodenburg et al., 2004; Kjaer, 2009; de Haas et al., 2010), foraging behaviour (de Haas et al., 2010), serotonin and dopamine transmission (Flisilowski et al., 2009; Kops et al., 2013, 2014), and production and immunological traits (Buitenhuis and Kjaer, 2008). Although selection against feather pecking has generally resulted in birds with the beneficial traits of lower fearfulness and activity and a calmer reaction to human handling, associations between feather pecking and neurotransmitters differ according to bird age or strain. Crucially, selection against feather pecking has had a mixed and sometimes negative effect on production parameters. Traditional selection of birds for individual high-production traits may therefore inadvertently also select for birds with a high pecking tendency, such that overall flock production is reduced on commercial farms (Rodenburg et al., 2010b).

In such a situation, where the behaviour of one bird impacts and adversely affects the performance of others, group selection techniques applied to outcome traits that are relatively simple to measure (e.g. mortality, feather loss) may be the best method for improving both welfare and production efficiency (Rodenburg et al., 2013). A notable programme in the USA has selected cage-housed birds based on their performance as a group (egg mass and mortality) over nine generations (Muir, 1996; Cheng and Muir, 2005). After just one generation, mortality due to injurious pecking had decreased from 60 to 10%. Even though this rate of progress could not be maintained, by the eighth generation, there was a significant reduction in mortality in the selected line compared with the unselected control, accompanied by reductions in plasma dopamine and serotonin (Cheng et al., 2001).

Based on the methods pioneered by Muir (1996), Ellen et al. (2007, 2008) developed a selection programme to reduce bird mortality in which information about the (individually housed) selection candidate was combined with information from its siblings housed in family groups. Including indirect effects (i.e. the effect of an individual on the survival of its group members) increased the total heritable variance for mortality in comparison with models that accounted for direct effects only (Ellen *et al.*, 2008). When the trait of feather condition score was used rather than mortality, a strong majority of estimated total heritable variation was due to indirect effects (Brinker *et al.*, 2014). Application of this group method of selection has led to progressive reductions in mortality, fear and stress sensitivity (Bolhuis *et al.*, 2009; Rodenburg *et al.*, 2009, 2013). It might seem that a straightforward next step would be to take this work and apply it in the commercial sphere, but considerable hurdles remain, not least difficulties relating to genetic × environment interactions. A programme of genetic selection conducted in one environment may produce birds that perform more poorly than expected when housed in a different commercial environment. In particular, selection programmes instigated in caged birds may have greatly reduced relevance for birds that are housed commercially in non-cage systems. Selection against feather pecking is currently being included by the major companies as a breeding goal, alongside more general traits for bird robustness and adaptability. However, the relative weight given to such welfare-friendly traits in current breeding models remains a closely guarded secret, making it difficult to know what the future holds. One clear conclusion arising from this exposition is that the genetic influences on feather pecking are only one part of a much larger picture. The way that chickens are housed, handled and managed has a substantial influence on their behaviour and welfare, for better or for worse. The following chapters will provide an introduction to some of the influences on their behavioural biology, before returning to the question of how best to safeguard their welfare.

References

Agnvall, B., Alis, A., Olby, S. and Jensen, P. (2014) Red junglefowl (*Gallus gallus*) selected for low fear of humans are larger, more dominant and produce larger offspring. *Animal* 8, 1498–1505.

Agnvall, B., Jöngren, M., Strandberg, E. and Jensen, P. (2012) Heritability and genetic correlations of fear-related behaviour in red jungle fowl – possible implications for early domestication. *PLoS One* 7, e35162.

Appleby, M.C., Mench, J.A. and Hughes, B.O. (2004) *Poultry Behaviour and Welfare*. CABI, Wallingford, UK.

Bassler, A.W., Arnould, C., Butterworth, A., Colin, L., De Jong, I.C., Ferrante, V., Ferrari, P., Haslam, S., Wemelsfelder, F. and Blokhuis, H.J. (2013) Potential risk factors associated with contact dermatitis, negative emotional state, and fear of humans in broiler chicken flocks. *Poultry Science* 92, 2811–2826.

Bessei, W., Reiter, K., Bley, T. and Zeep, F. (1999) Measuring pecking of a bunch of feathers in individual housed hens: first results of genetic studies and feeding related reactions. *Lohmann Information* 22, 27–31.

Bolhuis, J.E., Ellen, E.D., Van Reenan, C.G., De Groot, J., Ten Napel, J., Koopmanschap, R., De Vries Reilingh, G., Uitdehaag, K.A., Kemp, B. and Rodenburg, T.B. (2009) Effects of genetic group selection against mortality on behaviour and peripheral serotonin in domestic laying hens with trimmed and intact beaks. *Physiology and Behavior* 97, 470–475.

Bridle, B.W., Julian, R., Shewen, P.E., Vaillancourt, J.P. and Kaushik, A.K. (2006) T lymphocyte subpopulations diverge in commercially raised chickens. *Canadian Journal of Veterinary Research* 70, 183–190.

Bright, A. (2007) Plumage colour and feather pecking in laying hens, a chicken perspective? *British Poultry Science* 48, 253–263.

Brinker, T., Bijma, P., Visscher, J., Rodenburg, T.B. and Ellen, E.D. (2014) Plumage condition in laying hens: genetic parameters for direct and indirect effects in two purebred lines. *Genetics Selection Evolution* 46, 33.

Buitenhuis, A.J. and Kjaer, J.B. (2008) Long term selection for reduced or increased pecking behaviour in laying hens. *World's Poultry Science Journal* 64, 477–487.

Buitenhuis, A.J., Rodenburg, T.B., van Hierden, Y.M., Siwek, M., Cornelissen, S.J.B., Niewland, M.G.B., Crooijmans, R.P.M.A., Groenen, M.A.M., Koene, P., Korte, S.M., Bovenhuis, H. and van der Poel, J.J. (2003a) Mapping quantitative trait loci affecting feather pecking behaviour and stress response in laying hens. *Poultry Science* 82, 1215–1222.

Buitenhuis, A.J., Rodenburg, T.B., Siwek, M., Cornelissen, S.J.B., Niewland, M.G.B., Crooijmans, R.P.M.A., Groenen, M.A.M., Koene, P., Bovenhuis, H. and van der Poel, J.J. (2003b) Identification of quantitative trait loci for receiving pecks in young and adult laying hens. *Poultry Science* 82, 1661–1667.

Buitenhuis, A.J., Rodenburg, T.B., Siwek, M., Cornelissen, S.J.B., Nieuwland, M.G.B., Crooijmans, R.P.M.A., Groenen, M.A.M., Koene, P., Bovenhuis, H. and van der Poel, J.J. (2004) Identification of QTLs involved in open-field behaviour in young and adult laying hens. *Behavior Genetics* 34, 325–333.

Campler, M., Jongren, M. and Jensen, P. (2009) Fearfulness in red junglefowl and domesticated White Leghorn chickens. *Behavioural Processes* 81, 39–43.

Cavero, D., Schmutz, M., Philipp, H.C. and Preisinger, R. (2009) Breeding to reduce susceptibility to *Escherichia coli* in layers. *Poultry Science* 88, 2063–2068.

Cheng, H.W. and Muir, W.M. (2005) The effect of genetic selection for survivability and productivity on chicken physiological homeostasis. *World's Poultry Science Journal* 61, 383–397.

Cheng, H.W., Dillworth, G., Singleton, P., Chen, Y. and Muir, W.M. (2001) Effects of group selection for productivity and longevity on blood concentrations of serotonin, catecholamines, and corticosterone of laying hens. *Poultry Science* 80, 1278–1285.

Darwin, C. (1858) *On the Origin of Species*. John Murray, London.

Darwin, C. (1868) *The Variation of Animals and Plants under Domestication*. John Murray, London.

Dawkins, M.S. and Layton, R. (2012) Breeding for better welfare: genetic goals for broiler chickens and their parents. *Animal Welfare* 21, 147–155.

de Haas, E.N., Nielsen, B.L., Buitenhuis, A.J. and Rodenburg, T.B. (2010) Selection on feather pecking affects response to novelty and foraging behaviour in laying hens. *Applied Animal Behaviour Science* 124, 90–96.

Duncan, I.J.H., Savory, C.J. and Wood-Gush, D.G.M. (1978) Observations on the reproductive behaviour of domestic fowl in the wild. *Applied Animal Ethology* 4, 29–42.

Dunn, I.C., Fleming, R.H., McCormack, H.A., Morrice, D., Burt, D.W., Preisinger, R. and Whitehead, C.C. (2007) A QTL for osteoporosis detected in an F2 population derived from White Leghorn chicken lines divergently selected for bone index. *Animal Genetics* 38, 45–49.

Eklund, B. and Jensen, P. (2011) Domestication effects on behavioural synchronization and individual distances in chickens (*Gallus gallus*). *Behavioural Processes* 86, 250–256.

Elferink, M.G., Megens, H.-J., Vereijken, A., Hu, X.X., Crooijmans, R.P.M.A. and Groenen, M.A.M. (2012) Signatures of selection in the genomes of commercial and non-commercial chicken breeds. *PLoS One* 7, e32720.

Ellen, E.D., Muir, W.M. and Bijma, P. (2007) Genetic improvement of traits affected by interactions among individuals: sib selection schemes. *Genetics* 176, 489–499.

Ellen, E.D., Visscher, J., van Arendonk, J.A.M. and Bijma, P. (2008) Survival of laying hens: genetic parameters for direct and associative effects in three purebred layer lines. *Poultry Science* 87, 233–239.

Eriksson, J., Larson, G., Gunnarsson, U., Bed'hom, B., Tixier-Boichard, M., Stromstedt, L., Wright, D., Jungerius, A., Vereijken, A., Randi, E., Jensen, P. and Andersson, L. (2008) Identification of the *yellow skin* gene reveals a hybrid origin of the domestic chicken. *PLoS Genetics* 4, e1000010.

Flisilowski, K., Schwarzenbacher, H., Wysocki, M., Weigend, S., Preisinger, R., Kjaer, J.B. and Fries, R. (2009) Variation in neighbouring genes of the dopaminergic and serotonergic systems affects feather pecking behaviour of laying hens. *Animal Genetics* 40, 192–199.

Fumihito, A., Miyake, T., Takada, M., Shingu, R., Endo, T., Gojobori, T., Kondo, N. and Ohno, S. (1996) Monophyletic origin and unique dispersal patterns of domestic fowls. *Proceedings of the National Academy of Sciences USA* 93, 6792–6795.

International Chicken Genome Sequencing Consortium (2004) A genetic variation map for the chicken with 2.8 million single-nucleotide polymorphisms. *Nature* 432, 717–722.

Jensen, P. (2014) Behaviour epigenetics – the connection between environment, stress and welfare. *Applied Animal Behaviour Science* 157, 1–7.

Jie, H. and Liu, Y.P. (2011) Breeding for disease resistance in poultry: opportunities with challenges. *World's Poultry Science Journal* 67, 687–695.

Kanginakudru, S., Metta, M., Jakati, R.D. and Nagaraju, J. (2008) Genetic evidence from Indian red jungle fowl corroborates multiple domestication of modern day chicken. *BMC Evolutionary Biology* 8, 174.

Kapell, D.N.R.G., Hill, W.G., Neeteson, A.M., McAdam, J., Koerhuis, A.N.M. and Avendano, S. (2012) Twenty-five years of selection for improved leg health in purebred broiler lines and underlying genetic parameters. *Poultry Science* 91, 3032–3043.

Keeling, L., Andresson, L., Schuts, K.E., Kerje, S., Fredriksson, R., Carlborg, O., Cornwallis, C.K., Pizzari, T. and Jensen, P. (2004) Chicken genomics: feather-pecking and victim pigmentation. *Nature* 7009, 645–646.

Keeling, L.J. and Duncan, I.J.H. (1991) Social spacing in domestic fowl under seminatural conditions – the effect of behavioural activity and activity transitions. *Applied Animal Behaviour Science* 32, 205–217.

Kjaer, J.B. (2009) Feather pecking in domestic fowl is genetically related to locomotor activity levels: implications for a hyperactivity disorder model of feather pecking. *Behavior Genetics* 39, 564–570.

Kjaer, J.B. and Guemene, D. (2009) Adrenal reactivity in lines of domestic fowl selected on feather pecking behaviour. *Physiology and Behavior* 96, 370–373.

Kjaer, J.B. and Jorgensen, H. (2011) Heart rate variability in domestic chicken lines genetically selected on feather pecking behaviour. *Genes Brain and Behavior* 10, 747–755.

Kjaer, J.B. and Sorensen, P. (1997) Feather pecking behaviour in White Leghorns, a genetic study. *British Poultry Science* 38, 333–341.

Kjaer, J.B. and Sorensen, P. (2002) Feather pecking and cannibalism in free-range laying hens as affected by genotype, dietary level of methionine plus cysteine, light intensity during rearing and age at first access to the range area. *Applied Animal Behaviour Science* 76, 21–29.

Kjaer, J.B., Sorensen, P. and Su, G. (2001) Divergent selection on feather pecking behaviour in laying hens (*Gallus gallus domesticus*). *Applied Animal Behaviour Science* 71, 229–239.

Kjaer, J.B., Su, G., Nielsen, B.L. and Sorensen, P. (2006) Foot pad dermatitis and hock burn in broiler chickens and degree of inheritance. *Poultry Science* 85, 1342–1348.

Knowles, T.G., Kestin, S.C., Haslam, S.M., Brown, S.N., Green, L.E., Butterworth, A., Pope, S.J., Pfeiffer, D. and Nicol, C.J. (2008) Leg disorders in broiler chickens: prevalence, risk factors and prevention. *PLoS One* 3, e1545.

Kops, M.S., de Haas, E.N., Rodenburg, T.B., Ellen, E.D., Korte-Bouws, G.A.H., Olivier, B., Gunturkun, O., Bolhuis, J.E. and Korte, S.M. (2013) Effects of feather pecking phenotype (sever feather peckers, victims and non-peckers) on serotonergic and dopaminergic activity in four brain areas of laying hens (*Gallus gallus domesticus*). *Physiology and Behavior* 120, 77–82.

Kops, M.S., Kjaer, J.B., Gunturkun, O., Westphal, K.G.C., Korte-Bouws, G.A.H., Olivier, B., Bolhuis, J.E. and Korte, S.M. (2014) Serotonin release in the caudal nidopallium of adult laying hens genetically selected for high and low feather pecking behavior: an *in vivo* microdialysis study. *Behavioural Brain Research* 268, 81–87.

Labouriau, R., Kjaer, J.B., Abreu, G.C.G., Hedegaard, J. and Buitenhuis, A.J. (2009) Analysis of severe feather pecking behavior in a high feather pecking selection line. *Poultry Science* 88, 2052–2062.

Liu, Y.-P., Wu, G.-S., Yao, Y.-G., Miao, Y.-W., Luikart, G., Baig, M., Beja-Pereira, A., Ding, Z.L., Palanichamy, M.G. and Zhang, Y.P. (2006) Multiple maternal origins of chickens: out of the Asian jungles. *Molecular Phylogenetics and Evolution* 38, 12–19.

Muir, W.M. (1996) Group selection for adaptation to multiple-hen cages: selection program and direct responses. *Poultry Science* 75, 447–458.

Nabholz, B., Glémin, S. and Galtier, N. (2009) The erratic mitochondrial clock: variations of mutation rate, not population size, affect mtDNA diversity across birds and mammals. *BMC Evolutionary Biology* 9, 54.

Nicol, C.J. and Pope, S.J. (1996) The maternal feeding display of domestic hens is sensitive to perceived chick error. *Animal Behaviour* 52, 767–774.

Nicol, C.J., Bestman, M., Gilani, A.-M., de Haas, E.N., de Jong, I.C., Lambton, S., Wagenaar, J.P., Weeks, C.A. and Rodenburg, T.B. (2013) The prevention and control of feather pecking: application to commercial systems. *World's Poultry Science Journal* 69, 775–789.

Nordqvist, R.E., Heerkens, J.L.T., Rodenburg, T.B., Boks, S., Ellen, E.D. and van der Staay (2011) Laying hens selected for low mortality: behaviour in tests of fearfulness, anxiety and cognition. *Applied Animal Behaviour Science* 131, 110–122.

Oden, K., Keeling, L.J. and Algers, B. (2002) Behaviour of laying hens in two types of aviary systems on 25 commercial farms in Sweden. *British Poultry Science* 43, 169–181.

Park, T.S., Lee, H.C., Rengara, D. and Han, J.Y. (2014) Germ cell, stem cell and genomic modification in birds. *Stem Cell Research and Therapy* 4, 201.

Preisinger, R. and Icken, W. (2013) Challenges for breeding layers to work in different housing systems. In: *Proceedings of the 9th European Symposium on Poultry Welfare*, Uppsala, Sweden, pp. 76–81.

Rekaya, R., Sapp, R.L., Wing, T. and Aggrey, S.E. (2013) Genetic evaluation for growth, body composition, feed efficiency and leg soundness. *Poultry Science* 92, 923–929.

Rodenburg, T.B., Buitenhuis, A.J., Ask, B., Uitdehaag, K.A., Koene, P., van der Poel, J.J. and Bovenhuis, H. (2003) Heritability of feather pecking and open-field response of laying hens at two different ages. *Poultry Science* 82, 861–867.

Rodenburg, T.B., Buitenhuis, A.J., Ask, B., Uitdehaag, K.A., Koene, P., van der Poel, J.J., van Arendonk, J.A.M. and Bovenhuis, H. (2004) Genetic and phenotypic correlations between feather pecking and open-field response in laying hens at two different ages. *Behaviour Genetics* 34, 407–415.

Rodenburg, T.B., Bolhuis, J.E., Koopmanschap, R.E., Ellen, E.D. and Decuypere, E. (2009) Maternal care and selection for low mortality affect post-stress corticosterone and peripheral serotonin in laying hens. *Physiology and Behavior* 98, 519–523.

Rodenburg, T.B., de Haas, E.N., Nielsen, B.L. and Buitenhuis, A.J. (2010a) Fearfulness and feather damage in laying hens divergently selected for high and low feather pecking. *Applied Animal Behaviour Science* 128, 91–96.

Rodenburg, T.B., Bijma, P., Ellen, E.D., Bergsma, R., de Vries, S., Bolhuis, J.E., Kemp, B. and van Arendonk, J.A.M. (2010b) Breeding amiable animals? Improving farm animal welfare by including social effects in breeding programmes. *Animal Welfare* 19, 77–82.

Rodenburg, T.B. van Krimpen, M.M., de Jong, I.C., de Haas, E.N., Kops, M.S., Riedstra, B.J., Nordquist, R.E., Wagenaar, J.P., Bestman, M. and Nicol, C.J. (2013) The prevention and control of feather pecking in laying hens: identifying the underlying principles. *World's Poultry Science Journal* 69, 361–373.

Rubin, C.-J., Zody, M.C., Eriksson, J., Meadows, J.R.S., Sherwood, E., Webster, M.T., Jiang, L., Ingman, M., Sharpe, T., Ka, S., Hallböök, F., Besnier, F., Carlborg, O., Bed'hom, B., Tixier-Boichard, M., Jensen, P., Siegel, P., Lindblad-Toh, K. and Andersson, L. (2010) Whole-genome resequencing reveals loci under selection during chicken domestication. *Nature* 464, 587–591.

Sawai, H., Kim, H.L., Kuno, K., Suzuki, S., Gotoh, H., Takada, M., Satta, Y. and Akishinonomiya, F. (2010) The origin and genetic variation of domestic chickens with special reference to junglefowls *Gallus g. gallus* and *G. varius*. *PLoS One* 5, e10639.

Schreiweis, M.A., Hester, P.Y. and Moody, D.E. (2005) Identification of quantitative trait loci associated with bone traits and body weight in an F2 resource population. *Genetics Selection Evolution* 37, 677–698.

Schütz, K., Forkman, B. and Jensen, P. (2001) Domestication effects on foraging strategy, social behaviour and different fear responses: a comparison between the red junglefowl (*Gallus gallus*) and a modern layer strain. *Applied Animal Behaviour Science*,74, 1–14.

Schütz, K.E., Kerje, S., Jacobsson, L., Forkman, B., Carlborg, O., Andersson, L. and Jensen, P. (2004) Major growth QTLs in fowl are related to fearful behavior: possible genetic links between fear responses and production traits in a red jungle fowl × White Leghorn intercross. *Behavior Genetics* 34, 121–130.

Su, G., Kjaer, J.B. and Sorensen, P. (2005) Variance components and selection response for feather-pecking behaviour in laying hens. *Poultry Science* 84, 14–21.

Su, G., Kjaer, J.B. and Sorensen, P. (2006) Divergent selection on feather pecking behaviour in laying hens has caused differences between lines in egg production, egg quality, and feed efficiency. *Poultry Science* 85, 191–197.

Sykes, N. (2012) A social perspective on the introduction of exotic animals: the case of the chicken. *World Archaeology* 44, 158–169.

Tarlton, J.F., Wilkins, L.J., Toscano, M.J., Avery, N.C. and Knott, L. (2013) Reduced bone breakage and increased bone strength in free range laying hens fed omega-3 polyunsaturated fatty acid supplemented diets. *Bone* 52, 578–586.

Tiemann, I. and Rehkämper, G. (2012) Evolutionary pets: offspring numbers reveal speciation process in domesticated chickens. *PLoS One* 7, e41453.

Uitdehaag, K., Komen, H., Rodenburg, T.B., Kemp, B. and Van Arendonk, J. (2008) The novel object test as a predictor of feather damage in cage-housed Rhode Island Red and White Leghorn laying hens. *Applied Animal Behaviour Science* 109, 292–305.

Väisänen, J., Lindqvist, C. and Jensen, P. (2005) Co-segregation of behaviour and production related traits in an F3 intercross between red junglefowl and White Leghorn laying hens. *British Poultry Science* 46, 156–158.

West, B. and Zhou, B.-X (1988) Did chickens go North? New evidence for domestication. *Journal of Archaeological Science* 14, 515–533.

Wilkins, L.J., McKinstry, J.L., Avery, N.C., Knowles, T.G., Brown, S.N., Tarlton, J. and Nicol, C.J. (2011) Influence of housing system and design on bone strength and keel bone fractures in laying hens. *Veterinary Record* 169, 414–417.

Wood-Gush, D.G.M. (1959) A history of the domestic fowl from antiquity to the 19th century. *Poultry Science* 38, 321–326.

Zekarias, B., Ter Huume, A.A.H.M., Landman, W.J.M., Rebel, J.M.J., Pol, J.M.A. and Gruys, E. (2002) Immunological basis of differences in disease resistance in the chicken. *Veterinary Research* 33, 109–125.

2

Sensory Biology

Understanding the perceptual world of other species is a combined exercise in science and imagination. The scientific approach allows us to examine the physiological properties of sense organs and discover which stimuli are registered by animals at a peripheral level. Behavioural tests can also be employed to reveal how these stimuli are detected centrally by the brain, against a background of competing influences and in a variety of contexts. Some behavioural tests are based on classical conditioning, as in Pavlov's famous experiments where dogs showed signs of anticipation when they heard a bell that reliably signalled the arrival of food. Pavlov was interested in the process by which dogs learned that the bell was a good predictor of food. Conveniently, this same procedure can be used in sensory biology to detect the threshold at which the predictive stimulus is detected. The predictive stimulus might be a bell or another sound, but equally it could be a visual or an olfactory stimulus. To detect hearing thresholds, a sound could be presented at increasingly high frequencies or at progressively decreasing sound pressure levels until the point is reached where the stimulus is beyond the reach of the animal's sensory capability. This will be apparent to the experimenter as the point at which the animal ceases to show signs of reward anticipation. More active behavioural tests can also be used in sensory biology, for example requiring the animal to make an obvious response (e.g. pressing a panel) when it detects a stimulus or, in more sophisticated tests, when it distinguishes one stimulus from others.

All of these scientific methods have been used to investigate the sensory world of the chicken. We can also learn which sights, sounds and odours chickens find attractive or repellent by observing their behaviour and preferences. We should not forget some of the very early scientific studies on chicken senses that showed, for example, that chickens are deceived by some of the same optical illusions as humans (Revesz, 1924, cited by Wood-Gush, 1971; Winslow, 1933, cited by Wood-Gush, 1971), something that has recently been confirmed (Rosa Salva *et al.*, 2013).

However, it is our imaginations that come into play when we try to understand what the chicken's sensory world *feels* like. We can try and imagine what it might feel like to have a beak, to detect a broader range of colours, to see UV light or to detect the Earth's magnetic field. Nagel's elegant 1974 essay 'What is it like to be a bat?' prompts us to wonder what it might feel like to be an individual of any other species. In the end, however, we have to acknowledge that our imaginations are based on our own sensory experiences and we can only speculate and wonder about others, as there is no scientific tool that can reveal the quality of conscious subjective experience in another being, and tell us what it is like to be a chicken.

Vision

Vision is arguably the most important sense for a ground-living bird that must forage for concealed food while simultaneously being aware of terrestrial and aerial predators. Chickens' eyes are relatively large compared with the overall size of their head and brain, and these precious organs are lubricated and protected by a transparent nictitating membrane, or third eyelid, that moves horizontally across the eye, affording a degree of protection without obscuring sight. Like most birds, the eyes of chickens are fixed within bony sockets and cannot really be moved. Instead, a chicken must move its head to attend to a particular scene, and therein lies a potential problem. Photoreceptors in the retina need to be stimulated continuously for around 20 ms; any less than this and the image is degraded. Thus, when a chicken is moving through its environment, there is a risk that its vision will be blurred and of poor quality. The chicken's solution is its characteristic bobbing head movement shown in walk, a mechanism that keeps the bird's eyes as still as possible for short bursts of time, allowing a good image to form before the head is moved again. Indeed, if you pick up a chicken and move its body gently from side to side or up and down, you will notice how its head stays surprisingly motionless (http://www.youtube.com/watch?v=_dPlkFPowCc).[1]

Because the eyes of chickens are on each side of the head, rather than placed frontally as in humans or some other bird species such as owls, they have a visual field of at least 300°, but the majority of their vision is monocular. The binocular field is relatively small, less than 30° and may not support depth perception (or stereopsis) as the necessary cell types are rare (Wilson, 1980, cited by Martin, 2009). Rather, the main function of binocular vision in chickens is in the fine control of beak movements during foraging and to some degree in control of locomotion. It may even have a role in social recognition, which we will explore later. For now, let us consider how chickens respond to different light wavelengths before exploring other aspects of their visual function.

Spectral sensitivity

Light can be defined according to its physical properties, such as the amount of radiant energy emitted per second arriving at a given surface area, known as its irradiance. However, light detection is a biological process, and the relative efficiency with which light of different frequencies or wavelengths is detected depends upon characteristics of light-sensitive cells and neural processes. In biology therefore, we rarely speak of irradiance but instead apply a conversion to take account of the sensory biology of the species of interest. For humans, irradiance at given wavelengths is multiplied by the corresponding spectral sensitivity of the human perceptual system at those same wavelengths. This provides a biological measure of perceived brightness or illuminance, measured in units called lux (lx).

If the spectral sensitivity of chickens was identical to that of humans, we could use lux as a measure of brightness for both species, but there are differences. Chickens, like many other birds, possess rods, cones and double cones as photoreceptors, and they have four light-reactive pigments associated with their cone cells compared with just three pigments in humans. The maximal sensitivity of these pigments occurs at wavelengths of 415, 455, 508 and 571 nm. Chickens also possess coloured oil droplets in their cone cells, which act as filters (Bowmaker and Knowles, 1977). Spectral sensitivity curves have been derived for chickens using electrophysiology (Wortel et al., 1987), where direct recordings of the electrical responses of rod and cone cells in the retina are made by placing electrodes on the cornea and skin near the eye. During recording, a light stimulus is applied and the electric potentials contributed by the different cell types result in an electroretinogram, which gives baseline information about spectral sensitivity. To account for the additional role of central brain processing in perception, a useful adjunct is to use behavioural tests of the kind mentioned earlier. Prescott and Wathes (1999), for example, trained chickens to discriminate between a lit and an unlit panel for a food reward. The ability of chickens to perform the discrimination provided conclusive proof that they could detect the training wavelengths, and the sensitivity of their detection

---

[1] Video links have been included within this book. Videos form a useful learning resource, and I hope you find them enlightening. Online materials can be removed from the internet, however, or links break over time; if this happens, I would recommend a Google search to find similar resources.

was then examined by systematic alterations of wavelength and intensity.

It turns out that chickens have a slightly broader response to light than humans and are more sensitive to the blue (~480 nm) and red (~650 nm) parts of the spectrum. Behavioural data obtained by Prescott and Wathes (1999) suggested, for example, that chickens perceive UV radiation below 400 nm, and there is evidence that this can improve their detection of rapidly moving objects (Rubene *et al.*, 2010). Simple light detection tasks show a strong overlap in the visual capacities of chickens and humans (Fig. 2.1), and it is now considered unlikely that the spectral response curves of humans and chickens differ as much as was suggested previously by the narrow Commission Internationale de l'Eclairage (CIE) curve for humans, which was produced under demanding task conditions (Saunders *et al.*, 2008). Although chickens have a slightly different spectral sensitivity to humans, such that lux is an imperfect measure of perceived illuminance, it is not a bad approximation, and it may not be necessary to devise entirely new units of chicken light perception (such as the 'clux'; Prescott *et al.*, 2003).

Even so, it is still important to recognize the implications of the differences in spectral sensitivity between humans and birds. For example, artificial lighting generally does not contain light in the UV range, and this may have some effects on chicken behaviour and welfare (Saunders *et al.*, 2008). Studies so far have produced mixed results. Under controlled conditions, supplementary UV light does not significantly influence ovulation (Lewis *et al.*, 2000) or lead to profound behavioural effects. When 14 flocks of commercial pullets were reared with UV light and compared with 14 matched controls, no differences were found in the birds' behaviour or welfare (Defra report on AW1134, http://randd.defra.gov.uk/). In contrast, a laboratory study found that chicks reared with additional UV light had lower basal corticosterone concentrations than controls (Maddocks *et al.*, 2001), although no differences were found when similar experiments were conducted with quail (Smith *et al.*, 2005). There is some evidence

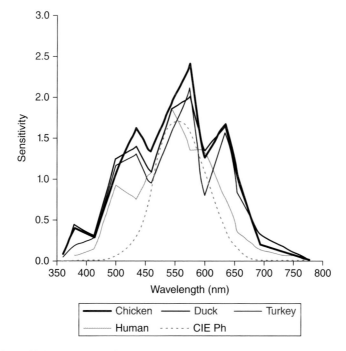

Fig. 2.1. Visual sensitivity in chickens. Saunders *et al.* (2008) plotted the spectral sensitivity of chicken, turkey, duck and human subjects, which had been measured under similar conditions by Prescott and Wathes (1999) and Barber *et al.* (2006). The sensitivities are based on threshold measures for the detection of a simple field against a light background. The CIE photopic function (CIE Ph) scaled to match the human data at 555 nm is included for comparison. With permission from Cambridge University Press.

that UV light plays a role in the sexual behaviour of broiler breeders, increasing the attention paid by hens to cockerels and resulting in a higher rate of mating attempts (Jones et al., 2001). Studies on other birds suggest that UV light can be an important influence on individual recognition and social behaviour, areas that remain underinvestigated for chickens.

Flicker sensitivity

A potential problem for chickens kept under artificial light sources is that they may detect these as flickering rather than continuous, something that has been raised as a potential welfare issue (Widowski and Duncan, 1996). Conventional and compact fluorescent light sources flicker at 100 or 120 Hz, and traditional television and computer screens flicker at between 50 and 85 Hz. To most humans, this flicker is imperceptible, as it is above their critical flicker fusion (CFF) threshold. The CFF threshold is the lowest frequency at which a flickering light source is seen as continuous, and is reported to be between 50 and 60 Hz in humans but may be considerably higher for chickens.

Like spectral sensitivity, CFF thresholds can also be investigated using both physiological and behavioural methods including electroretinograms (100–118 Hz; Lisney et al., 2012), discrimination tasks (71.5 Hz; Jarvis et al., 2002; 87–100 Hz; Lisney et al., 2011), conditional reward response (70–105 Hz; Nuboer et al., 1992) and both simultaneous and conditional presentation methods (68–95 Hz; Railton et al., 2009). Using behavioural methods, CFF thresholds are generally lower than those detected using physiology alone, suggesting that central brain processing mediates the signal produced by the eye. As detection of flicker depends on the spectrum and brightness of the light source used, as well as individual characteristics of different birds, there is no one absolute value for the chicken CFF threshold. However, under most conditions, most chickens do not perceive flicker above 95 Hz. This suggests that, contrary to earlier worries, chickens are probably not aware of the flicker of artificial lighting in commercial housing. Despite this, Lisney et al. (2012) cautioned that there could still be detrimental effects of subconscious physiological reactions to 'invisible flicker'.

Spatial acuity

Although chickens detect a broader light spectrum than humans and are more sensitive to flicker, they are nowhere near as good as us in the level of detail they detect in images. The ability to detect image detail is called acuity and is strictly a measure of spatial resolution. Tests of acuity present a repeated number of fuzzy dark and light bars over a stated visual angle (a sine wave grating). The resolution limit is defined as the minimum grating fineness that can just be distinguished from a uniform grey stimulus at the same illuminance, and the unit of measurement is the number of 'cycles' (black and white line pairs) per degree of visual angle. The visual acuity of the chicken using a variety of behavioural tests has been reported to vary between 1.5 and 8.6 cycles per degree, depending on the age of bird and the illuminance under which the test was conducted (Demello et al., 1992; Prescott et al., 2003; Gover et al., 2009). If a higher effort is required by the chicken, then their acuity appears to be greater, suggesting that they can (up to a point) adjust the amount of effort they invest in visual acuity (Demello et al., 1993). Using a wider range of target sizes and simultaneous discrimination techniques, Jarvis et al. (2009) reported that chickens had an acuity of about seven cycles per degree under bright light. Under dimmer light, their acuity dropped to five (Jarvis et al., 2009) or even three cycles (Gover et al., 2009) per degree. In contrast, humans can detect dark and light line pairs at up to 50 or 60 per degree, allowing much greater perception of image detail. In addition, the spatial acuity of chickens decreases more rapidly than that of humans as luminance levels fall.

Obtaining a clear and memorable image

The process by which an object can be viewed as a clearly focused image across varying distances is called accommodation. Accommodation is achieved by adjusting the shape of the lens and cornea to refract light, and the degree to which an eye can adjust its optical power in this way is called the accommodative range. Overall, the refractive power of the chicken eye is greater than that of humans and their accommodative range is good. For effective foraging, chickens must focus on small objects that are only a few centimetres distance from their eye. The necessary refraction of the eye is achieved by the lens thickening and the cornea bulging under the action of the ciliary muscles. The trick to being an effective chicken is maintaining a focus on potential food items on the ground in the lower visual field via a lateral extension of the area centralis, while simultaneously focusing on more distant stimuli in the upper visual field via a central extension of this same area. However, Dawkins (1995) noticed that chickens also appear to use the myopic lower visual field in social recognition, a theory she subsequently tested by examining the head positions of test chickens presented with familiar or unfamiliar companion birds. If the companion was at a distance of more than 0.7 m, the test bird used its lateral visual field, but at distances of less than 0.2 m, the test bird viewed the companion head on using the binocular lower visual field. The hens also displayed their general preference to be near familiar companions only once the companions had been inspected at close range (Dawkins, 1995, 1996).

The degree and type of light stimulation during rearing can affect the development of the eye, increasing intraocular pressure and affecting the shape and weight of the eye. Dim light, very short or long photoperiods, and continuous illumination all adversely affect the development of the eye and its ability to focus (Lewis and Gous, 2009). Ambient light of low illuminance is a particular risk factor for axial elongation and myopia in chicks (Cohen et al., 2008, 2011). It is interesting to note that the worldwide prevalence of human myopia has risen greatly during the past 30 years, in parallel with rapid urbanization and increased time spent under artificial light. Time spent outdoors during childhood is a protective factor against refractive problems in children (Rose et al., 2008) and may also have beneficial effects for chickens.

Clear visual images may be obtained by accommodation of the eye, but object recognition seems to depend also on the ability of hens to compare images taken from the same viewpoint. We have already described how chickens stabilize their heads to enable image fixation, but the approach behaviour, orientation and head movements of chickens all interact functionally in object inspection. Individual birds fixate objects at particular distances and angles, which they repeat during subsequent reinspections, building up a simplified set of views that can be compared to determine the nature of the object (Dawkins and Woodington, 2000). The head movements made by chickens may be a way of ensuring that all relevant aspects of an object are processed by the different specialized areas of the retina (Dawkins, 2002).

Use of polarized light

It has been suggested that some bird species may be able to detect polarized light and use this for navigation purposes. Sunlight is a form of unpolarized light comprising a sum of randomly oriented wave trains, resulting in a wave whose orientation also changes randomly. Skylight polarization arises when this unpolarized sunlight is scattered or bounced by matter in the Earth's atmosphere such that an overall spatial orientation can be discerned. The polarization pattern depends on the position of the sun. The maximum degree of polarization occurs in a circular band 90° from the sun (viewed from Earth). When the sun is at its highest point, this circle wraps around the horizon, and at twilight the circle wraps around the meridian. Polarized light is visible on partially cloudy days when the sun itself is hidden and can be used for navigation by bees and other insects. Despite earlier suggestions, it now seems that homing pigeons and other birds do not use information from skylight polarization directly for orientation, although they can use polarized light cues at twilight to calibrate their normal sun compass (Muheim, 2011). The ability of chickens to use polarized light has not been examined, although one of their relatives, the Japanese quail, failed to discriminate stimuli differing in their polarization pattern in a foraging task (Greenwood et al., 2003). However, chickens do possess a retinal double cone, which has been suggested to be a possible receptor for polarized light.

Pineal gland or 'third eye'

The pineal gland is a small endocrine gland found in most vertebrate species and is of interest here because the pinealocyte cells share similarities with the retinal photoreceptor cells of the eye. In birds, including chickens, the direct light sensitivity of pinealocytes is one way in which circadian rhythms and seasonal breeding patterns are regulated via melatonin release. Melatonin release can be detected as early as E13 (embryonic day 13) in the chick embryo, and circadian rhythms start to develop at this point (Hill et al., 2004). Broiler chicks have higher melatonin concentrations in the pineal gland than layer chicks, something that may be associated with their more confident early neonatal behaviour (Furuse, 2007).

Light preferences: spectrum, source and illuminance

Chickens are not just passive receivers of sensory stimulation. The importance of vision in their daily lives together with their refined sensory capacities make it likely that they will seek out or avoid certain types of light stimulation. Their ability to do so may be greatly constrained in the artificial conditions in which they are housed commercially, but evidence of their light preferences is important and may contribute to future improved lighting regimes.

Light spectrum preferences are highly sensitive to early developmental exposure. Broilers reared in white, red, green or blue light initially prefer the colour with which they are most familiar, although many birds later develop a preference for blue or green light (Prayitno et al., 1997a; Khosravinia, 2007). By 6 weeks of age, broilers that are initially indifferent to the light source show a clear preference for biolux fluorescent and warm-white fluorescent light over other fluorescent or incandescent sources (Kristensen et al., 2007). The biolux and warm-white sources provide an approximately similar contribution of all wavelengths between 350 and 700 nm, as does natural daylight. The preference for these daylight-mimicking sources persisted across a range of illuminance values adjusted for chicken vision.

The importance of early experience is exemplified by a study showing that adult hens subsequently exhibited a strong preference for incandescent light over natural light if they had been reared under incandescent light (Gunnarsson et al., 2008a). The junglefowl ancestry of chickens might lead them to prefer a relatively low light intensity rather than bright sunlight, but in this study, the Scandinavian daylight was received through a window and produced a relatively low light intensity of between 20 and 150 lx in the test environment. In contrast, Widowski et al. (1992) found that, even though they were reared under incandescent light, hens preferred fluorescent to incandescent light as adults.

It seems that while young (2-week-old) broiler and layer birds prefer high light intensities (200 lx), older birds prefer lower intensities (6 lx rather than 20, 60 or 200 lx) (Davis et al., 1999). The diminishing preference of older birds for bright light may cause particular problems in free-range or organic production where birds are expected to range outside. Rearing young birds with natural light can partially counteract the tendency of older birds to avoid bright daylight, with adult hens showing a slight preference for natural light over incandescent light during the afternoon (Gunnarsson et al., 2008b). A sensible commercial strategy followed by many organic producers is to give young birds experience of natural light during rearing.

Many new light sources (e.g. light-emitting diodes (LEDs)) are now being trialled in commercial production systems in an attempt to minimize energy inputs, with promising results from a production viewpoint (Rozenboim et al., 1998; Min et al., 2012). When housed under LED light, broiler chickens preferred cooler-white (blueish hue) to warmer-white hues (Riber, 2013). However, little else is known about chickens' preferences for new types of light source.

Influence of light spectrum and illuminance on broiler behaviour and welfare

The lighting of broiler chicken houses was traditionally directed towards the goal of providing sufficient light to encourage birds to feed (and hence grow) as fast as possible while minimizing electricity costs. This led to a system where birds were housed under near-continuous photoperiods but at low light intensities, both factors having some deleterious effects on bird behaviour and welfare. However, the EU Broiler Directive (Council Directive 2007/43/CE) now specifies that broilers must have a light intensity of at least 20 lx at bird eye level, across at least 80% of the useable area. Some of the evidence that led to this decision is reviewed next.

Generally, higher light intensities result in broiler chickens showing a more pronounced circadian activity rhythm and greater activity during the photoperiod (e.g. broilers reared at 50 or 200 lx compared with 5 lx (Blatchford *et al.*, 2009); broilers reared at 200 lx compared with 1 lx (Blatchford *et al.*, 2012)). The increased activity seen under higher light intensities can reduce the incidence of leg problems (Prayitno *et al.*, 1997b). Higher light intensities are also associated with increased behavioural synchrony in the flock and reduced interruptions to resting behaviour during the dark phase (Alvino *et al.*, 2009a). Broilers reared under illuminance levels of 1 lx (Deep *et al.*, 2012) or 5 lx (Alvino *et al.*, 2009b) tend to perform less preening and foraging than birds reared under higher light intensities. In a set-up where birds could move freely between compartments of differing light intensity, Davis *et al.* (1999) found that broilers performed more active behaviour, such as foraging, under the brightest light available (200 lx). Because of the often inverse relationship between activity and production, producers may not wish to increase activity too much. Another approach has therefore been to investigate the potential of providing a light intensity that varies across time. Kristensen *et al.* (2006) examined the effects of switching between 5 and 100 lx intensities every 2 or 4 h during a 16 h photoperiod. The broilers responded rapidly to a step-up in light intensity, with many birds running and flapping their wings as the light intensity increased. However, they also became less active during the 5 lx periods in comparison with birds kept at 5 lx all the time, so the overall activity levels were not greatly changed. Sherlock *et al.* (2010) also assessed the effect of alternating light intensity between 10 and 200 lx but found no differences in bird activity or leg health in comparison with a control group reared at a constant 10 lx. More positively, commercial flocks of broilers reared with natural light provided via windows at an average 85 lx performed less lying behaviour and had improved leg health compared with birds housed under artificial lighting at 11 lx (Bailie *et al.*, 2013). This effect may not be due entirely to the greater illuminance provided by natural light, as additional influences could be the more variable ambience and/or the broader daylight wavelength spectrum.

In studies where illuminance levels are equated (albeit usually in relation to human spectral sensitivity, not bird sensitivity), it seems that light wavelength has its own effects on bird behaviour and production. Broilers reared under red or white light, for example, show more activity than those reared under blue or green light (Prayitno *et al.*, 1997a) and their growth may be slightly reduced. There is increasing interest in maximizing production by rearing broilers under monochromatic green light until approximately day 10 and then switching to monochromatic blue light thereafter (Rozenboim *et al.*, 2004; Cao *et al.*, 2012).

It is clear that strategic adjustments in light intensity and wavelength can be used to influence broiler production, behaviour and welfare. However, a fundamental choice seems to be whether to maximize production by limiting bird activity or to prioritize welfare by encouraging activity and reducing leg health problems.

Influence of light spectrum and illuminance on laying hen behaviour and welfare

The sexual maturity of laying hens is stimulated by red or white light more than by blue or green, and particularly by a change from blue or green to red or white (Lewis and Morris, 2000). Laying hens preen more under the fluorescent light they prefer than under incandescent light (Widowski *et al.*, 1992), and they generally do more pecking and preening under high-intensity light (Vandenberg and Widowski, 2000; O'Connor *et al.*, 2011). Housing under 5 lx compared with 150 lx did not affect stress levels in hens during the early laying period (O'Connor *et al.*, 2011). Pullets housed under natural light prepare for night-time roosting earlier than pullets housed under artificial light of the same photoperiod (Gunnarsson *et al.*, 2008b).

Work on other birds suggests that light is likely to be an important influence for social discrimination in domestic fowl, although it is a little-studied subject. The fact that hens generally prefer to be in close proximity to familiar birds was used as a starting point for one study. D'Eath and Stone (1999) found that hens selected familiar feeding companions over unfamiliar birds only in bright white light (77 lx). In bright red or blue light, or dim light of any colour (adjusted according to spectral sensitivity so that illuminance was equivalent for all colours), little apparent social discrimination occurred. Light intensities of 1 lx or below may interfere more drastically with the recognition of individual birds and with assessment of relative rank. In an experiment where hens were able to inspect each other closely before competing for food, competitive behaviour was partially disrupted at 1 lx but not at illuminances of 5, 20 or 100 lx (Kristensen *et al.*, 2009).

Photoperiod and welfare: broilers

Broilers have traditionally been reared under long daylengths (photoperiods) with short dark periods (scotoperiods) in an effort to maximize their feeding and growth during their short lives. This practice has many adverse welfare consequences (Classen *et al.*, 1991; Sorensen *et al.*, 1999) and EU legislation now requires that, excluding the first 7 days of life and last 3 days before slaughter, lighting for broilers must 'follow a 24-hour rhythm and include periods of darkness lasting at least six hours in total, with at least one uninterrupted period of darkness of at least four hours, excluding dimming periods' (EC Council Directive 2007/43/EC). However, in many other countries of the world, the practice of housing broilers with photoperiods of 23 h or more continues.

The detrimental effects of a long photoperiod include increased bird mortality and lameness (Sanotra *et al.*, 2002; Knowles *et al.*, 2008). Recent Canadian studies reveal linear relationships between photoperiod (varying from 14 to 23 h), mortality and lameness (Schwean-Lardner *et al.*, 2013). Long photoperiods also result in increased eye size (Lewis and Gous, 2009; Schwean-Lardner *et al.*, 2013) due to a disruption of the normal circadian rhythm of eye growth. Paradoxically, reducing broiler growth by including a longer dark period can result in overall increased productivity, once mortality and losses due to disease are accounted for (Schwean-Lardner *et al.*, 2012a). During the dark period, birds are able to rest and sleep in an uninterrupted fashion, melatonin production is increased and healing can take place. More pronounced circadian activity rhythms can improve leg health and overall activity levels (Schwean-Lardner *et al.*, 2012b).

If a scotoperiod of at least 4 h is provided, then illuminance appears to have a greater effect on broiler behaviour and welfare than a further increase in the dark period (Blatchford *et al.*, 2012). There is no reason why the compulsory (under EU legislation) 6 h dark period must be provided continuously. In the first few weeks of life, chicks reared by broody hens make frequent excursions to forage and explore, returning for relatively short bursts of rest under the hen. There is increasing interest in the broiler industry in partially mimicking this pattern by providing a number of patterned dark periods totalling 6 h in every 24 h period. The effects of such intermittent lighting programmes have yet to be fully evaluated.

Object colour preferences and influences

When transmitted light hits a surface, and depending on the characteristics of that surface, some wavelengths are selectively absorbed and others are selectively reflected back. Humans categorize the reflected spectrum into groupings that we call colours, with qualities that depend on hue, saturation and brightness. Domestic chicks generally prefer to peck at food particles that we name as orange or red (reviewed by Ham and Osorio, 2007). The chicks' general preference for red grains or fruits is highly context specific. If odour or visual cues suggest that the food might be an insect, then the colour red is interpreted as a colour warning of potential toxicity and is strongly avoided (Rowe and Guilford, 1996; Gamberale-Stille and Tullberg, 2001). Obviously, the chicks do not use language to name colours, but there is some evidence that they categorize reflected wavelengths in similar ways to humans (Ham and Osorio, 2007). Once chicks are trained that yellow and red food particles (for example) are palatable, they also regard intermediate colours (orange in this case) as food. The chicks thus show interpolation, and prefer the interpolated colour more than would be predicted by normal generalization processes. This preference is not due simply to novelty, as the same categorization is not extended to colours beyond the limits of the rewarded colours, i.e. the chicks do not extrapolate, even if the spectral distance from the initial red or yellow is no greater than for the interpolated colour. Kelber and Osorio (2010) suggest that the known rewarded colours form a category in the birds' colour space.

The reflected colours of encountered objects can influence other aspects of chicken behaviour. Nest choice, for example, is influenced by exposure to specific colours in early life. Chicks exposed to the colours blue, green and red preferred yellow nests when in lay, whereas exposure to yellow resulted in an indifference towards the colour of the nest (Huber-Eicher, 2004; Zupan *et al.*, 2007). However, no preference for white, brown or black perches was discovered for hens kept in test cages (Chen and Bao, 2012).

Detection of artificial images

All manner of behavioural, social and cognitive research can be facilitated if animals react to repeatable visual stimuli presented as photographs or on TV or computer screens. There is no doubt that chickens react to some stimuli presented in this way. Video images of overhead hawk predators elicit alarm calling (Evans and Marler, 1991), and young chicks are attracted to video images of other chicks feeding (Clarke and Jones, 2001). Impressively, if junglefowl watch training videos of their companions finding food in coloured bowls, they are more likely to approach and feed from similarly coloured bowls themselves (McQuoid and Galef, 1992). In McQuoid and Galef's experiment, watching a film of another bird eating was more effective than simply observing the other bird standing close to the food bowl. There is also some evidence of social facilitation in response to video images, with an increasing chance that observers will show the same feeding or dust-bathing behaviour as the on-screen demonstrator (Keeling and Hurnik, 1993; Lundberg and Keeling, 2003). All of these results suggest that, under some circumstances, chickens perceive video images presented on traditional TV screens as equivalent to the real thing. However, chickens appear unable to discriminate familiar from unfamiliar birds using photographs (Dawkins, 1996) or video images (D'Eath and Dawkins, 1996).

To provide more information about the extent to which chickens perceive on-screen images as equivalent to real-world stimuli, formal discrimination tests can be used. In such tests, birds are required to discriminate two different images presented on a screen and then, without any further training, apply the learned discrimination to the equivalent real-world stimuli. In the reverse procedure, they first learn to discriminate between real stimuli and are then tested to see whether they can transfer this to screen images of those same stimuli. In early studies using conventional screens, simple real stimuli (red versus green objects) were treated as equivalent to their screen images, but this was not always the case for more complex stimuli. For example, hens did not show transfer between real and screen images in a hen versus no-hen discrimination task (Patterson-Kane *et al.*, 1997). It was also much quicker and easier to train a group of hens to discriminate between a real hen and a ball than to train a second group to discriminate between video images of these, something that occurred only after 500 trials or so (Patterson-Kane *et al.*, 1997). Transfer from real stimuli to screen image stimuli was poor in both groups, suggesting that complex video images presented on conventional screens are not regarded as fully equivalent to the real stimuli.

The flicker rate of the conventional cathode-ray screen may be partially responsible for poor transfer, as recent work shows that discrimination accuracy increases with a higher flicker rate (Railton *et al.*, 2010). Researchers today are likely to use high-definition plasma screens with high flicker rates, or flicker-free liquid crystal screens, to present images in behavioural studies with chickens. It is also important to ensure that chickens can view the images at a distance of less than 30 cm or thereabouts (Dawkins, 1996; Dawkins and Woodington, 1997). Following these guidelines seems to give good results, even in studies of detailed social communication. For example, hens react appropriately to playback of male tit-bitting sequences (a social behaviour described in full in Chapter 6) (Smith and Evans, 2009, 2011) and manipulations of male ornamentation (Smith *et al.*, 2009).

Hearing

Until very recently, there had been far more research on chicken vision than on their hearing capacities. However, this situation is now changing as there is increasing recognition of the role of sound in chicken behaviour and also some concern about the effects of the commercial sound environment on their welfare.

The point at which a minimum sound level can be detected against a silent background is known as the absolute threshold and it varies for different pure tones. Hearing thus depends on both the frequency (tone) and the pressure level of the sound. Other attributes of the sound, such as its duration, contribute to a subjective impression of loudness. In human hearing tests, the joint influences of frequency and sound pressure level are plotted together on an audiogram, and tone duration can also be varied by the tester. Often, the pressure level is A-weighted to account for the reduced sensitivity of the human ear at low frequencies, and is then reported as db(A).

Audiograms can be prepared for animals too, often using positive reinforcement techniques where the animal performs a response to receive a reward when it hears a tone. Temple *et al.* (1984) used a positive discrimination task to assess auditory thresholds in chickens using A-weighted sound pressure levels of between 45 and 80 dB, and frequencies of 260–8000 Hz. Hens could discriminate the lowest-frequency tones at around 40 dB and most other tones at about 20–40 dB. However, the 8000 Hz tones were too high to be detected at sound pressure levels that were bearable by the experimenters. Saunders and Salvi (1993) also used a positive reinforcement method with three adult white leghorn chickens. Their results were similar in that the birds were rather sensitive to pure tones within the range 400–4000 Hz, which could be detected at low sound pressure levels of between 11 and 20 dB (similar to a very quiet whisper). Frequencies outside this range could only be detected at higher sound pressure levels. Detection threshold also varied with stimulus duration.

The abilities of chickens in comparison with other avian species have been reviewed recently by Gleich and Langemann (2011). Chickens are less attuned to high frequency sounds than, say, owls or canaries, but they have a flatter threshold curve than many other galliform species. Figure 2.2, for example, shows that chickens can detect high- and low-frequency sounds at lower pressure levels than quail or turkeys, while being slightly less sensitive than quail at their best frequency.

The adeptness of chickens in detecting low-frequency sounds, including a capacity to detect sounds that humans cannot hear (hence called infra-sound at below 20 Hz; Warchol and Dallos, 1989) suggests that low-frequency perception is important. This may relate especially to the importance of low-frequency sound in the communication

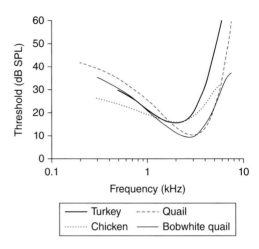

Fig. 2.2. Hearing sensitivity in chickens. Gleich and Langemann (2011) plotted the absolute hearing threshold sensitivities of chickens alongside those of three other galliform species. SPL, sound pressure level.

between hen and chicks. Sounds that attract and comfort small chicks are generally of low frequency below 800 Hz. The roosting call of the broody hen is perhaps the lowest-frequency vocalization of the chicken, described as a musical note of 'very low and slowly falling pitch in the range from 90 down to 70 c.p.s (in the second-lowest octave of the piano), and with all harmonics about equally strong up to 2000 c.p.s' (Collias and Joos, 1953). Today, cycles per second (c.p.s.) are generally reported as Hz.

Sound preferences and aversions

There is growing evidence that the typical ambient sound levels of commercial hen and broiler houses are both harmful and aversive to poultry. Chronic exposure to 80 dB(A) compared with 60 dB(A) in young laying hens led to more resting behaviour and reduced egg production (O'Connor *et al.*, 2011). In hens, higher sound levels of 80 or 90 dB increase indices of stress (Campo *et al.*, 2005) and are a risk factor for the early onset of feather pecking (Drake *et al.*, 2010). In broilers, both chronic and acute exposure to loud noises can increase indices of stress, such as the ratio of blood heterophils to lymphocytes (Bedanova *et al.*, 2010) and corticosterone concentrations (Chloupek *et al.*, 2009). Although lower noise levels appear less stressful (Bedanova *et al.*, 2010), chronic exposure to 70 dB is none the less associated with reduced weight gain (Voslarova *et al.*, 2011). Extensive cochlear hair-cell damage and loss occurs in commercial chicken flocks. The pathological signs of damage are far greater for broiler chickens kept in a commercial housing (average 90 dB) than for age-matched controls kept in a quieter environment (Smittkamp *et al.*, 2002; Smittkamp and Durham, 2004). Remarkably, broiler chickens also increase the volume of their own calls as the level of background noise increases (Brumm *et al.*, 2009). The understandable attempt by each individual chicken to be heard may aggravate the background noise level in commercial housing.

Evidence that loud sounds are aversive comes from work showing that hens will perform key-pecking tasks to turn off high-intensity pure tones of 105–110 dB(A) and a variety of taped mechanical and animal sounds at 90 dB(A) (McKenzie *et al.*, 1993). A slightly different test presents the chicken with two keys, each producing a food reward but with one key also initiating a test sound. If this sound is perceived neutrally, the chickens will peck both keys, but if the sound is perceived to be aversive, a pecking bias towards the non-sound key is observed. Using this method, McAdie *et al.* (1993) found that hens strongly avoided pecking a key that reproduced sound from a commercial poultry house at 100 dB(A). More recently, it has been found that chickens will avoid a chamber that they associate with previous exposure to white noise (Jones *et al.*, 2012).

Of course, not all sounds are aversive. Newly hatched chicks, like young human babies, find harmonic consonant music preferable to dissonant sound intervals (Chiandetti and Vallortigara, 2011), and young pullets reared with exposure to classical music appear to be less stressed than controls (Dávila *et al.*, 2011). However, this does not mean that constantly playing a radio to adult flocks, a common commercial practice, is always beneficial. Classical music at 75 dB led to increased fearfulness in old Spanish breeds (Campo *et al.*, 2005), and broiler chicks responded to music with head flicking (Christensen and Knight, 1975), more of an alerting response than an indication of relaxation. It may be important to acclimatize flocks to certain sounds when they are young, and to avoid excessive sound pressure levels.

Many of the investigations described so far have focused on sounds that are, whether generated from musical instruments, machinery or from the high stocking densities of commercial production, fundamentally anthropogenic. But what about the reaction of chickens to more natural sounds, including those with which domestic fowl may have co-evolved? As chicks find the warning colours, patterns and odours of some insect species repellent, it is reasonable to ask whether certain natural noises might also serve a warning function. In one study, the buzzing sound of a wasp seemed to facilitate avoidance of a striped warning pattern in chicks (Haglund *et al.*, 2006), but another study found no evidence that bumblebee buzzing affected the chicks' feeding decisions or food-related memory (Siddall and Marples, 2011).

Chemical Senses: Taste and Olfaction

Contrary to some reports, chickens have well-developed senses of smell and of taste (gustation), as well as a lesser-known chemical sense that responds to airborne or soluble irritants via the trigeminal nerve. The trigeminal nerve branches into sensitive free nerve endings distributed throughout the mucous membranes of the eyes, mouth and nasal cavity. The trigeminal system has a primary role in generating rapid protective responses to harmful or irritating chemical stimuli. Many behavioural studies of chicken olfaction may depend to an unacknowledged extent on a degree of participation of gustatory and trigeminal chemoreception systems, and vice versa, as there are strong inter-relationships among all of them. However, olfaction appears to be the dominant chemical sense, with the chicken genome containing at least 229 genes coding for olfactory receptors, but far fewer coding for genes related to human bitter or sweet taste receptors (Lagerstrom *et al.*, 2006).

Chicks are capable of detecting and reacting to odours on the day before hatching, once their beak has penetrated the egg sac. Once hatched, even sleeping chicks will react to odours presented on a cotton bud by head shaking and beak clapping (Porter *et al.*, 1999). The varied behavioural contexts in which chickens use their sense of olfaction were reviewed by Jones and Roper (1997). There is a reasonable body of evidence that chicks develop attachments to familiar odours when placed in an otherwise novel environment, whether the familiar odours are natural (soiled litter) or artificial (orange or geranium oil). In one experiment, the presence of a familiar odour increased chick activity and vocalization in a novel environment (Jones and Gentle, 1985), although whether this was due to increased or reduced fear is less certain. Chicks also use their olfactory sense to avoid predatory (cat) odour and, in some circumstances, the odour of conspecific blood (Fluck et al., 1996). A comprehensive set of experiments found that domestic fowl of an old-fashioned Swedish breed increased their vigilance behaviour more after exposure to the odour of faecal material from a predator (tiger or dhole) than after exposure to herbivore faecal material (elephant or antelope) (Zidar and Lovlie, 2012).

There is good evidence that chickens have individual body odours, deriving from the preen gland (uropygial gland), situated dorsally just in front of the tail feathers. These individual chicken odours can be clearly distinguished by trained mice (Karlsson et al., 2012). Chickens too can detect odours from the uropygial gland, as male birds with an intact sense of smell show preferential courtship behaviour towards females with an intact preen gland, whereas male birds who have lost their sense of smell show no preference in their courtship behaviour (Hirao et al., 2009). However, olfaction does not appear to be used by younger chicks in the development of social familiarity with their companions – visual cues seem precedent in this context (Porter et al., 2005). Jones and Roper (1997) speculated that chickens may emit body odours when they are stressed or frightened but were unable to find solid evidence in support at the time of their review.

More certainly, chickens use their sense of olfaction to guide their feeding behaviour. Young chicks with a blocked sense of smell are slower to gain weight than controls, for example (Porter et al., 2002). Experiments show that if vanilla or almond odour is reliably paired with bitter but odourless quinine water, the detectable cues are used by chicks to select the untainted drink. Novel odours are particularly effective in deterring or slowing chicks from eating food of a novel appearance (Marples and Roper, 1996), while some novel odours (orange oil) will even dissuade chicks from eating their normal familiar feed (Turro et al., 1994). Although especially effective when novel, the aversive properties of orange oil continue to

reduce feed consumption in chicks even after it has become familiar (Dixon et al., 2006). A functional explanation for this aversion to orange is lacking as there is no known association of orange oil with harmful or toxic prey items.

Unlike orange oil, odours that are reliably associated with harmful or toxic prey items could be perceived as innately aversive. Alternatively, they could act as part of a suite of information to influence the feeding behaviour of chickens. Experiments on chicks have examined the effects of odorifous compounds called pyrazines that are produced naturally by damaged, and often toxic, insects. Pyrazines seem relatively innocuous in the presence of familiar food, but they trigger hesitation or avoidance of feeding by young chicks presented with yellow or red food, suggesting a synergy in warning colour and warning odour effects (Rowe and Guilford, 1996). More recently, it has been found that pyrazines increase the speed of avoidance learning and the duration of avoidance memory for chicks presented with unpalatable yellow food (Siddall and Marples, 2008).

Chemical preferences and aversions

Just as the commercial poultry house may present chickens with a light and sound environment that is far from their preferred milieu, so it also presents a challenging chemical environment containing gases such as hydrogen sulphide and ammonia. Chickens react spontaneously to the odour of hydrogen sulphide and to the odour (or possibly the irritant properties) of ammonia (McKeegan et al., 2005). Humans should not be exposed to ammonia concentrations greater than 25 p.p.m. (parts per million) but there are no legal limits for farm animals. Assurance schemes suggest their own limits (e.g. 20 p.p.m. for broiler chickens by the RSPCA Freedom Food scheme), but in reality, levels often exceed this, sometimes greatly.

In an ingenious feat of agricultural engineering, Wathes et al. (2002) constructed a ring-shaped exposure chamber with multiple chambers each containing a different concentration of ammonia within a range of approximately 3–40 p.p.m. The chickens were free to move between compartments via plastic curtains and the number and duration of visits made to each compartment were recorded. Broilers and laying hens both preferred fresh air and spent less time in compartments containing ammonia at concentrations of approximately 10 p.p.m. or above. The aversion to ammonia was not influenced by familiarity, as birds reared with or without 20 p.p.m. of ammonia showed similar patterns of ammonia avoidance (Jones et al., 2005). A general finding was that the aversion to ammonia was not instantaneous, making it unlikely that the smell of this gas was responsible for the birds' dislike. An idea proposed by the authors was that detrimental effects of ammonia exposure accumulated over a period of up to an hour, inducing after this time a possible sensation of malaise and a desire to seek fresh air.

Another context in which chickens' responses to the chemical environment are highly relevant is the question of stunning using lethal gases prior to slaughter at commercial abattoirs. Because handling is avoided, gas stunning could potentially improve chicken welfare but only if the gases used are not experienced as highly aversive. McKeegan et al. (2006) examined the birds' initial responses to a range of potential stunning gases by allowing them to feed and then observing their behaviour before, during and after a 10 s administration of either carbon dioxide in air or nitrogen, or a mix of inert argon with nitrogen. There was little reaction to the inert gases, but broiler chickens reacted to carbon dioxide, at concentrations of 10% or more, with head shaking and increased respiration effort. Despite this, although the experimental set-up enabled birds to move away to avoid the gas, such avoidance was rarely observed. Sandilands et al. (2011) also examined the aversion of chickens to gas mixtures including carbon dioxide in air, argon in carbon dioxide or nitrogen in carbon dioxide, each at low, medium or high concentrations, by seeing how much each gas interrupted the birds' feeding behaviour. All the gas mixtures were aversive, but the inert mixtures were less so than the carbon dioxide in air. Decisions about whether to use gas mixtures to stun or kill poultry, whether for routine or emergency purposes, cannot be made solely on the basis of the chicken's initial sensory reaction. Factors such as the speed with which the gas renders the chicken unconscious and the extent to which chickens react in ways that may cause physical damage are also relevant.

Taste preferences and aversions in chickens are difficult to detect, as food can be accepted or rejected based on its physical characteristics or novelty, or by post-ingestive assessment of nutritional value. Broiler chicks can detect and prefer diets containing a balance of amino acids, and this may partly be accomplished by oral sensing of the umami taste (Roura et al., 2013). Adult hens eat a very wide range of vegetable and animal-based foodstuffs and have few marked taste aversions. They have a specific appetite for calcium-rich foods (a mineral needed in abundance for eggshell formation), but the relative role of calcium taste perception (Tordoff et al., 2008) versus assessment of post-ingestive consequences in mediating this food preference is not known. Chickens possess fewer taste buds than most mammals, and show little preference for sugars in feed, and a preference for moderate but an aversion to strong salt solutions (McKeegan, 2004). The absence of the human gene that detects sweetness suggests that chickens may be a sweet-insensitive species (Roura et al., 2013). Sometimes the lack of aversion shown by chickens is surprising, for example, their apparent indifference to the properties of chilli peppers (Furuse et al., 1994). However, taste aversions can be demonstrated experimentally, particularly for bitter substances. Quinine solution, for example, has no odour, and food that is artificially adulterated with this substance is readily ingested by naïve chickens. Typically, on first tasting this bitter substance, chickens shake their heads, wipe their beaks and avoid the same food on future occasions. As quinine is harmless, this provides a useful way of training chickens in learning experiments to eat some foods but not others (Nicol and Pope, 1996). Hens are attracted to peck at and eat feathers that are coated with oil from the uropygial gland (McKeegan and Savory, 2001). To counteract this tendency, it has been suggested that spraying feathers with quinine solution could be a way of training hens not to peck at the feathers of their conspecifics (Harlander-Matauschek et al., 2009; Harlander-Matauschek and Rodenburg, 2011).

Magnetoreception

The possibility that some bird species might use the Earth's magnetic field for orientation and navigation has long been discussed, but, despite circumstantial evidence in favour of this hypothesis in homing pigeons, the usual behavioural techniques used in sensory biology have provided limited evidence of a magnetic sense. Operant experiments showed that pigeons could react to a magnetic anomaly, but more subtle discrimination of magnetic field direction would be required if this sense was to be used to provide compass or map information. Given that chickens are not known for their great migratory or navigational powers, it is surprising that the first demonstration of sensitivity to the direction of a magnetic field came from young layer-strain chicks (Freire *et al.*, 2008). If chicks are imprinted on a red ball, they will try to follow and seek that ball when it is hidden. Layer-strain chicks can use magnetic cues to find the ball, making correct responses on a significant proportion of tests and shifting their choices by 90° when the magnetic field is rotated (Fig. 2.3). Their magnetic ability appears to develop sometime during the second week after hatching (Denzau *et al.*, 2013; Munro *et al.*,

2014) and they seem particularly able to use field intensities that approximate those found in nature (Wiltschko *et al.*, 2007). Clearly, this ability is not possessed by humans, but more surprisingly it is either not present, or not utilized, by broiler-strain chicks (Freire *et al.*, 2008).

Exactly how chickens detect magnetic field and intensity is an area of active investigation. Some detection may take place using a protein called cryptochrome, which is present in the retina and which depends also on the presence of visible light. However, iron-containing subcellular particles of maghemite and magnetite have been found in the skin lining the upper beak of homing pigeons, migratory birds and chickens (Falkenberg *et al.*, 2010), so a magnetic sense could reside here. However, it remains unclear exactly how these particles are used in functional short-distance location, which may depend on the ability

Fig. 2.3. (a) A chick, imprinted on a red ball, is placed in a transparent start box. (b) The chick is trained to locate the ball by heading north. (c) The fact that the chick is using magnetic cues is demonstrated when the magnetic field is shifted by 90° by running an electric current through 2 m diameter copper coils that surround the white test box. (d) The chick now searches behind screens on the east–west axis. (Photos courtesy of Dr Raf Freire, Charles Sturt University, Australia.)

to make use of cues relating to magnetic field intensity, direction and inclination. Anaesthesia of the beak does not disrupt chick orientation (Wiltschko *et al.*, 2007), but beak trimming (as practised routinely in many agricultural systems) did reduce the time chicks spent with a magnetic stimulus that had previously been associated with food (Freire *et al.*, 2011).

References

Alvino, G.M., Archer, G.S. and Mench, J.A. (2009a) Behavioural time budgets of broiler chickens reared in varying light intensities. *Applied Animal Behaviour Science* 118, 54–61.

Alvino, G.M., Blatchford, R.A., Archer, G.S. and Mench, J.A. (2009b) Light intensity during rearing affects the behavioural synchrony and resting patterns of broiler chickens. *British Poultry Science* 50, 275–283.

Bailie, C.L., Ball, M.E.E. and O'Connell, N.E. (2013) Influence of the provision of natural light and straw bales on activity levels and leg health in commercial broiler chickens. *Animal* 7, 618–626.

Bedanova, I., Chloupek, J., Chloupeck, P., Knotkova, Z., Voslarova, E., Pistekova, V. and Vecerek, V. (2010) Responses of peripheral blood leukocytes to chronic intermittent noise exposure in broilers. *Berliner and Munchener Tierartztliche Wochenschrift* 123, 186–191.

Blatchford, R.A., Klasing, K.C., Shivaprasad, H.L., Wakenell, P.S., Archer, G.S. and Mench, J.A. (2009) The effect of light intensity on the behaviour, eye and leg health and immune function of broiler chickens. *Poultry Science* 88, 20–28.

Blatchford, R.A., Archer, G.S. and Mench, J.A. (2012) Contrast in light intensity, rather than day length, influences the behaviour and health of broiler chickens. *Poultry Science* 91, 1768–1774.

Bowmaker, J.K. and Knowles, A. (1977) Visual pigments and oil droplets of chicken retina. *Vision Research* 17, 755–764.

Brumm, H., Schmidt, R. and Schrader, L. (2009) Noise-dependent vocal plasticity in domestic fowl. *Animal Behaviour* 78, 741–746.

Campo, J.L., Gil, M.G. and Dávila, S.G. (2005) Effects of specific noise and music stimuli on stress and fear levels of laying hens of several breeds. *Applied Animal Behaviour Science* 91, 75–84.

Cao, J., Wang, Z., Dong, Y., Zhang, Z., Li, J., Li, F. and Chen, Y. (2012) Effect of combinations of monochromatic lights on growth and productive performance of broilers. *Poultry Science* 91, 3013–3018.

Chen, D.H. and Bao, J. (2012) General behaviors and perching behaviors of laying hens in cages with different colored perches. *Asian-Australasian Journal of Animal Sciences* 25, 717–724.

Chiandetti, C. and Vallortigara, G. (2011) Chicks like consonant music. *Psychological Science* 22, 1270–1273.

Chloupek, P., Voslarova, E., Chloupeck, J., Bedanova, I., Pistekova, V. and Vecerek, V. (2009) Stress in broiler chickens due to acute noise exposure. *Acta Veterinaria Brno* 78, 93–98.

Christensen, A.C. and Knight, A.D. (1975) Observations on the effects of music exposure to growing performance of meat-type chicks. *Poultry Science*, 54, 619–621.

Clarke, C.H. and Jones, R.B. (2001) Domestic chicks' runway responses to video images of conspecifics. *Applied Animal Behaviour Science* 70, 285–295.

Classen, H.L., Riddell, C. and Robinson, F.E. (1991) Effects of increasing photoperiod length on performance and health of broiler chickens. *British Poultry Science* 32, 21–29.

Cohen, Y., Belkin, M., Yehezkel, O., Avni, I. and Polat, U. (2008) Light intensity modulates corneal power and refraction in the chick eye exposed to continuous light. *Vision Research* 48, 2329–2335.

Cohen, Y., Belkin, M., Yehezkel, O., Solomon, A.S. and Polat, U. (2011) Dependency between light intensity and refractive development under light-dark cycles. *Experimental Eye Research* 92, 40–46.

Collias, N. and Joos, M. (1953) The spectrographic analysis of sound signals of the domestic fowl. *Behaviour* 5, 175–188.

Dávila, S.G., Campo, J.L., Gil, M.G., Prieto, M.T. and Torres, O. (2011) Effects of auditory and physical enrichment on 3 measurements of fear and stress (tonic immobility duration, heterophil to lymphocyte ratio, and fluctuating asymmetry) in several breeds of layer chicks. *Poultry Science* 90, 2459–2466.

Davis, N.J., Prescott, N.B., Savory, C.J. and Wathes, C.M. (1999) Preferences of growing fowls for different light intensities in relation to age, strain and behaviour. *Animal Welfare* 8, 193–203.

Dawkins, M.S. (1995) How do hens view other hens – the use of lateral and binocular visual-fields in social recognition. *Behaviour* 132, 591–606.

Dawkins, M.S. (1996) Distance and social recognition in hens: implications for the use of photographs as social stimuli. *Behaviour* 133, 663–680.

Dawkins, M.S. (2002) What are birds looking at? Head movements and eye use in chickens. *Animal Behaviour* 63, 991–998.

Dawkins, M.S. and Woodington, A. (1997) Distance and the presentation of visual stimuli to birds. *Animal Behaviour* 54, 1019–1025.

Dawkins, M.S. and Woodington, A. (2000) Pattern recognition and active vision in chickens. *Nature* 403, 652–655.

D'Eath, R.B. and Dawkins, M.S. (1996) Laying hens do not discriminate between video images of conspecifics. *Animal Behaviour* 52, 903–912.

D'Eath, R.B. and Stone, R.J. (1999) Chickens use visual cues in social discrimination: an experiment with coloured lighting. *Applied Animal Behaviour Science* 62, 233–242.

Deep, A., Schwean-Lardner, K., Crowe, T.G., Fancher, B.I. and Classen, H.L. (2012) Effect of light intensity on broiler behaviour and diurnal rhythms. *Applied Animal Behaviour Science* 136, 50–56.

Demello, L.R., Foster, T.M. and Temple, W. (1992) Discriminative performance of the domestic hen in a visual-acuity task. *Journal of the Experimental Analysis of Behavior* 58, 147–157.

Demello, L.R., Foster, T.M. and Temple, W. (1993) The effect of increased response requirements on discriminative performance of the domestic hen in a visual-acuity task. *Journal of the Experimental Analysis of Behavior* 60, 595–609.

Denzau, S., Niessner, C., Rogers, L.J. and Wiltschko, W. (2013) Ontogenetic development of magnetic compass orientation in domestic chickens (*Gallus gallus*). *Journal of Experimental Biology* 216, 3143–3147.

Dixon, G., Green, L.E. and Nicol, C.J. (2006) Effect of diet change on the behaviour of chicks of an egg-laying strain. *Journal of Applied Animal Welfare Science* 9, 41–58.

Drake, K.A., Donnelly, C.A. and Dawkins, M.S. (2010) Influence of rearing and lay risk factors on propensity for feather damage in laying hens. *British Poultry Science* 51, 725–733.

Evans, C.S. and Marler, P. (1991) On the use of video images as social stimuli in birds – audience effects on alarm calling. *Animal Behaviour* 41, 17–26.

Falkenberg, G., Fleissner, G., Schuchardt, K., Kuehbacher, M., Thalau, P., Mouritsen, H., Heyers, D., Wellenreuther, G. and Fleissner, G. (2010) Avian magnetoreception: elaborate iron mineral containing dendrites in the upper beak seem to be a common feature of birds. *PLoSOne* 5, e9231.

Fluck, E., Hogg, S., Mabbutt, P.S. and File, S.E. (1996) Behavioural and neurochemical responses of male and female chicks to cat odour. *Pharmacology Biochemistry and Behavior* 54, 85–91.

Freire, R., Munro, U., Rogers, L.J., Sagasser, S., Wiltschko, R. and Wiltshko, W. (2008) Different responses in two strains of chickens in a magnetic orientation test. *Animal Cognition* 11, 547–552.

Freire, R., Eastwood, M.A. and Joyce, M. (2011) Minor beak trimming in chickens leads to a loss of mechanore-ception and magnetoreception. *Journal of Animal Science* 89, 1201–1206.

Furuse, M. (2007) Behavioral regulators in the brain of neonatal chicks. *Animal Science Journal* 78, 218–232.

Furuse, M., Nakajima, S.-I., Miyagawa, S., Nakagawa, J. and Okumura, J.-I. (1994) Feeding behaviour, abdominal fat and laying performance in laying hens given diets containing red pepper. *Japanese Poultry Science* 31, 45–52.

Gamberale-Stille, G. and Tullberg, B.S. (2001) Fruit or aposematic insect? Context-dependent colour preferences in domestic chicks. *Proceedings of the Royal Society B – Biological Sciences* 268, 2525–2529.

Gleich, O. and Langemann, U. (2011) Auditory capabilities of birds in relation to the structural diversity of the basilar papilla. *Hearing Research* 273, 80–88.

Gover, N., Jarvis, J.R., Abeyesinghe, S.M. and Wathes, C.M. (2009) Stimulus luminance and the spatial acuity of domestic fowl. *Vision Research* 49, 2747–2753.

Greenwood, V.J., Smith, E.L., Church, S.C. and Partridge, J.C. (2003) Behavioural investigation of polarisation sensitivity in the Japanese quail (*Coturnix coturnix japonica*) and the European starling (*Sturnus vulgaris*). *Journal of Experimental Biology* 206, 3201–3210.

Gunnarsson, S., Heikkila, M., Hultgren, J. and Valros, A. (2008a) A note on light preferences in layer pullets reared in incandescent or natural light. *Applied Animal Behaviour Science* 112, 395–399.

Gunnarsson, S., Heikkila, M. and Valros, A. (2008b) Effect of day length and natural versus incandescent light on perching and the diurnal rhythm of feeding behaviour in layer chicks (*Gallus g. domesticus*). *Acta Agriculturae Scandinavica Section A – Animal Science* 58, 93–99.

Ham, A.D. and Osorio, D. (2007) Colour preferences and colour vision in poultry chicks. *Proceedings of the Royal Society B – Biological Sciences* 274, 1941–1948.

Harlander-Matauschek, A., Beck, P. and Piepho, H.-P. (2009) Taste aversion learning to eliminate feather pecking in laying hens, *Gallus gallus domesticus*. *Animal Behaviour* 78, 485–490.

Harlander-Matauschek, A. and Rodenburg, T.B. (2011) Applying chemical stimuli on feathers to reduce feather pecking in laying hens. *Applied Animal Behaviour Science* 132, 146–151.

Haug, K., Hagen, S.B. and Lampe, H.M. (2006) Responses of domestic chicks (*Gallus gallus domesticus*) to multi-modal aposematic signals. *Behavioral Ecology* 17, 392–398.

Hill, W.L., Bassi, K.L., Bonaventura, L. and Sacus, J.E. (2004) Prehatch entrainment of circadian rhythms in the domestic chick using different light regimes. *Developmental Psychobiology* 45, 174–186.

Hirao, A., Aoyama, M. and Sugita, S. (2009) The role of uropygial gland on sexual behaviour in domestic chicken, *Gallus gallus domesticus*. *Behavioural Processes* 80, 115–120.

Huber-Eicher, B. (2004) The effect of early colour preference and of a colour exposing procedure on the choice of nest colours in laying hens. *Applied Animal Behaviour Science* 86, 63–76.

Jarvis, J.R., Taylor, N.R., Prescott, N.B., Meeks, I. and Wathes, C.M. (2002) Measuring and modelling the photopic flicker sensitivity of the chicken (*Gallus g. domesticus*). *Vision Research* 42, 99–106.

Jarvis, J.R., Abeyesinghe, S.M., McMahon, C.E. and Wathes, C.M. (2009) Measuring and modelling the spatial contrast sensitivity of the chicken (*Gallus g.domesticus*). *Vision Research* 49, 1448–1454.

Jones, A.R., Bizo, L.A. and Foster, T.M. (2012) Domestic hen chicks' conditioned place preferences for sound. *Behavioural Processes* 89, 30–35.

Jones, E.K.M., Prescott, N.B., Cook, P., White, R.P. and Wathes, C.M. (2001) Ultraviolet light and mating behaviour in domestic broiler breeders. *British Poultry Science* 42, 23–32.

Jones, E.K.M., Wathes, C.A. and Webster, A.J.F. (2005) Avoidance of atmospheric ammonia by domestic fowl and the effect of early experience. *Applied Animal Behaviour Science* 90, 293–308.

Jones, R.B. and Gentle, M.J. (1985) Olfaction and behavioural modification in domestic chicks (*Gallus domesticus*). *Physiology and Behavior* 34, 917–924.

Jones, R.B. and Roper, T.J. (1997) Olfaction in the domestica fowl: a critical review. *Physiology and Behavior* 62, 1009–1018.

Karlsson, A.C., Jensen, P., Elgland, M., Laur, K., Fyrner, T., Konradsson, P. and Laska, M. (2012) Red junglefowl have individual body odors. *Journal of Experimental Biology* 213, 1619–1624.

Keeling, L.J. and Hurnik, J.F. (1993) Chickens show socially facilitated feeding behaviour in response to a video image of a conspecific. *Applied Animal Behaviour Science* 36, 223–231.

Kelber, A. and Osorio, D. (2010) From spectral information to animal colour vision: experiments and concepts. *Proceedings of the Royal Society B – Biological Sciences* 277, 1617–1625.

Khosravinia, H. (2007) Preference of broiler chicks for color of lighting and feed. *Journal of Poultry Science* 44, 213–219.

Knowles, T.G., Kestin, S.C., Haslam, S.M., Brown, S.N., Green, L.E., Butterworth, A., Pope, S.J., Pfeiffer, D. and Nicol, C.J. (2008) Leg disorders in broiler chickens: prevalence, risk factors and prevention. *PLoS One* 3, e1545.

Kristensen, H.H., Aerts, J.M., Leroy, T., Wathes, C.M. and Berckmans, D. (2006) Modelling the dynamic activity of broiler chickens in response to step-wise changes in light intensity. *Applied Animal Behaviour Science* 101, 125–143

Kristensen, H.H., Prescott, N.B., Perry, G.C., Ladewig, J., Ersbøll, A.K., Overvad, K.C. and Wathes, C.M. (2007) The behaviour of broiler chickens in different light sources and illuminances. *Applied Animal Behaviour Science* 103, 75–89.

Kristensen, H.H., White, R.P. and Wathes, C.M. (2009) Light intensity and social communication between hens. *British Poultry Science* 50, 649–656.

Lagerstrom, M.C., Hellstrom, A.R., Gloriam, D.E., Larsson, T.P., Schioth, H. and Fredriksson, R. (2006) The G protein-coupled receptor subset of the chicken genome. *PLoS Computational Biology* 2, 493–507.

Lewis, P.D. and Gous, R.M. (2009) Photoperiodic conditions of broilers. II. Ocular development. *British Poultry Science* 50, 667–672.

Lewis, P.D. and Morris, T.R. (2000) Poultry and coloured light. *World's Poultry Science Journal* 56, 189–207.

Lewis, P.D., Perry, G.C. and Morris, T.R. (2000) Ultraviolet radiation and laying pullets. *British Poultry Science* 41, 131–135.

Lisney, T.J., Rubene, D., Rozsa, J., Lovlie, H., Hastad, O. and Odeen, A. (2011) Behavioural assessment of flicker fusion frequency in chicken *Gallus gallus domesticus*. *Vision Research* 51, 1324–1332.

Lisney, T.J., Ekesten, B., Tauson, R., Hastad, O. and Odeen, A. (2012) Using electroretinograms to assess flicker fusion frequency in domestic hens *Gallus gallus domesticus*. *Vision Research* 62, 125–133.

Lundberg, A.S. and Keeling, L.J. (2003) Social effects on dustbathing behaviour in laying hens: using video images to investigate effect of rank. *Applied Animal Behaviour Science* 81, 43–57.

Maddocks, S.A., Cuthill, I.C., Goldsmith, A.R. and Sherwin, C.M. (2001) Behavioural and physiological effects of absence of ultraviolet wavelengths for domestic chicks. *Animal Behaviour* 62, 1013–1019.

Marples, N.M. and Roper, T.J. (1996) Effects of novel colour and smell on the response of naïve chicks towards food and water. *Animal Behaviour* 51, 1417–1424.

Martin, G.R. (2009) What is binocular vision for? A birds' eye view. *Journal of Vision* 9, article 14.

McAdie, T.M., Foster, T.M., Temple, W. and Matthews, L.R. (1993) A method for measuring the aversiveness of sounds to domestic hens. *Applied Animal Behaviour Science* 37, 223–238.

McKeegan, D.E.F. (2004) Sensory perception: chemoreception. In: Perry, G.C. *Welfare of the Laying Hen*. CABI, Wallingford, UK.

McKeegan, D.E.F. and Savory, C.J. (2001) Feather eating in individually caged hens which differ in their propensity to feather peck. *Applied Animal Behaviour Science* 73, 131–140.

McKeegan, D.E.F., Smith, F.S., Demmers, T.G.M., Wathes, C.M. and Jones, R.B. (2005) Behavioral correlates of olfactory and trigeminal gaseous stimulation in chickens, *Gallus domesticus. Physiology and Behavior* 84, 761–768.

McKeegan, D.E.F., McIntyre, J., Demmers, T.G.M., Wathes, C.M. and Jones, R.B. (2006) Behavioural responses of broiler chickens during acute exposure to gaseous stimulation. *Applied Animal Behaviour Science* 99, 271–286.

McKenzie, J.G., Foster, T.M. and Temple, W. (1993) Sound avoidance by hens. *Behavioural Processes* 30, 143–156.

McQuoid, L.M. and Galef, B.G. (1992) Social stimuli influencing feeding behaviour of Burmese fowl – a video analysis. *Animal Behaviour* 46, 13–22.

Min, J.K., Hossan, M.S., Nazma, A., Jae, C.N., Han, T.B., Hwan, K.K., Dong, W.K., Hyun, S.C., Hee, C.C. and Ok, S.S. (2012) Effect of monochromatic light on sexual maturity, production performance and egg quality of laying hens. *Avian Biology Research* 5, 69–74.

Muheim, R. (2011) Behavioural and physiological mechanisms of polarized light sensitivity in birds. *Philosophical Transactions of the Royal Society B – Biological Sciences* 366, 763–771.

Munro, U., Luu, P., DeFilippis, L. and Freire, R. (2014) Ontogeny of magnetoreception in chickens (*Gallus gallus domesticus*). *Journal of Ethology* 32, 69–74.

Nagel, T. (1974) What is it like to be a bat? *Philosophical Review* 83, 435–450.

Nicol, C.J. and Pope, S.J. (1996) The maternal feeding display of domestic hens is sensitive to perceived chick error. *Animal Behaviour* 52, 767–774.

Nuboer, J.F.W., Coemans, M.A.J.M. and Vos, J.J. (1992) Artificial lighting in poultry houses: do hens perceive the modulation of fluorescent lamps as flicker? *British Poultry Science* 33, 123–133.

O'Connor, E.A., Parker, M.O., Davey, E.L., Grist, H., Owen, R.C., Szladovits, B., Demmers, T.G.M., Wathes, C.M. and Abeyesinghe, S.M. (2011) Effect of low light and high noise on behavioural activity, physiological indicators of stress and production in laying hens. *British Poultry Science* 52, 666–674.

Patterson-Kane, E., Nicol, C.J., Foster, T.M. and Temple, W. (1997) Limited perception of video images by domestic hens. *Animal Behaviour* 53, 951–963.

Porter, R.H., Hepper, P.G., Bouchot, C. and Picard, M. (1999) A simple method for testing odor detection and discrimination in chicks. *Physiology and Behavior* 67, 459–462.

Porter, R.H., Picard, M., Arnould, C. and Tallet, C. (2002) Chemosensory deficits are associated with reduced weight gain in newly hatched chicks. *Animal Research* 51, 337–345.

Porter, R.H., Roelofsen, R., Picard, M. and Arnould, C. (2005) The temporal development and sensory mediation of social discrimination in domestic chicks. *Animal Behaviour* 70, 359–364.

Prayitno, D.S., Phillips, C.J.C. and Omed, H. (1997a) The effects of color of lighting on the behavior and production of meat chickens. *Poultry Science* 76, 452–457.

Prayitno, D.S., Phillips, C.J.C. and Stokes, D.K. (1997b) The effects of color and intensity of light on behaviour and leg disorders in broiler chickens. *Poultry Science* 76, 1674–1681.

Prescott, N.B. and Wathes, C.M. (1999) Spectral sensitivity of the domestic fowl (*Gallus g. domesticus*). *British Poultry Science* 40, 332–339.

Prescott, N.B., Wathes, C.M. and Jarvis, J.R. (2003) Light, vision and the welfare of poultry. *Animal Welfare* 12, 269–288.

Railton, R.C.R., Foster, T.M. and Temple, W. (2009) A comparison of two methods for assessing critical flicker fusion frequency in hens. *Behavioural Processes* 80, 196–200.

Railton, R.C.R., Foster, T.M. and Temple, W. (2010) Transfer of stimulus control from a TFT to CRT screen. *Behavioural Processes* 85, 111–115.

Riber, A. (2013) Colour temperature of LED lighting used in broiler housing – preference and effects on performance. In: *Proceedings of the 9th European Symposium on Poultry Welfare*, Uppsala, Sweden, p. 99.

Rosa Salva, O., Rugani, R., Cavazzana, A., Regolin, L. and Vallortigara, G. (2013) Perception of the Ebbinghaus illusion in four-day-old domestic chicks (*Gallus gallus*). *Animal Cognition* 16, 895–906.

Rose, K.A., Morgan, I.G., Ip, J., Kifley, A., Huynh, S., Smith, W. and Mitchell, P. (2008) Outdoor activity reduces the prevalence of myopia in children. *Opthalmology* 115, 1279–1287.

Roura, E., Guzman-Pino, S.A. and Fu, M. (2013) The taste system from chickens to humans: a common link in search for a nutritionally balanced diet. In: *Proceedings of the 34th Western Nutrition Conference – Processing, Performance & Profit*, Saskatoon, Saskatchewan, pp. 29–41.

Rowe, C. and Guilford, T. (1996) Hidden colour aversions in domestic chicks triggered by pyrazine odours of insect warning displays. *Nature* 383, 520–522.

Rozenboim, I., Zilberman, E. and Gvaryahu, G. (1998) New monochromatic light source for hens. *Poultry Science* 77, 1695–1698.

Rozenboim, I., Biran, I., Chaiseha, Y., Yahav, S., Rosenstrauch, A., Sklan, D. and Halevy, O. (2004) The effect of a green and blue monochromatic light combination on broiler growth and development. *Poultry Science* 83, 842–845.

Rubene, D., Håstad, O., Tauson, R., Wall, H. and Ödeen, A. (2010) Presence of UV wavelengths improves the temporal resolution of the avian visual system. *Journal of Experimental Biology* 213, 3357–3363.

Sandilands, V., Raj, A.B.M., Baker, L. and Sparks, N.H.C. (2011) Aversion of chickens to various lethal gas mixtures. *Animal Welfare* 20, 253–262.

Sanotra, G.S., Lund, J.D. and Vestergaard, K.S. (2002) Influence of light-dark schedules and stocking density on behaviour, risk of leg problems and occurrence of chronic fear in broilers. *British Poultry Science* 43, 344–354.

Saunders, J.E., Jarvis, J.R. and Wathes, C.M. (2008) Calculating luminous flux and lighting levels for domesticated mammals and birds. *Animal* 2, 921–932.

Saunders, S.S. and Salvi, R.J. (1993) Psychoacoustics of normal adult chickens: thresholds and temporal integration. *Journal of the Acoustical Society of America* 94, 83–90.

Schwean-Lardner, K., Fancher, B.I. and Classen. H.L. (2012a) Impact of daylength on behavioural output in commercial broilers. *Applied Animal Behaviour Science* 137 43–52.

Schwean-Lardner, K., Fancher, B.I. and Classen H.L. (2012b) Impact of daylength on the productivity of two commercial broiler strains. *British Poultry Science* 53, 7–18.

Schwean-Lardner, K., Fancher, B.I., Gomis, S., Van Kessel, A., Dalal, S. and Classen, H.L. (2013) Effect of day length on cause of mortality, leg health and ocular health in broilers. *Poultry Science* 92, 1–11.

Sherlock, L., Demmers, T.G.M., Goodship, A.E., Mccarthy, I.D. and Wathes, C.M. (2010) The relationship between physical activity and leg health in the broiler chicken. *British Poultry Science* 51, 22–30.

Siddall, E.C. and Marples, N.M. (2008) Better to be bimodal: the interaction of color and odor on learning and memory. *Behavioral Ecology* 19, 425–432.

Siddall, E.C. and Marples, N.M. (2011) Hear no evil: the effect of auditory warning signals on avian innate avoidance, learned avoidance and memory. *Current Zoology* 57, 197–207.

Smith, C.L. and Evans, C.S. (2009) Silent tidbitting in male fowl, *Gallus gallus*: a referential visual signal with multiple functions. *Journal of Experimental Biology* 212, 835–842.

Smith, C.L. and Evans, C.S. (2011) Exaggeration of display characteristics enhances detection of visual signals. *Behaviour* 148, 287–305.

Smith, C.L., Van Dyk, D.A., Taylor, P.W. and Evans, C.S. (2009) On the function of an enigmatic ornament: wattles increase the conspicuousness of visual displays in male fowl. *Animal Behaviour* 78, 1433–1440.

Smith, E.L., Greenwood, V.J., Goldsmith, A.R. and Cuthill, I.C. (2005) Effect of supplementary ultraviolet lighting on the behaviour and corticosterone levels of Japanese quail chicks. *Animal Welfare* 14, 103–109.

Smittkamp, S.E. and Durham, D. (2004) Contributions of age, cochlear integrity and auditory environment to avian cochlear nucleus metabolism. *Hearing Research* 195, 79–89.

Smittkamp, S.E., Colgan, A.L., Park, D.L., Girod, D.A. and Durham, D. (2002) Time course and quantification of changes in cochlear integrity observed in commercially raised broiler chickens. *Hearing Research* 170, 139–154.

Sorensen, P., Su, G. and Kestin, S.C. (1999) The effect of photoperiod:scotoperiod on leg weakness in broiler chickens. *Poultry Science* 78, 336–342.

Temple, W., Foster, T. and Odonnell, C.S. (1984) Behavioral estimates of auditory thresholds in hens. *British Poultry Science* 25, 487–493.

Tordoff, M.G., Shao, H., Alarcon, L.K., Margolskee, R.F., Mosinger, B., Bachmanov, A.A., Reed, D.R. and McCaughey, S.A. (2008) Involvement of T1R3 in calcium-magnesium taste. *Physiological Genomics* 34, 338–348.

Turro, I., Porter, R.H. and Picard, M. (1994) Olfactory cues mediate food selection by young chicks. *Physiology and Behavior* 55, 761–767.

Vandenberg, C. and Widowski, T.M. (2000) Hens' preferences for high-intensity high-pressure sodium or low-intensity incandescent lighting. *Journal of Applied Poultry Research* 9, 172–178.

Voslarova, E., Chloupek, P., Chloupek, J., Bedanova, I., Pistekova, V. and Vecerek, V. (2011) The effects of chronic intermittent noise exposure on broiler chicken performance. *Animal Science Journal* 82, 601–606.

Warchol, M.E. and Dallos, P. (1989) Neural response to very low-frequency sound in the avian cochlear nucleus. *Journal of Comparative Physiology A – Sensory Neural and Behavioral Physiology* 166, 83–95.

Wathes, C.M., Jones, J.B., Kirstensen, H.H., Jones, E.K.M. and Webster, A.J.F. (2002) Aversion of pigs and domestic fowl to atmospheric ammonia. *Transactions of the ASAE* 45, 1605–1610.

Widowski, T.M. and Duncan, I.J.H. (1996) Laying hens do not have a preference for high-frequency versus low-frequency compact fluorescent light sources. *Canadian Journal of Animal Science* 76, 177–181.

Widowski, T.M., Keeling, L.J. and Duncan, I.J.H. (1992) The preferences of hens for compact fluorescent over incandescent lighting. *Canadian Journal of Animal Science* 72, 203–211.

Wiltschko, W., Freire, R., Munro, U., Ritz, T., Rogers, L., Thalau, P. and Wiltschko, R. (2007) The magnetic compass of domestic chickens, *Gallus gallus. Journal of Experimental Biology* 210, 2300–2310.

Wood-Gush, D.G.M. (1971) *The Behaviour of the Domestic Fowl.* Heinemann Educational Books, London.

Wortel, J.F., Rugenbrink, H. and Nuboer, J.F.W. (1987) The photopic spectral sensitivity of the dorsal and ventral retinae of the chicken. *Journal of Comparative Physiology A – Sensory Neural and Behavioral Physiology* 160, 151–154.

Zidar, J. and Lovlie, H. (2012) Scent of the enemy: behavioural responses to predator faecal odour in the fowl. *Animal Behaviour* 84, 547–554.

Zupan, M., Kruschwitz, A. and Huber-Eicher, B. (2007) The influence of light intensity during early exposure to colours on the choice of nest colours by laying hens. *Applied Animal Behaviour Science* 105, 154–164.

3

Development of the Brain and Behaviour

The Chicken Brain

Views about the structure and function of bird brains have changed fundamentally in the past 10 years, not so much in terms of essential structure but certainly in terms of our understanding of how their brain structure supports diverse neural functions. The basic structure of the bird brain is shared with other vertebrates and is formed from five divisions. The spinal cord ascends to the hindbrain (myelencephalon and metencephalon), the midbrain (mesencephalon) and the forebrain (diencephalon and telencephalon). The functions of the hindbrain and midbrain (e.g. in controlling respiration, reflex movements, blood pressure and water balance) are relatively conserved among all vertebrate groups. In most vertebrates, including the chicken, the diencephalon develops into the thalamus and hypothalamus regions. The hypothalamus secretes hormones and regulates autonomic processes, while the thalamus is best characterized as a relay station involved in the transfer of information between brain regions. Every sensory system except for the olfactory system includes a thalamic nucleus (a concentration of neurons) that receives sensory signals and then forwards them to the various higher forebrain regions. In these respects, the avian brain does not differ fundamentally from the brains of other vertebrates, but substantial and significant differences are seen between birds and other vertebrates in the extent and nature of differentiation of the telencephalon. Our understanding of the

chicken brain continues to advance but has not been helped by considerable historical confusion. A clear understanding about how the anatomy of the bird brain contributes to its many functions has been clouded by an old and misconceived view of evolution as a ladder of progress or a 'scala naturae'. This view portrays fish with their primitive forebrains languishing at the bottom of the ladder, while amphibians, reptiles and birds have achieved higher positions by accumulating new forebrain structures (given names such as the paleocortex and archicortex). In this hierarchy, these vertebrates merely herald the ascent of the mammals, with their massively expanded and differentiated telencephalon. Humans, with their complex, layered and folded neocortex supporting higher thought and reason, are always placed at the top of this hierarchy.

Appealing as this view may be, it ran into increasing difficulties, especially as applied to the evolution of the avian brain. By the second half of the 20th century, the problems were brought into ever sharper focus as gene expression, pathway tracing and behavioural studies proliferated. It became apparent that certain brain regions in birds received and processed visual, auditory and somatosensory input from the thalamus, in a manner very similar to the mammalian neocortex. Additionally, regions of the avian brain once thought to be unique to birds (e.g. the so-called hyperstriatum) turned out not to have uniquely avian functions but rather were involved in control processes supporting motor actions and both simple and complex learning in ways that closely paralleled the mammalian brain.

The mismatch between the historic terminology describing the brain structure of birds and the growing evidence of functional similarities with mammalian brains led, in 2004, to a completely new terminology being adopted. The Avian Brain Nomenclature Consortium concluded that the avian telencephalon is organized into three main, developmentally distinct domains that are homologous in fish, amphibians, reptiles, birds and mammals, and proposed that these should be described as the pallial, striatal and pallidal domains (Fig. 3.1).

The striatal and pallidal domains and their derivatives (the basal ganglia and amygdala) share common structural, developmental and functional features among all extant vertebrates. GABAergic neurons are the predominant and defining neuron type of these two domains. In contrast to the conserved structure and function of the striatal and pallidal domains, the pallial domains of the different vertebrate classes are far more diverse. The avian pallium is organized into four main subdivisions, the hyperpallium and the dorsal ventricular ridge, comprising the mesopallium, nidopallium and arcopallium. The hyperpallium has a unique semi-layered organization that has been found only in birds, while the compact clusters of neurons (nuclei) that typify the grey matter composition of the dorsal ventricular ridge are shared by birds and reptiles. A different form of layered neuronal organization and a unique six-layered cortex is unique to the mammalian telencephalon.

In birds, like mammals, the adult pallium comprises about 75% of the telencephalic volume. The nidopallium in particular supports an array of higher cognitive functions in birds feted for their intelligence (e.g. parrots, crows and ravens) but also in the humble chicken, as we will explore in later chapters.

One interesting question is whether birds, and chickens in particular, possess the neural

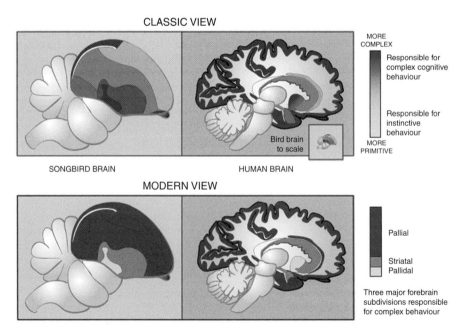

Fig. 3.1. Diagrams of classic and modern views of the bird brain. The traditional view of the avian brain (top row) was that it lacked the types of structure that could support complex cognition. The diagram of the bird brain on the left seems to suggest that birds lack brain structures that might be equivalent to the mammalian brain cortex. The modern view is presented in the diagrams at the bottom. Here, large areas of the avian pallium are seen to have some functional equivalence with the mammalian cortex (Jarvis *et al.*, 2005). (From http://avianbrain.org/new_terminology.html; Erich Jarvis, Duke University Medical Center, NC, USA, and Zina Deretsky, National Science Foundation, VA, USA.)

substrates that might support consciousness. After all, if chickens are nothing more than feathered robots, reacting to their environments without experiencing any sensations or feelings, then we probably do not need to worry about their welfare. The topic of conscious experience is far too broad to consider in any detail here, but in humans at least, primary conscious experience (the feeling of seeing something, for example) appears to depend on a rapid relay of information between the thalamus and cortical regions. All healthy mammals and birds (at least those beyond a certain stage of embryonic development) possess the neural circuit patterns that should support similar types of experience. In contrast, the corresponding circuits in reptiles lack some components, and in amphibians and fish there is an absence of the major circuits and cell clusters responsible for the rapid relay of information between the thalamus and pallium (Butler and Cotterill, 2006; Butler, 2008). In short, based on our current understanding of neural function, there is no reason to rule out the possibility of primary conscious experience in chickens – even if we cannot know its subjective nature.

The complex chicken brain forms during a short pre-hatching period. The speed, just 21 days, and predictability of the development process and the fact that chick embryos can be examined far more easily than mammalian embryos or fetuses, means that a great deal of our general knowledge about vertebrate brain development comes from studies of domestic chicks. We will consider some of the findings from this large body of research in the next section, focusing particularly on those aspects of brain development that are important for post-hatching behaviour.

Pre-hatching Development

The development of the chicken brain is best regarded as a complex interaction between genetic and environmental influences. The egg is far from a sealed unit, and multiple environmental stimuli exert their effects. The nature and timing of these sensory inputs result in an integration of the chick's senses by the time it hatches. Certain forms of stimulation during embryonic development induce changes in behaviour, while others alter the timing of development, leading for example to synchronized hatching. It is important to appreciate therefore that development proceeds according to many self-correcting processes that are continually modified by environmental influences.

With that caveat in mind, there remains an essentially predictable pattern to the development of the chick's nervous system. Four key processes are involved: proliferation of cells in the anterior part of the neural tube forming the rapidly expanding telencephalon, differentiation of neurons from their stem-cell precursors, migration of immature neurons to their final destination and synapse formation. In terms of timing, peak neuron formation occurs in the chick at embryonic day 8 (E8), and there is a peak in the formation of synaptic connections between neurons in the forebrain at day E15. Spontaneous movements are seen in the embryo during this period, and increased co-ordination of these movements signal, by day E16, the onset of pre-hatching behaviour. By day E17, sensory connections are more established, neurotransmission begins and the embryo acquires the ability to react and even learn from sensory inputs and their consequences. On day E18, the chick gets into a tucked position with its head turned to the left side of the body, its beak positioned against the air sac membrane, its left eye hidden by the body and its right eye adjacent to the air sac membrane. Far more detail on the timings and interactions of these developmental processes has been given by Rogers (1996).

The egg has a permeable structure that allows volatile compounds to penetrate the shell and other membranes. Such compounds could potentially influence the behaviour of the incubating chicks. Bertin et al. (2010) found that prenatal exposure to low (but not high) concentrations of a blended orange and vanilla odour between days E12 and E20 resulted in chicks preferring food containing the same odour when they were approximately 1 week old. The timing of this influence was subsequently examined in more detail by presenting the odour either before day E16 (when the chicks are surrounded by and swallowing amniotic fluid) or after day E17 (when the chicks have penetrated the air sac membrane with their beaks and have begun to breathe with their lungs and vocalize). This work showed that olfactory stimulation towards the end of incubation had the greater effect (Bertin et al., 2012).

Exposure to sounds during the embryonic stages increases the responsiveness of chicks to the same sounds after hatching and assists in the integration of the entire sensory system. Falt (1981) found that the pre-hatch experience of an individual maternal call facilitated discrimination of the mother but was not absolutely essential to later auditory imprinting. Auditory stimulation is increasingly effective in promoting sensory development from day E10 onwards (Roy *et al.*, 2013).

Brain Lateralization

Some of the most far-reaching and profound discoveries that have emerged from studies of chick development relate to the lateralization of the chicken brain such that specialized functions are accomplished by each hemisphere. Lateralization was once considered a unique feature of the human brain but is now known to be widespread among birds and mammals, and is a particularly strong guiding influence on chicken behaviour. In mammals, a connecting structure called the corpus callosum mediates communication and information transfer between the hemispheres. No such structure is present in birds. Although a degree of visual information transfer may occur via crossing optic nerve pathways, it is likely that the two hemispheres function relatively independently in chickens. In this section, we will briefly cover some of the mechanisms that lead to brain lateralization in the pre-hatch period and then discuss some of the functional implications for later chicken behaviour.

Chicks hatched from eggs incubated in darkness have symmetrical neuron development (from the retina to a collection of nuclei in the thalamus), but chicks exposed to light during the latter phases of incubation have asymmetrical neuron development that persists for 2–3 weeks post-hatching. Light penetrating the eggshell affects the anatomical development of the eye. As chicks lie with their right eye uppermost, the right eye receives more light stimulation and becomes larger and heavier than the left. After hatching, the refractive ability of the two eyes differs. However, the most profound effect of light exposure is the legacy of lateralized brain function, even after the structural asymmetry has largely disappeared. The process of lateralization has been examined in many experiments where the embryo is deliberately exposed to light, for as

little as 2 h, on day E19. Of course, in the natural situation, a broody hen would be sitting on the eggs for the majority of the time, but even in this situation there are opportunities for light to penetrate the eggshell. Sometimes the hen stands and turns the eggs, and broody hens also occasionally leave the nest, to drink or eat briefly. As some lateralization is observed even in chicks that are incubated in total darkness (Andrew *et al.*, 2004; Chiandetti *et al.*, 2005), it is likely that the natural incubation process provides sufficient light to support a greater degree of lateralization, although to exactly what extent is not fully known. Some findings relating to hemispheric specialization may depend on experimentally imposed light exposure, while in other cases experimental pre-hatching light exposure has no influence on lateralized function (Wichman *et al.*, 2009a).

This discrepancy is explained partly by the fact that light is not the only factor influencing brain lateralization. The hormonal environment of the embryo also has an important influence. If levels of testosterone or oestrogen are relatively low, greater lateralization is observed (Rogers and Rajendra 1993), and lower corticosterone concentrations also result in greater lateralization (Freire *et al.*, 2006). As female embryos have higher levels of oestrogen, and there is little difference in testosterone levels between male and female embryos pre-hatching, the brain and eye lateralization of male chicks is generally more pronounced than that of females. This should be remembered when perusing research papers, which generally focus on male birds, especially if trying to apply results to adult female laying hens.

The principle advantage of a lateralized brain appears to be the ability to perform specialized foraging while also remaining attentive to environmental cues signalling danger or novelty. The multi-tasking abilities of chicks can be studied when chicks are using just one eye by fitting a dark patch over the other eye during experiments. Chicks using their right eye learn rapidly to discriminate grains of food from small pebbles, while chicks using the left eye continue to peck indiscriminately at both grain and pebbles (Rogers, 1997). Chicks using their left eye are more easily distracted by novel events and stimuli, and react faster and more strongly to predatory stimuli appearing on this side (Dharmaretnam and Rogers, 2005). This is partly because chicks using their right eye are more able to inhibit premature pecks and other inappropriate responses, but also because the left eye (right hemisphere) is more involved in the response to novelty (Rogers *et al.*, 2007). The division of labour between hemispheres means that lateralized chicks can better perform simultaneous feeding and vigilance (Rogers *et al.*, 2004). The more pronounced lateralization of male chicks is highlighted by their performance in observational learning tasks. Male chicks trained with both eyes but tested with one eye covered succeed at an observational learning task (avoiding pecking at a bitter-tasting bead) only when using their left eye, whereas female chicks are competent using either eye (Rosa-Salva *et al.*, 2009).

Although the importance of lateralization was first framed in terms of its enabling effect for simultaneous foraging and vigilance, the full effects of lateralization may be more profound and general. The right hemisphere appears to be specialized to detect and respond to novel stimuli and to control a range of fear, flight and avoidance responses. One way of characterizing this division of labour is to argue that the right hemisphere is responsive to global cues in a reactive manner, while the left hemisphere is more involved in proximal cues during proactive (internally motivated or scheduled) behaviour (Tommasi *et al.*, 2003). The two hemispheres also take differing roles in the control of spatial behaviour, such that in a working memory task domestic chicks generally use both hemispheres to process object and position information. However, in a variety of food location tasks, and particularly when there is a conflict between different cues, the right hemisphere is specialized for processing geometric spatial information including distal and positional cues from the wider environment (Rashid and Andrew, 1989; Regolin *et al.*, 2005). Meanwhile, chicks using their right eye and left hemisphere pay particular attention to object-specific cues and proximal or landmark cues (Tommasi and Vallortigara, 2001; Freire and Rogers, 2005; Regolin *et al.*, 2005; Della Chiesa *et al.*, 2006a,b).

An interesting aspect of lateralization is that changes in the chick's behaviour coincide with changes in the hemispheric dominance. On day 10 post-hatching, there is a shift towards control of behaviour by the right hemisphere (Rogers and Ehrlich, 1983), and this is also the age at which domestic chicks reared in semi-natural conditions begin to move out of sight of the mother hen.

In the laboratory, providing domestic chicks with the ability to move out of sight of an imprinting stimulus, by using opaque screens, has been found to lead to improved orientation in a detour test and better retrieval of a visually displaced goal (Freire *et al.*, 2004). When a conflict is established between the use of proximal cues (e.g. markings on a screen) and distal cues (outside the apparatus), chicks with experience of opaque screens on days 10–11 show an improvement in their ability to make use of the distal information. The same experience of opaque screens provided at day 8, when the left hemisphere is dominant, does not lead to an improvement in the use of distal spatial cues (Freire and Rogers, 2007).

Social behaviour is also influenced by pre-hatch light exposure and lateralization. It has long been known that chicks using the left eye are better, for example, at distinguishing familiar from unfamiliar conspecifics (Vallortigara and Andrew, 1994). Male chicks hatched from eggs incubated in darkness form a more flexible group structure, with a less stable social hierarchy than light-exposed chicks (Rogers and Workman, 1989; Wichman *et al.*, 2009b), such that lower-ranked chicks are more likely to gain access to food in a competitive situation. Possibly, a hierarchy is more stable in a group composed of individuals all having the same light-induced asymmetry.

Before we get too carried away by the array of functions that are influenced by a lateralized brain, it is worth noting that pre-hatching light exposure has not been found to affect the ability of chicks to imprint (Andrew *et al.*, 2004) or their preference for familiar versus unfamiliar conspecifics (Deng and Rogers, 2002), and initial reports that light exposure increases a tendency to feather peck in young birds (Riedstra and Groothuis, 2004) have not been confirmed (Wichman *et al.*, 2009b).

The relative ease of investigating and manipulating the degree of lateralization of visual processing in the chick may partly obscure the fact that the processing of information from other sensory modalities can also be lateralized. For example, chicks respond more to olfactory cues processed by their right nostril (left hemisphere) than their left (Burne and Rogers, 2002). Their magnetic sense, too, is affected by eye use. As we have seen in Chapter 2, chicks are able to locate a hidden red ball or similar object on which they have been imprinted, using cues from experimentally imposed magnetic fields. Their ability to use magnetic cues is greater when they use their right eye, as shown in experiments where the direction of the magnetic field is shifted. Chicks using their right eye oriented according to magnetic cues, whereas chicks using the left eye appeared to prioritize distal geographical cues and continued to choose the training direction (Rogers *et al.*, 2008). Whether other bird species do the same is a subject of current debate (Hein *et al.*, 2011).

Hormone-mediated Maternal Effects on Development

There has been much interest in possible epigenetic influences in vertebrate development, whereby gene expression and phenotype are affected by mechanisms other than the direct contribution of genetic material. In chickens and the closely related quail (which have been more studied in this context), there is good evidence that mother hens affect the development of their chicks by transferring hormones into the follicles that develop into egg yolk and nourish the chicks. Such hormones influence embryonic development and can produce long-lasting effects on subsequent adult patterns of behaviour.

Corticosterone

Corticosterone is an important regulatory steroid hormone, released from the adrenal gland as part of a complex stress response. The possibility that maternal stress could influence chick development is theoretically interesting and of practical importance, but it is a difficult area to study. One problem has been technical issues whereby some assays for corticosterone have cross-reacted with other steroid hormones such as progesterone (Rettenbacher *et al.*, 2009; Quillfeldt *et al.*, 2011). This problem means that a cautious interpretation must be applied to studies that estimate yolk or albumen corticosterone using radio-immunoassay or high-performance liquid chromatography techniques, for example the finding that white layer birds deposit more corticosterone in their eggs than brown birds (Navara and Pinson, 2010) or the finding that direct injection of corticosterone into hens increases albumen corticosterone levels (Downing and Bryden, 2008).

However, rather more specific enzyme immunoassays confirm that corticosterone is present in the yolk of eggs laid by commercial chickens (Singh *et al.*, 2009) and in the yolk of quail eggs (Okuliarova *et al.*, 2010). Experimentally increasing corticosterone levels at E1 *reduces* chick growth rate and body symmetry (Eriksen *et al.*, 2003), so it is slightly unexpected that the yolks of the fastest-growing broiler strains contain significantly more corticosterone than the yolks of slower-growing breeds (Ahmed *et al.*, 2013). Possibly, the fast-growth broiler-breeder females are more stressed due to chronic food restriction and this raises the corticosterone levels of their eggs (by either an active or a passive process, which we shall explore later). If so, the artificial selection for growth rate must have been so strong in these strains that it has overcome the usual growth-inhibiting effects of corticosterone.

Heightened corticosterone levels do not simply influence growth; they can have far-reaching effects on the developing chick. Early embryonic exposure to heightened corticosterone levels can increase fearfulness, reduce competitive ability and inhibit learning (Janczak *et al.*, 2006; Davis *et al.*, 2008). Experimental increases in corticosterone at later stages of embryonic development can also have profound effects, with one study reporting reduced neuronal asymmetry and brain lateralization (Rogers and Deng, 2005) and another reporting that chicks subsequently gave more distress vocalizations in response to isolation in a novel arena, showed reduced sleep and took longer to detect a predatory stimulus in vigilance tests (Freire *et al.*, 2006). The same effects were not seen if the chicks were directly injected with corticosterone on day 1 post-hatching, emphasizing the potency of the effects of corticosterone before chicks hatch.

The injection of steroids into an egg by an experimenter does not necessarily mimic the natural process by which mother hens influence their chicks' development. Such processes will have been influenced by natural selection, and there are good evolutionary arguments to suspect that maternal strategies that prepare chicks for the post-hatch environment they will face will have been favoured. If the post-hatch environment is risky and dangerous, then more fearful chick phenotypes may do better than less fearful ones, for example. The quality of the post-hatch environment may be most clearly signalled to the developing embryos by the stress status of their mother, perhaps via the deposition of steroid hormones in the yolk (Groothuis *et al.*, 2005). There are potential costs as well as benefits in this strategy, the main risk being that the environment may change again before the chicks hatch. However, maternal stress level does seem to have some influence on developing chicks. A fascinating question is whether this process is passive (in which case, we would expect a direct relationship between maternal stress level and yolk corticosterone concentration) or active (with the mother bird playing a role in adjusting the corticosterone levels of the egg somewhat independently of her own hormone levels).

Evidence for a passive relationship in birds comes from studies that have experimentally increased the concentration of maternal plasma corticosterone directly (via implants) or indirectly (by subjecting mother birds to stressful conditions) and then looked at the resultant changes in yolk corticosterone concentrations. For example, Okuliarova *et al.* (2010) subjected quail to chronic restraint stress and found that maternal corticosterone entered the yolk during the earliest stages of egg formation. In the chronically stressed birds but not in the controls, additional corticosterone was passed to the yolk during the last stages of egg formation. A separate study of barn owls (*Tyto alba*) found that implants that increased the concentration of maternal plasma corticosterone within a normal physiological range also increased the corticosterone in incubated eggs (Almasi *et al.*, 2012). However, not all studies have found a clear relationship between maternal and egg corticosterone concentrations (Henriksen *et al.*, 2011a). For example, Janczak *et al.* (2007) reported that unpredictable feed restriction increased hens' faecal corticosterone metabolite concentrations (as expected) but had no significant effect on the concentrations of corticosterone detected in their eggs. However, when eggs from feed-restricted and control hens were artificially incubated, the chicks from feed-restricted hens subsequently became more fearful adults, suggesting effects mediated by other hormonal pathways. As corticosterone can downregulate gonadal hormones and reduce the progesterone and testosterone concentrations in eggs (Henriksen *et al.*, 2011b), there is more than one mechanism by which maternal stress could influence offspring behaviour. Inconsistent results in this relatively new field of

research could also arise from the intriguing possibility that hens may be able (partially) to control the amount of corticosterone that they deposit in the egg. Such active adjustment would eliminate any simple relationship between maternal stress, egg corticosterone levels and subsequent chick behaviour, opening up a far more complex area of research.

Overall, at this early stage, the balance of evidence suggests that there is often a positive relationship between avian plasma hormone concentrations and egg yolk hormone levels (Groothuis and Schwabl, 2008), but we are not yet able to conclude whether these are regulated independently. In addition, recent work shows that direct manipulation of maternal corticosterone level has an effect on offspring behaviour, even when chicks are not hatched or reared by their mothers. Eggs produced while maternal corticosterone was elevated were smaller, and the chicks that hatched were less successful at stealing a mealworm from a companion. Surprisingly, however, these same chicks were less fearful in a tonic immobility test (Henriksen *et al.*, 2013), showing that it is difficult to predict exactly how the prenatal environment may affect offspring behaviour. A variety of other indirect effects of maternal stress also remain to be investigated, for example whether corticosterone levels alter the incubation behaviour of female birds, and whether altered incubation patterns in turn affect chick behaviour. Research on rodents reveals that paternal stress can also be a significant factor influencing offspring development, via epigenetic mechanisms such as altered microRNA content in sperm (Rodgers *et al.*, 2013), and this will undoubtedly become an area for future chicken work.

Gonadal hormones

Corticosterone is far from the only hormone that has an influence on chicken development. Indeed, chick embryos develop within a complex hormonal environment containing a mix of maternal gonadal hormones. Unlike corticosterone, these hormones can be deposited directly into the egg and so circulating blood levels of maternal hormone are not likely accurately to reflect egg deposition levels. The exact concentrations of these hormones in the egg are influenced by the initial level of maternal deposition (which is in turn influenced by factors such as maternal age or experience; Guibert *et al.*, 2012), incubation time, the sex of the embryo and the time at which embryonic gonads start to produce their own hormones.

It has been proposed that mother hens could influence the sex of their offspring, possibly by yolk hormones influencing the segregation of sex-determining chromosomes (female chickens have one Z and one Y chromosome, while males have two Z chromosomes) at meiosis. Early studies failed to make much progress in investigating this possibility, as hormone concentrations were examined in eggs that had already started the incubation process. This is a potential problem because the hormone levels surrounding male and female embryos could change during incubation due, for example, to sex-specific differences in the rate of usage of androgens in the yolk or, after some days of development, to embryonic gonadal activity. Thus, to examine the possible role of the initial maternal allocation of hormones in sex determination, recent studies have examined unincubated eggs. However, no differences in initial gonadal hormone levels have been detected in eggs containing male or female embryos (quail: Pilz *et al.*, 2005; chickens: Aslam *et al.*, 2013). It seems, therefore, that maternal hormone allocation has no role in chromosomal sex determination, but it may be involved in sexual differentiation, which starts at around day E4. Even though initial maternal hormone levels do not appear to differ for male and female embryos, maternal hormones are likely to affect male and female embryos in different ways.

During a short critical period in chick development, it is the oestrogens that play a determining and irreversible role in triggering the development of female chicks and feminized patterns of brain development. Male chickens exposed to oestradiol as embryos show greatly reduced copulatory, aggressive (Clifton and Andrew, 1989) and even crowing (Marx *et al.*, 2004) behaviours as adults, while inhibition of oestradiol synthesis in female embryos encourages the development of masculine behaviour in female birds. Maternal oestrogens rather than maternal androgens seem to be the main hormones involved in sex differentiation (Eising *et al.*, 2003).

Later in development, maternal hormones, particularly the androgens, influence the manner in which the embryonic brain develops and is organized. Maternal hormones continue to influence development and subsequent chick behaviour, but this influence is increasingly constrained by the preceding neural organization.

The interacting and competing effects of the complex hormonal environment on the development of chick behaviour are difficult to disentangle. There is no doubt that steroidal hormones influence chick behaviour. Direct intramuscular injection of testosterone increased persistence of search behaviour but reduced the ability of young chicks to adjust to new situations, for example (Andrew and Rogers, 1972). Androgenic effects on behaviour also arise via early embryonic exposure, for example, experimental injection of testosterone into the yolk produces a short post-hatch increase in fear in quail chicks (Okuliarova *et al.*, 2007). By the time testosterone-exposed chicks are a few days of age, they seem to show a bolder profile, with a faster approach to novel objects, fewer distress calls and faster performance in a detour test than untreated controls (Daisley *et al.*, 2005). The effects of yolk manipulation experiments can produce longer-lasting effects with altered patterns of male social behaviour apparent when quail chicks were 8 weeks of age (Schweitzer *et al.*, 2013). In wild birds, there are competing costs and benefits of varying testosterone yolk concentration on offspring survival. Higher testosterone in young birds can stimulate growth but suppress immune function, a balance of functional costs and benefits that may explain the wide variation in androgen levels within and among clutches under natural conditions (Groothuis *et al.*, 2005).

Increasing the steroid hormone levels of her eggs may be of more value to some mothers than others. Heavier mothers deposit more androgens in the yolks of their eggs (Aslam *et al.*, 2013), and socially dominant mothers selectively deposit androgens in the yolks of male chicks (Muller *et al.*, 2002). The behavioural traits associated with higher androgen levels may be of most benefit to male offspring that will, as adults, have more variable reproductive success than females. However, unequivocal and comprehensive evidence that mother hens allocate steroid hormones strategically in this way is not currently available.

The stress level of the mother hen is another factor that influences the allocation of gonadal hormones to the egg, but again results are contradictory. Stress in maternal quail sometimes increases and sometimes decreases egg yolk androgen levels, possibly because mild stressors increase egg testosterone while more serious or longer-term stressors decrease levels of this hormone. Providing female quail with a place to hide from regular disturbances (presumably decreasing maternal stress) decreased egg testosterone concentrations and resulted in offspring that were subsequently less fearful (Guesdon et al., 2011). In contrast, Bertin et al. (2008) found that the provision of regular handling (presumably decreasing maternal stress once the birds had habituated) increased egg testosterone levels and resulted in offspring that were subsequently more fearful. Perhaps the most interesting finding in this area to date comes from work that genetically selected for quail that deposited more testosterone in their eggs. This was achieved despite no rise in the mothers' own plasma levels, suggesting a degree of active transfer (Okuliarova et al., 2011).

Maternal effects and commercial production

There are potential practical applications in considering maternal effects in the context of commercial poultry production. The parent birds of both broilers and laying hens are housed in large flocks but often in environments that differ substantially from those in which the offspring will be reared. The presence of male birds in the breeding flocks but their absence in commercial laying hen flocks is just one example of a potential mismatch between the environment in which the parental strains produce the fertile eggs and the environment in which the offspring are kept. Fear and stress levels in both white- and brown-strain parent flocks vary greatly, and a negative relationship has been detected between maternal stress level and egg size (de Haas et al., 2013). In addition, if the female breeding birds are stressed by their living environment, then, as we have seen above, this will affect the hormonal environment in which the embryos develop, whether for better or worse.

The extent to which parental stress levels are relevant to commercial practice requires investigation. A promising and comprehensive approach to this has been taken by de Haas et al. (2014).

These authors examined the long-term influences of maternal stress levels on the development of commercial laying hens. Chicks originating from ten different parent flocks were separated into 47 different rearing flocks and exposed to housing conditions that differed in litter availability. Although paternal stress levels and concentrations of hormones in eggs were not examined in this study, the behaviour and physiology of the mother hens were studied in detail. Maternal plasma corticosterone concentration varied from less than 0.5 to over 1.4 ng ml^{-1} and was positively correlated with maternal plumage damage. Overall, in white-strain but not in brown-strain birds, maternal stress was significantly and positively associated with the fearfulness and feather pecking activities of young chicks up to 5 weeks of age. In this study, the maternal influence seemed to fade as the chicks grew older. None the less, this important work demonstrates the relevance of fundamental studies to commercial practice and will surely stimulate further research on the longevity and relative importance of maternal influence.

If future research confirms that hens actively deposit hormones in the egg as part of a strategy that 'guesses' that chicks will hatch in a similar environment, this has important implications for the commercial environment, where chicks are reared separately from hens.

Hatching

Embryos communicate with each other, making clicks and beak clapping that synchronize hatching. The development of an isolated egg can be accelerated by exposing it to clicking sounds from a loudspeaker (Vince, 1973). Vigorous head movements produce the first cracks in the eggshell and then the chick appears to rest for around 24 h. Hatching comprises a series of back thrusts of the beak, with the chick using its egg tooth (a small, sharp protuberance that develops from the edge of the material of the top beak) to enlarge the hole, before the chick finally emerges from the egg. The egg tooth drops off a few days after hatching. Incidentally, it is sometimes stated that artificial selection for chickens with a reduced tendency for injurious pecking might have the unwanted side-effect of reducing their overall pecking strength and ability to hatch. However, as the neck muscles and egg tooth used to break the egg are highly specialized for that purpose alone, little credence can be given to this idea.

The chick hatches as a fully functioning precocial animal, although synapse development and patterns of neurotransmission continue to mature until approximately 60 days after hatching. In commercial egg production, only females are needed. Male layer-strain chicks have no commercial value, which means that at least 50% of the chicks hatched have to be killed. Given the brain function of the newly hatched chick, this raises practical and ethical issues. Many chicks are fed into macerating machines or are gassed. Methods are available to sex eggs before hatching (Steiner *et al.*, 2011), but with a general lack of consumer pressure calling for a change in practice, these methods are not currently cost-effective to implement.

Filial Imprinting

In the natural environment, the most important thing that a chick must learn after hatching is the identity of its mother. Once this learning task has been accomplished, the chick will approach and follow its mother, receiving many benefits (e.g. warmth, protection and instruction) that we will consider in the section on brooding behaviour. Because all chicks have a very strong predisposition to recognize and remember their mother, and because the imprinting process takes place during a predictable few days after hatching, it has been extensively studied by neurobiologists interested in general biochemical and neural aspects of learning. As a process, imprinting has much in common with other forms of recognition learning and memory, although the relatively narrow time window in which imprinting can take place is a special and functional feature to ensure chick survival.

Learning must take place quickly if the chick is to prosper, and a sensitive period during the first few days after hatching is often described. The alternative term 'critical period' is no longer widely used, as it suggests a very narrow window of opportunity when the identity of the mother is somehow pressed (imprinted) on to the developing brain, in a manner that does not match current biological knowledge. The reality of the imprinting process and its timing is far more interesting.

Some degree of auditory imprinting will already have occurred in the few days before hatching, when chicks listen to maternal calls and begin to treat these as familiar. Chicks respond to their mother's call by beak clapping and, in between maternal calling, they give both contentment and distress calls, which increase as hatching approaches. After hatching, any initial auditory imprinting is strengthened and accompanied by visual (Bolhuis, 1991) and olfactory (Burne and Rogers, 1999) imprinting. Provided that the chick experiences sufficient exposure to its mother (or another complex, naturalistic imprinting stimulus), the process draws to a natural close (Bateson, 1979; Bolhuis, 1991). However, where the quantity or quality of exposure to the mother or other imprinting stimulus is insufficient, the chick retains the potential to imprint for longer, i.e. the sensitive period is extended. Normally, even in the absence of stimulation, the sensitive period draws to a close around 3 days after hatching, but it can be extended for up to 8 days by certain pharmacological treatments (McCabe, 2013).

Chicks hatch with a predisposition to learn about the visual, and to learn more about the auditory, characteristics of their mother. In other words, they possess initial biases that influence their attention. Chicks listen to maternal clucks (which have slightly longer durations and intervals than maternal food calls), which are stable and characteristic features of individual hens but vary somewhat among hens, aiding maternal recognition (Kent, 1993). Chicks approach loud maternal clucks rapidly but are then calmed by more intense maternal calls, which, under natural conditions, would signal the close proximity of the mother (Evans, 1975; Robinson-Guy and Schulman, 1980). In the visual domain, young chicks have an innate preference for attending to social and moving cues (Bolhuis, 1991), particularly stimuli associated with head and face regions (Rosa-Salva *et al.*, 2010). This is functionally significant, as it is subtle details that differentiate one adult hen from another and permit individual recognition. The predisposition to attend to stimuli from the head region is intensified in stressed chicks and may be a mechanism by which an isolated chick increases its chances of returning to the safe base of its mother (McCabe, 2013).

A short period of exposure with the mother or other imprinting stimulus is necessary for a full social attachment to develop. Many features of the hen's appearance play a role at this stage, with the chick consolidating its representation of her as a familiar individual. However, the chicks do not necessarily have to be exposed to all possible features for imprinting to proceed normally (Bateson and Horn, 1994). If, in short succession, chicks are exposed to two different imprinting stimuli, they will classify them together, a mechanism that may aid the chicks in learning that different images or sounds emanating from the mother are all part of a greater whole. There is also a degree of stimulus generalization in the imprinting process (Bolhuis and Horn, 1992). We can infer that, in the natural environment, these mechanisms allow for a degree of change in the appearance of the attachment object. For example, chicks should recognize their mother even if she loses a few feathers or becomes bedraggled in a rainstorm. In the same way, the attachment of chicks to each other is maintained as their appearance changes with age.

Chicks also become familiar with, and somewhat attached to, each other after just their first day together, a process that depends primarily on visual cues (Porter et al., 2005). If the natural mother is not present during the imprinting period, then chicks will imprint on less naturalistic stimuli (such as red boxes or balls used extensively in learning studies). A red stimulus that moves and is accompanied by exposure to a recorded maternal call is particularly effective in the experimental situation.

A conceptual model of the imprinting process was developed by Bateson and Horn (1994), which simulated many aspects of the real-life phenomenon. The model is conceived as a set of connections between three separate systems concerning: (i) the initial analysis of stimuli; (ii) the categorization of relevant stimuli as familiar; and (iii) the output or execution of attachment behaviour. Exposure to relevant stimuli strengthens connections among all systems and promotes an increasing aversion to novelty, signalling the end of the sensitive period of imprinting. Since the publication of this model, a great deal more information has become available about the real biological substrates that support the imprinting process.

The type of recognition learning that is involved in the imprinting process is of great interest in its own right. Unlike associative learning, where neural connections are made between stimuli and reinforcers, the process of learning that

a particular individual is familiar appears to be rewarding by itself, and no additional reinforcement is necessary. Also, once the memory is formed it can be used not only in the original learning context (e.g. in approach and following behaviour) but in other contexts too. The brain region responsible for recognition and memory of the mother is located within a part of the chick forebrain called the intermediate and medial mesopallium (previously termed the intermediate and medial hyperstriatum ventrale), particularly in the left hemisphere (McCabe, 2013). Exposure to the mother, or other imprinting stimulus, during the sensitive period is associated with increases in neuronal size and activity, gene expression and connectivity in this brain region. Signs of increased neuronal activity start to increase in response to the imprinting stimulus after as little as 15 min of exposure (Suge and McCabe, 2004) and firing rate rises after just 1 h of exposure (Horn et al., 2001), but it is important that chicks are able to sleep for a period of about 6 h to consolidate and retain their memory for the imprinting stimulus 24 h later (Jackson et al., 2008). A supplementary storage area outside the intermediate and medial mesopallium permits information acquired during imprinting to be used in other learning contexts (McCabe, 2013).

The process of learning the identity of the mother (or other imprinting stimulus) is accompanied simultaneously by the development of a social attachment. After imprinting has taken place, chicks will approach, follow and show a preference for the attached object compared with other conspicuous objects in the environment. Indeed, they gradually become fearful of novel objects, reducing the chances of imprinting on anything else. When chicks are in close proximity to their mother or another attachment object, they give characteristic twitter calls, indicating contentment, and when separated they give high-intensity distress ('peep') calls. In the commercial hatchery, in the absence of a mother hen, chicks will imprint on each other.

Chick Distress and Fearfulness

There has been considerable independent interest in studying the distress calls made by chicks when isolated or separated from their mother

(Collias, 1952; Andrew, 1964; McBride *et al.*, 1969). As we have seen, the chick distress call attracts the mother and stimulates maternal clucking and brooding, which in turn calms the chick and reduces its distress calling. The rate and intensity of distress calling by the chick separated from its mother is positively correlated with increased circulatory levels of corticosterone and with behavioural measures of fearfulness (Jones and Williams, 1992). Distress calling is increased significantly by a range of factors including the administration of adrenocorticotropic hormone (Panksepp and Normansell, 1990), unfamiliar handling (Sufka and Hughes, 1991), separation from broodmates (Sufka and Hughes, 1991; Jones and Williams 1992; Marx *et al.*, 2001) and brief exposure to cold (Bugden and Evans, 1997). The emotional component of the chicks' experience is highlighted by the fact that distress calling can be decreased by the use of certain drugs that have anti-anxiety effects in people (Feltenstein *et al.*, 2004). If isolation persists beyond an hour or so, the rate of distress calling declines. This has been termed a depression-like phase (Kim and Sufka, 2011). The severity of this phase is reduced when chicks are housed in an enriched environment (Kim and Sufka, 2011) or when they are given antidepressant drugs (Hymel and Sufka, 2012). The isolation procedure is sometimes termed the 'chick separation stress paradigm' or the 'chick anxiety–depression model'. The simplicity of the procedure and the predictability of the chicks' responses mean that they may be increasingly used in the search for new anxiolytic and anti-depressive drugs (Feltenstein *et al.*, 2004; Sufka and White, 2013).

Learning to Feed

As chicks are not actively fed by their mothers, they must also learn to eat within a few days of hatching. Oils in the maternal diet are conveyed to the yolk with lasting effects, evidenced by the preferential approach of chicks to diets eaten previously by their mothers (Bertin *et al.*, 2010; Aigueperse *et al.*, 2013). Young chicks tend to peck at small, round objects and small moving stimuli, but during the first few days of life, their pecking behaviour is not particularly sensitive to any rewarding properties of food arising from ingestion. They are therefore likely to peck at edible

and inedible particles alike without some guidance. In the natural situation, this guidance is provided by the mother. Junglefowl and domestic hens attract their young to food with short bursts of repetitive food calls and by directing pecking movements towards food items on the ground (Sherry, 1977; Nicol and Pope, 1996). With higher-quality food items, the mother's food calls are longer and more intense, attracting chicks to the area more rapidly (Moffatt and Hogan, 1992).

The flexibility of the hens' response is noteworthy as she does not simply perform this food-calling display in a fixed manner whenever she encounters a suitable food particle. Rather, her display is enhanced when the chicks move away or fail to respond (Stokes, 1971) and diminished when the young chicks are feeding properly (Sherry, 1977). In experimental conditions, hens vocalize for longer and give more food calls when chicks are visible but are physically separated from the hen than when the chicks are free to interact with the hen (Wauters *et al.*, 1999), suggesting that hens note the proximity and behaviour of their chicks during this early learning phase. We will consider the complexity of the maternal feeding display further in Chapter 6.

The mother is not the only influence on early feeding behaviour. After a few days of age, chicks find it rewarding to peck at small grains and objects that can be manipulated with the beak. In the commercial environment, food is made readily available, and commonly, for the first few days after hatching, grains are spread on a paper or another smooth surface, maximizing the opportunities for chicks to learn the characteristics of their feed. Commercial chicks will also influence each other, perhaps to a greater extent than under natural conditions. Social facilitation plays a role, and it has long been known that the amount of food ingested is increased by the mere presence of a companion chick and more so by the active feeding behaviour of a companion. The stimulus conditions that evoke pecking in young chicks were examined in detail in a series of papers by Tolman in the 1960s (reviewed by Nicol, 1995). He found that an increase in pecking rate by a real or a model companion resulted in a corresponding increase in subject pecking rate, and concluded that visual exposure to the companion's pecking movement was the stimulus that evoked pecking in the subject. The tapping sound of pecking had a further enhancing effect when presented in conjunction with visual exposure. Watching a companion chick pecking at an aversive substance (and showing its disgust by beak wiping and head shaking) assists avoidance learning in the observer chick (Johnston et al., 1998). Social influences seem to be more important in the acquisition of food preferences in young chicks than in older birds, which are more able to monitor the rewarding effects of food ingestion.

After the first week of life, chicks tend to move away from their mother and explore on their own or in the company of their broodmates. The social influences on feeding shift towards more flexible forms of learning where both social transmission and individual associative learning are involved. Some individual chicks take on a leadership role (Collias, 2000). Domestic chicks are attracted to slightly older birds that can guide their own feeding behaviour (Adret-Hausberger and Cumming, 1987). Older birds may be easier to observe, but it is their prior knowledge that is of most importance. Gajdon et al. (2001) studied 2- to 8-day-old chicks kept in small groups, some with a slightly older knowledgeable demonstrator (pre-trained to find food at marked locations) and some with a naïve demonstrator. Chicks paired with a knowledgeable demonstrator ate more, both when their demonstrator foraged with them and on subsequent trials when they foraged alone.

Effects of Maternal Brooding

The role of oxytocin in inducing maternal care in mammalian mothers has been widely studied, but the hormonal regulation of brooding behaviour in birds is not well understood. It has recently been shown that neurons sensitive to mesotocin (the avian homologue of oxytocin) increase in number when native Thai chickens enter a broody state (Chokchaloemwong et al., 2013). In addition, if chicks are removed from broody hens, the number of these neurons declines again, indicating that chick presence has an active role in maintaining levels of this maternal hormone. However, there is much still to discover about the factors controlling incubation and rearing behaviour in broody hens. In the natural situation, the hen learns to recognize her own chicks, partly on the basis of their colour (Kent, 1992),

and her brooding behaviour is essential to their survival, guiding their feeding behaviour as just described. Her brooding behaviour also keeps the chicks warm, aids the development of their own thermoregulatory abilities (Sherry, 1981) and facilitates sleep (Malleau et al., 2007). Brooding occurs in regular bouts, whereby the hen alternately sits with the chicks resting underneath her, performs a range of active behaviours and sits again. The average duration of each brooding bout is around 30 min, although there is a great deal of variation among hens. As the chicks age, the brooding bout duration does not change much but the overall proportion of time spent being active increases (Hogan et al., 1998).

From the moment of hatching, there is within-day (ultradian) rhythmicity in the sleep/wake and active/inactive behavioural cycles of young chicks (Nielsen et al., 2008). In groups of unbrooded chicks, these independent rhythms become synchronized, most likely due to mutual social influences, but only for a short time. In the absence of a broody hen, by around 3 days of age, group synchronization disappears (Nielsen et al., 2008). The presence of a broody mother hen thus has a major role in prolonging and strengthening behavioural synchronization and organization of chicks' behaviour. Although brooded and non-brooded chicks perform essentially the same behaviours (and for the same overall amount of time), brooded chicks do so in longer, less-interrupted bouts (Wauters et al., 2002; Formanek et al., 2009). In older chicks, up to 16 days of age, the increased behavioural bout length of brooded chicks is associated with a significantly greater synchrony of active or resting behaviours within the brood than occurs in unbrooded chicks (Riber et al., 2007a).

Another fundamental role of the broody hen is in reducing the fear levels of her brood and thereby promoting exploration (Rodenburg et al., 2009a; de Margerie et al., 2013) and further behavioural development (Bertin and Richard-Yris, 2005). This is something of a paradox, as we have seen that an increased fear of novelty occurs as the sensitive phase of imprinting draws to a close. However, once chicks have learned to keep in close contact with their mother, they have more of a safe base from which to explore as they age. Chicks reared by a broody hen to the age of approximately 3 weeks showed more dustbathing and foraging in their home pens and lower fearfulness in behavioural tests (Shimmura et al., 2010).

Chicks reared by a broody hen to the age of approximately 8 weeks showed reduced fearfulness during the time they were with the hen (Roden and Wechsler, 1998) and subsequently reacted with less fear to a novel object as young independent adults (Perre et al., 2002). Finally, when birds reared by a broody hen were given manual restraint tests as adults, their blood serotonin levels were higher than those of non-brooded controls, although there were no differences in corticosterone levels (Rodenburg et al., 2009b). In several species, low serotonin levels are associated with fearfulness and anxiety and with sleep disorders, so this again suggests a positive influence of maternal brooding. Finally, a mother-hen odour derived from the uropygial gland (the preen gland, situated on the hen's back just in front of her tail but normally covered by feathers) can act in isolation to reduce fearfulness in chicks (Madec et al., 2008). The reduced fearfulness of brooded chicks, acting in tandem with visual cues and vocalizations from the hen, encourages chicks to begin perching at least 3 days earlier than non-brooded chicks (Riber et al., 2007b).

However, a mother hen will reduce fear in her brood only if she is relatively confident herself. In experiments where quail chicks were raised by adoptive mothers (ruling out genetic influences), a fear of humans was transmitted by mothers to their brood and was detectable even after the end of maternal contact when the quail were 90 days of age. Chicks raised by mothers who had received prior habituation showed reduced specific fear responses to humans, without any apparent differences between the two groups in general fearfulness (Bertin and Richard-Yris, 2004). Cross-fostering experiments show that maternal fear behaviour is a more important influence on genetically fearful chicks than on genetically calmer chicks (Houdelier et al., 2011).

Although natural brooding of chicks does not occur on commercial farms, studying early behavioural development can have commercial applications. For example, it might be sensible to expose young chicks to a variety of visual and auditory stimuli in commercial settings, as this has been shown to reduce fearfulness (Jones and Waddington, 1992). Alternatively, it could be beneficial to play hen food calls to young chicks, if small-scale experimental results showing improved growth and feed-conversion efficiency (Woodcock et al., 2004) could be replicated on a larger scale. The lighting patterns under which broiler chicks

are reared is another area where knowledge of more natural behaviour could be used to suggest improvements. Currently, in the EU, broiler chicks are usually kept under continuous light with the intention of stimulating feeding. Once they reach 7 days of age, they must be provided with at least 6 h darkness, including at least one continuous 4 h dark period. During the first week, and possibly thereafter, broiler chicks may not get sufficient sleep due to the combination of bright light and disturbance from other chicks. Buijs et al. (2010) found that broiler chicks rest up against pen walls, rather than staying in the centre of the flock, to minimize this type of disturbance. Malleau et al. (2007) examined the effects of providing chicks with an alternating 40 min light:40 min dark cycle, mimicking the natural brooding cycle of the domestic fowl (Wood-Gush et al., 1978). Both broiler and layer chicks rested more when kept under these conditions, with no adverse effects on growth or feeding behaviour.

One of the most interesting commercial applications, deriving from an appreciation of the importance of the broody hen in directing chick behaviour, is the use of 'dark brooders' for the commercial rearing of layer-strain chicks. A dark brooder, shown in Fig. 3.2, provides a conventional heat source for the young chicks but differs from normal practice in that the heat is distributed within a sheltered dark place, screened by a fringed curtain that chicks can pass through at will.

There have been mixed findings regarding the influence of a broody hen on feather pecking behaviour in chicks and older birds. Riber et al. (2007b) reported reduced mortality due to feather pecking in brooded birds, but other studies have failed to show that birds reared by a broody hen perform less feather pecking behaviour (Roden and Wechsler, 1998; Angevaare et al., 2012). A more robust result is that less feather pecking is observed in small experimental groups of chicks reared with dark brooders than in chicks reared under conventional heat lamps (Johnsen and Kristensen, 2001; Jensen et al., 2006). This may be because dark brooders provide the benefits of a warm resting place and synchronization of activity without the more variable influence of the mother hen on chick fearfulness. Gilani et al. (2012) recently reported significantly less severe feather pecking and plumage loss in commercial flocks of birds that had been reared with dark brooders, an effect that persisted into lay. The

Fig. 3.2. Chicks in a small commercial rearing house find it easy to move through the black plastic curtains to reach a warm, dark area where they can rest and sleep. This 'dark brooder' mimics some features of the broody hen and keeps active chicks separate from inactive ones, reducing the risk of early feather pecking. (Photo courtesy of Dr Anne-Marie Gilani, University of Bristol, UK.)

mechanism by which brooding reduces feather pecking is not known, but possibilities include the effects of brooding in lowering fear levels, or the fact that resting under the dark brooder places inactive chicks (potential victims) out of the sight of more active chicks. There does not appear to be any current intention to use dark brooders on commercial broiler farms, although such systems could potentially provide benefits here too.

Development of Maintenance and Spatial Behaviour

After hatching, chicks spend a considerable amount of time asleep, although this time decreases gradually from hatching to around 2 weeks of age (Mascetti *et al.*, 2004). Chicks usually sleep with both eyes closed, but this is interspersed with brief periods of monocular sleep where only one eye is shut and only one half of the brain is asleep. Highly lateralized brain activity (learning a new task, for example) can promote subsequent sleep in the hemisphere that has been working hard. However, closing both eyes appears to be more restful, and sleep deprivation is compensated

with a relative increase in binocular sleep (Bobbo *et al.*, 2008).

The early formation of non-social preferences, such as for foraging or pecking substrates, shares some characteristics with the development of social preferences reviewed above. For example, Vestergaard and Lisborg (1993) argued that chicks 'imprinted' on dustbathing substrates during the first 10 days of life, with a sensitive period peaking at around day 3 (Vestergaard and Baranyiova, 1996), and there is evidence that development of a full structural sequence of dustbathing movements is facilitated by exposure to a suitable substrate (Larsen *et al.*, 2000). However, the parallels with the filial imprinting process are not strong. Although the 'sensitive period' for substrate imprinting is said to end by about day 10, birds can revise their substrate preferences as they age (Sanotra *et al.*, 1995). Adult foraging behaviour is little influenced by early substrate exposure and depends most on the current substrate available (Nicol *et al.*, 2001). In addition, early preferences for dustbathing substrates appear to be revised by experience gained at around 60 days (Nicol *et al.*, 2001), and adult birds prefer to dustbathe in peat, irrespective of their early experience of this substrate (Wichman and Keeling, 2008).

This strongly suggests that adult foraging and dustbathing behaviour is flexible and not unduly governed by early experience.

As we have seen, the precocial chick hatches with finely tuned visual, auditory and olfactory abilities. However, its ability to use its magnetic sense appears to develop only in the second week of life (Denzau et al., 2013). This may relate to the maturation of other spatial abilities in young birds. In the first week of life, the left brain hemisphere dominates behaviour, facilitating food and social learning, but at around day 10, there is a shift to right-hemisphere dominance as chicks begin to explore their wider environment. Between days 10 and 12, chicks voluntarily start to move out of sight of their mother for short periods before returning to regain contact. Experimentally provided experience of such occlusion at around this age enhances subsequent spatial and searching skills (Freire et al., 2004) and increases neural development in the hippocampus (Freire and Cheng, 2004). At around this age, chicks also start to frolic and perch. The amount of daytime perching increases steadily until by 6 weeks or so chicks will spend around a quarter of the light period on perches (Heikkila et al., 2006).

Night-time roosting is a later development, only initiated when chicks are around 6 weeks of age (McBride et al., 1969; Heikkila et al., 2006). There is no strong evidence that the development of perching behaviour is associated with two-dimensional spatial ability (Wichman et al., 2007). However, the early opportunity to perch is not always available to commercially reared chicks, and this can adversely affect their later spatial and navigational abilities (Gunnarsson et al., 2000). Early experience with perches also results in adult birds that are better able to use three-dimensional space to escape from attack (Yngvesson, 2002), less likely to lay eggs on the floor and less likely to have problems with feather pecking and cannibalism (Gunnarsson et al., 1999; Lambton et al., 2013).

Development of Social and Sexual Behaviour

The early social behaviours of chicks cluster into three rather distinct groups – running/sparring, leaping and aggression (Rushen, 1982). Running and sparring take place between male chicks but rarely between females or between males and females. As development progresses, leaps between males are 'increasingly not reciprocated' (Rushen, 1982) and the first male–male threats and head pecks are observed, supporting earlier observations by Kruijt (1964) that sparring is replaced by aggression. Female chicks develop head pecking shortly after their male counterparts.

Pronounced aggressive behaviour does not develop until chicks are 2–3 weeks of age and peaks between 6 and 12 weeks (Estevez et al., 2003). Aggressive behaviour is a key determinant of social status, but it may not be the only one. One report found that differences in performance in resource competition tests (e.g. competing for a worm-like object by picking it up and running) were correlated with social status in very young chicks (Rogers and Astiningsih, 1991). However, the function of this worm-running behaviour is not clear, although it may be a way for young birds to practise skills associated with prey capture and retention (Cloutier et al., 2004). Chick embryos exposed to corticosterone in the egg are less successful than their peers in this type of test (Janczak et al., 2006), while exposure to testosterone increases success (Rogers and Astiningsih, 1991). However, there are no clear links between worm-running success and aggression (Cloutier et al., 2004), and further confirmation would be needed to regard this as an indicator of future social status.

The development of sexual behaviour in the chicken has been well described by Wood-Gush (1971). Increases in crowing, waltzing and wing flapping in males are highly correlated, but actual success in mating depends also upon the separate timing of development of female receptivity and crouching behaviour, with juvenile females tending to avoid displaying males (Rushen, 1982, 1983). Wood-Gush (1958) was one of the first researchers to show the importance of early social experience in the development of normal courtship and reproductive behaviour, finding that isolated cockerels were more likely to show aggression towards females. The early social experience of the young chick also influences its subsequent preference for a reproductive partner, and this is sometimes described as sexual imprinting. For example, chicks can recognize and avoid mating with members of the same hatched brood (normally their siblings) while also seeking out other individuals that share species and breed characteristics similar to those of their early companions. In this way, adult birds seek out partners that differ, but only slightly, from their broodmates (Bateson, 1982). Although there is considerable information on the mate choice preferences of chickens, which we will consider later (see Chapter 6), there has been remarkably little experimental work examining how these sexual preferences *develop* in chickens, or whether the processes involved really bear comparison with those acting during filial imprinting. It is known that pairs of males and females with similar rearing experience (both reared with, or both reared without, the opposite sex) have greater mating success than pairs with differing rearing experiences, and there is some evidence that adult females select compatible sexual partners using visual cues (Leonard *et al.*, 1996), but beyond this the field is wide open for further study.

References

Adret-Hausberger, M. and Cumming, R.B. (1987) Social experience and selection of diet in domestic chickens. *Bird Behaviour* 7, 37–43.

Ahmed, A.A., Ma, W.Q., Guo, F., Ni, Y.D., Grossmann, R. and Zhao, R.Q. (2013) Differences in egg deposition of corticosterone and embryonic expression of corticosterone metabolic enzymes between slow and fast growing broiler chickens. *Comparative Biochemistry and Physiology A – Molecular and Integrative Physiology* 164, 200–206.

Aigueperse, N., Calandreau, L. and Bertin, A. (2013) Maternal diet influences offspring feeding behaviour and fearfulness in the precocial chicken. *PLoS One* 8, e77583.

Almasi, B., Rettenbacher, S., Mueller, C., Brill, S., Wagner, H. and Jenni, L. (2012) Maternal corticosterone is transferred into the egg yolk. *General and Comparative Endocrinology* 178, 139–144.

Andrew, R.J. (1964) Vocalization in chicks and the concept of "stimulus contrast". *Animal Behaviour* 12, 64–76.

Andrew, R.J. and Rogers, L.J. (1972) Testosterone, search behaviour and persistence. *Nature* 237, 343–345.

Andrew, R.J., Johnston, A.N.B., Robins, A. and Rogers, L.J. (2004) Light experience and the development of behavioural lateralisation in chicks: II. Choice of familiar versus unfamiliar model social partner. *Behavioural Brain Research* 155, 67–76.

Angevaare, M.J., Prins, S., van der Staay, F.J. and Nordquist, R.E. (2012) The effect of maternal care and infrared beak trimming on development, performance and behaviour of Silver Nick hens. *Applied Animal Behaviour Science* 140, 70–84.

Aslam, M.A., Hulst, M., Hoving-Bolink, R.A.H., Smits, M.A., de Vries, B., Weites, I., Groothuis, T.G.G. and Woelders, H. (2013) Yolk concentrations of hormones and glucose and egg weight and egg dimensions in unincubated chicken eggs, in relation to egg sex and hen body weight. *General and Comparative Endocrinology* 187, 15–22.

Bateson, P. (1979) How do sensitive periods arise and what are they for? *Animal Behaviour* 27, 470–486.

Bateson, P. (1982) Preferences for cousins in Japanese quail. *Nature* 295, 236–237.

Bateson, P. and Horn, G. (1994) Imprinting and recognition memory: a neural net model. *Animal Behaviour* 48, 695–715.

Bertin, A. and Richard-Yris, M.A. (2004) Mothers' fear of human affects the emotional reactivity of young in domestic Japanese quail. *Applied Animal Behaviour Science* 89, 215–231.

Bertin, A. and Richard-Yris, M.A. (2005) Mothering during early development influences subsequent emotional and social behaviour in Japanese quail. *Journal of Experimental Zooogy Part A – Comparative Experimental Biology* 303A, 792–801.

Bertin, A., Richard-Yris, M.-A., Houdelier, C., Lumineau, S., Mostl, E., Kuchar, A., Hirschenhauser, K. and Kotrschal, K. (2008) Habituation to humans affects yolk steroid levels and offspring phenotype in quail. *Hormones and Behavior* 54, 396–402.

Bertin, A., Calandreau, L., Arnould, C., Nowak, R., Levy, F., Noirot, V., Bouvarel, I. and Leterrier, C. (2010) *In ovo* olfactory experience influences post-hatch feeding behaviour in young chickens. *Ethology* 116, 1027–1037.

Bertin, A., Calandreau, L., Arnould, C. and Levy, F. (2012) The developmental stage of chicken embryos modulates the impact of *in ovo* olfactory stimulation on food preferences. *Chemical Senses* 37, 253–261.

Bobbo, D., Nelini, C. and Mascetti, G.G. (2008) Binocular and monocular/unihemispheric sleep in the domestic chick (*Gallus gallus*) after a moderate sleep deprivation. *Experimental Brain Research* 185, 421–427.

Bolhuis, J.J. (1991) Mechanisms of avian imprinting: a review. *Biology Reviews* 66, 303–345.

Bolhuis, J.J. and Horn, G. (1992) Generalization of learned preferences in filial imprinting. *Animal Behaviour* 44, 185–187.

Bugden, S.C. and Evans, R.M. (1997) Vocal solicitation of heat as an integral component of the developing thermo-regulatory system in young domestic chickens. *Canadian Journal of Zoology – Revue Canadienne de Zoologie* 75, 1949–1954.

Buijs, S., Keeling, L.J., Vangestel, C., Baert, J., Vangeyte, J. and Tuyttens, F.A.M. (2010) Resting or hiding? Why broiler chickens stay near walls and how density affects this. *Applied Animal Behaviour Science* 124, 97–103.

Burne, T.H.J. and Rogers, L.J. (1999) Changes in olfactory responsiveness by the domestic chick after early exposure to odorants. *Animal Behaviour* 58, 329–336.

Burne, T.H.J. and Rogers, L.J. (2002) Chemosensory input and lateralization of brain function in the domestic chick. *Behavioural Brain Research* 133, 293–300.

Butler, A.B. (2008) Evolution of brains, cognition and consciousness. *Brain Research Bulletin* 75, 442–449.

Butler, A.B. and Cotterill, R.M.J. (2006) Mammalian and avian neuroanatomy and the question of consciousness in birds. *Biological Bulletin* 211, 106–127.

Chiandetti, C., Regolin, L., Rogers, L.J. and Vallortigara, G. (2005) Effects of light stimulation of embryos on the use of position-specific and object-specific cues in binocular and monocular domestic chicks (*Gallus gallus*). *Behavioural Brain Research* 163, 10–17.

Chokchaloemwong, D., Prakobsaeng, N., Sartsoongnoen, N., Kosonsiriluk, S., El Halawani, M. and Chaiseha, Y. (2013) Mesotocin and maternal care of chicks in native Thai hens (*Gallus domesticus*). *Hormones and Behaviour* 64, 53–69.

Clifton, P.G. and Andrew, R.J. (1989) Contrasting effects of pre- and posthatch exposure to gonadal steroids on the development of vocal, sexual and aggressive behaviour of young domestic fowl. *Hormones and Behavior* 23, 572–589.

Cloutier, S., Newberry, R.C. and Honda, K. (2004) Comparison of social ranks based on worm-running and aggressive behaviour in young domestic fowl. *Behavioural Processes* 65, 79–86.

Collias, N.E. (1952) The development of social behaviour in birds. *Auk* 69, 127–159.

Collias, N.E. (2000) Filial imprinting and leadership among chicks in family integration of the domestic fowl. *Behaviour* 137, 197–211.

Daisley, J.N., Bromundt, V., Mostl, E. and Kotrschal, K. (2005) Enhanced yolk testosterone influences behavioural phenotype independent of sex in Japanese quail chicks *Coturnix japonica*. *Hormone and Behavior* 47, 185–194.

Davis, K.A., Schmidt, J.B., Doescher, R.M. and Satterlee, D.G. (2008) Fear responses of offspring from divergent quail stress response line hens treated with corticosterone during egg formation. *Poultry Science* 87, 1303–1313.

de Haas, E.N., Kemp, B., Bolhuis, J.E., Groothuis, T. and Rodenburg, T.B. (2013) Fear, stress and feather pecking in commercial white and brown laying hen parent-stock flocks and their relationships with production parameters. *Poultry Science* 92, 2259–2269.

de Haas, E.N., Bolhuis, J.E., Kemp, B., Groothuis, T.G.G. and Rodenburg, T.B. (2014) Parents and early life environment affect behavioural development of laying hen chickens. *PLoS One* 9, e90577.

de Margerie, E., Peris, A., Pittet, F., Houdelier, C., Lumineau, S. and Richard-Yris, M.A. (2013) Effects of mothering on the spatial exploratory behavior of quail chicks. *Developmental Psychobiology* 55, 256–264.

Della Chiesa, A., Peccia, T., Tommasi, L. and Vallortigara, G. (2006b) Multiple landmarks, the encoding of environmental geometry and the spatial logics of a dual brain. *Animal Cognition* 9, 281–293.

Della Chiesa, A., Speranza, M., Tommasi, L. and Vallortigara, G. (2006a) Spatial cognition based on geometry and landmarks in the domestic chick (*Gallus gallus*). *Behavioural Brain Research* 175, 119–127.

Deng, C. and Rogers, L.J. (2002) Social recognition and approach in the chick: lateralisation and effect of visual experience. *Animal Behaviour* 63, 697–706.

Denzau, S., Niessner, C., Rogers, L.J. and Wiltschko, W. (2013) Ontogenetic development of magnetic compass orientation in domestic chickens (*Gallus gallus*). *Journal of Experimental Biology* 216, 3143–3147.

Dharmaretnam, M. and Rogers, L.J. (2005) Hemispheric specialisation and dual processing in strongly versus weakly lateralised chicks. *Behavioural Brain Research* 162, 62–70.

Downing, J.A. and Bryden, W.L. (2008) Determination of corticosterone concentrations in egg albumen: a non-invasive indicator of stress in laying hens. *Physiology and Behavior* 95, 381–387.

Eising, C.M., Muller, W., Dijkstra, C. and Groothuis, T.G.G. (2003) Maternal androgens in egg yolks: relation with sex, incubation time and embryonic growth. *General and Comparative Endocrinology* 132, 241–247.

Eriksen, M.S., Haug, A., Torjesen, P.A. and Bakken, M. (2003) Prenatal exposure to corticosterone impairs embryonic development and increases fluctuating asymmetry in chickens. *British Poultry Science* 44, 690–697.

Estevez, I., Keeling, L.J. and Newberry, R.C. (2003) Decreasing aggression with increasing group size in young domestic fowl. *Applied Animal Behaviour Science* 84, 213–218.

Evans, R.M. (1975) Stimulus-intensity and acoustical communication in young domestic chicks. *Behaviour* 55, 73–80.

Falt, B. (1981) Development of responsiveness to the individual maternal clucking by domestic chicks (*Gallus gallus domesticus*). *Behavioural Processes* 6, 303–317.

Feltenstein, M.W., Warnick, J.E., Guth, A.N. and Sufka, K.J. (2004) The chick separation stress paradigm: a validation study. *Pharmacology Biochemistry and Behavior* 77, 221–226.

Formanek, L., Richard-Yris, M.A., Houdelier, C. and Lumineau, S. (2009) Epigenetic maternal effects on endogenous rhythms in precocial birds. *Chronobiology International* 26, 396–414.

Freire, R. and Cheng, H.W. (2004) Experience-dependent changes in the hippocampus of domestic chicks: a model for spatial memory. *European Journal of Neuroscience* 20, 1065–1068.

Freire, R. and Rogers, L.J. (2005) Experience-induced modulation of the use of spatial information in the domestic chick. *Animal Behaviour* 69, 1093–1100.

Freire, R. and Rogers, L.J. (2007) Experience during a period of right hemispheric dominance alters attention to spatial information in the domestic chick. *Animal Behaviour* 74, 413–41.

Freire R., Cheng, H.W. and Nicol, C.J. (2004) Development of spatial memory in occlusion-experienced chicks. *Animal Behaviour* 67, 141–150.

Freire, R., van Dort, S. and Rogers, L.J. (2006) Pre- and post-hatching effects of corticosterone treatment on behavior of the domestic chick. *Hormones and Behavior* 49, 157–165.

Gajdon, G.K., Hungerbuhler, N. and Stauffacher, M. (2001) Social influence on early foraging of domestic chicks (*Gallus gallus*) in a near-to-nature procedure. *Ethology* 107, 913–937.

Gilani, A.-M., Knowles, T.G. and Nicol, C.J. (2012) The effect of dark brooders on feather pecking on commercial farms. *Applied Animal Behaviour Science* 142, 42–50.

Groothuis, T.G.G. and Schwabl, H. (2008) Hormone-mediated maternal effects in birds: mechanisms matter but what do we know of them? *Philosophical Transactions of the Royal Society B – Biological Sciences* 363, 1647–1661.

Groothuis, T.G.G., Eising, C.M., Dijkstra, C. and Muller, W. (2005) Balancing between costs and benefits of maternal hormone deposition in avian eggs. *Biology Letters* 1, 78–81.

Guesdon, V., Bertin, A., Houdelier, C., Lumineau, S., Formanek, L., Kotrschal, K., Mostl, E. and Richard-Yris, M.A. (2011) A place to hide in the home-cage decreases yolk androgen levels and offspring emotional reactivity in Japanese quail. *PLoS One* 6, e23941.

Guibert, F., Richard-Yris, M.A., Lumineau, S., Kotrschal, K., Mostl, E. and Houdelier, C. (2012) Yolk testosterone levels and offspring phenotype correlate with parental age in a precocial bird. *Physiology and Behavior* 105, 242–250.

Gunnarsson, S., Keeling, L.J. and Svedberg, J. (1999) Effect of rearing factors on the prevalence of floor eggs, cloacal cannibalism and feather pecking in commercial flocks of loose housed laying hens. *British Poultry Science* 40, 12–18.

Gunnarsson, S., Yngvesson, J., Keeling, L.J. and Forkman, B. (2000) Rearing without early access to perches impairs the spatial skills of laying hens. *Applied Animal Behaviour Science* 67, 217–228.

Heikkila, M., Wichman, A., Gunnarsson, S. and Valros, A. (2006) Development of perching behaviour in chicks reared in enriched environment. *Applied Animal Behaviour Science* 99, 145–156.

Hein, C.M., Engels, S., Kishkinev, D. and Mouritsen, H. (2011) Robins have a magnetic compass in both eyes. *Nature* 471, E11–E12.

Henriksen, R., Rettenbacher, S. and Groothuis, T.G.G (2011a) Prenatal stress in birds: pathways, effects, function and perspectives. *Neuroscience and Biobehavioural Reviews* 35, 1484–1501.

Henriksen, R., Groothuis, T.G. and Rettenbacher, S. (2011b) Elevated plasma corticosterone decreases yolk testosterone and progesterone in chickens: linking maternal stress and hormone-mediated effects. *PLoS One* 6, e23824.

Henriksen, R., Rettenbacher, S. and Groothuis, T.G.G. (2013) Maternal corticosterone elevation during egg formation in chickens (*Gallus gallus domesticus*) influences offspring traits, partly via prenatal undernutrition. *General and Comparative Endocrinology* 191, 83–91.

Hogan, J.A., Hogan-Warburg, A.J., Panning, L. and Moffatt, C.A. (1998) Causal factors controlling the brooding cycle of broody junglefowl hens with chicks. *Behaviour* 135, 957–980.

Horn, G., Nicol, A.U. and Brown, M.W. (2001) Tracking memory's trace. *Proceedings of the National Academy of Sciences USA* 98, 5282–5287.

Houdelier, C., Lumineau, S., Bertin, A., Guibert, F., De Margerie, E., Augery, M. and Richard-Yris, M.A. (2011) Development of fearfulness in birds: genetic factors modulate non-genetic maternal influences. *PLoS One* 6, e14604.

Hymel, K.A. and Sufka, K.J. (2012) Pharmacological reversal of cognitive bias in the chick anxiety–depression model. *Neuropharmacology* 62, 161–166.

Jackson, C., McCabe, B.J., Nicol, A.U. and Grout, A.S. (2008) Dynamics of a memory trace: effects of sleep on consolidation. *Current Biology* 18, 393–400.

Janczak, A., Braastad, B.O. and Bakken, M (2006). Behavioural effects of embryonic exposure to corticosterone in chickens. *Applied Animal Behaviour Science* 96, 69–82.

Janczak, A.M., Toriesen, P., Palme, R. and Bakken, M. (2007) Effects of stress in hens on the behaviour of their offspring. *Applied Animal Behaviour Science* 107, 66–77.

Jarvis, E.D., Güntürkün, O., Bruce, L., Csillag, A., Karten, H., Kuenzel, W., Medina, L., Paxinos, G., Perkel, D.J., Shimizu, T., Striedter, G., Wild, J.M., Ball, G.F., Dugas-Ford, J., Durand, S.E., Hough, G.E., Husband, S., Kubikova, L., Lee, D.W., Mello, C.V., Powers, A., Siang, C., Smulders, T.V., Wada, K., White, S.A., Yamamoto, K., Yu, J., Reiner, A. and Butler, A.B. (2005) Avian brains and new understanding of vertebrate brain evolution. *Nature Neuroscience Reviews* 6, 151–159.

Jensen, A.B., Palme, R. and Forkman, B. (2006) Effects of brooders on feather pecking and cannibalism in domestic fowl (*Gallus gallus domesticus*). *Applied Animal Behaviour Science* 99, 287–300.

Johnsen, P.F. and Kristensen, H.H. (2001) Effect of brooder quality on the early development of feather pecking behaviour in domestic chicks. In: *Proceedings of the 6th European Symposium on Poultry Welfare*, Zollikofen, Switzerland, pp. 209–212.

Johnston, A.N.B., Burne, T.H.J. and Rose, S.P.R. (1998) Observation learning in day-old chicks using a one-trial passive avoidance learning paradigm. *Animal Behaviour* 56, 1347–1353.

Jones, R.B. and Waddington, D. (1992) Modification of fear in domestic chicks, *Gallus gallus domesticus*, via regular handling and early environmental enrichment. *Animal Behaviour* 43, 1021–1033.

Jones, R.B. and Williams, J.B. (1992) Responses of pair-housed male and female domestic chicks to the removal of a companion. *Applied Animal Behaviour Science* 32, 375–380.

Kent, J.P. (1992) The relationship between the hen and the chick, *Gallus gallus domesticus*: the hen's recognition of the chick. *Animal Behaviour* 44, 996–998.

Kent, J.P. (1993) The chick's preference for certain features of the maternal cluck vocalization in the domestic fowl (*Gallus gallus*). *Behaviour* 125, 177–187.

Kim, E.H. and Sufka, K.J. (2011) The effects of environmental enrichment in the chick anxiety-depression model. *Behavioural Brain Research* 221, 276–281.

Kruijt, J.P. (1964) *Ontogeny of Social Behaviour in Burmese Red Junglefowl (Gallus gallus spadiceus)*. Brill, Leiden, The Netherlands.

Lambton, S.L., Nicol, C.J., Friel, M., Main, D.C.J., McKinstry, J.L., Sherwin, C.M., Walton, J. and Weeks, C.A. (2013) A bespoke management package can reduce levels of injurious pecking in loose-housed laying hen flocks. *Veterinary Record* 172, 423–429.

Larsen, B.H., Vestergaard, K.S. and Hogan, J.A. (2000) Development of dustbathing behaviour sequences in the domestic fowl: the significance of functional experience. *Developmental Psychobiology* 37, 5–12.

Leonard, M.L., Zanette, L. and Clinchy, M. (1996) The effect of early exposure to the opposite sex on mate choice in White Leghorn chickens. *Applied Animal Behaviour Science* 48, 15–23.

Madec, I., Gabarrou, J.F. and Pageat, P. (2008) Influence of a maternal odorant on coping strategies in chicks facing isolation and novelty during a standardized test. *Neuroendocrinology Letters* 29, 507–511.

Malleau, A.E., Duncan, I.J.H., Widowski, T.M. and Atkinson, J.L. (2007) The importance of rest in young domestic fowl. *Applied Animal Behaviour Science* 106, 52–69.

Marx, G., Leppelt, J. and Ellendorf, F. (2001) Vocalisation in chicks (*Gallus gallus* dom.) during stepwise social isolation. *Applied Animal Behaviour Science* 75, 61–74.

Marx, G., Jurkevich, A. and Grossman, R. (2004) Effects of estrogens during embryonal development on crowing in the domestic fowl. *Physiology and Behavior* 82, 637–645.

Mascetti, G.G., Bobbo, D., Rugger, M. and Vallortigara, G. (2004) Monocular sleep in male domestic chicks. *Behavioural Brain Research* 153, 447–452.

McBride, G., Parer, I.P. and Foenander, F. (1969) The social organisation and behaviour of the feral domestic fowl. *Animal Behaviour Monographs* 2, 126–181.

McCabe, B.J. (2013) Imprinting. *Wiley Interdisciplinary Reviews – Cognitive Science* 4, 375–390.

Moffatt, C.A. and Hogan, J.A. (1992) Ontogeny of chick responses to maternal food calls in the Burmese red junglefowl (*Gallus gallus spadiceus*). *Journal of Comparative Psychology* 106, 92–96.

Muller, W., Eising, C.M., Dijkstra, C. and Groothuis, T.G.G. (2002) Sex differences in yolk hormones depend on maternal social status in Leghorn chickens (*Gallus gallus domesticus*). *Proceedings of the Royal Society B – Biological Sciences* 269, 2249–2255.

Navara, K.J. and Pinson, S.E. (2010) Yolk and albumen corticosterone concentrations in eggs laid by white versus brown caged laying hens. *Poultry Science* 89, 1509–1513.

Nicol, C.J. (1995) The social transmission of information and behaviour. *Applied Animal Behaviour Science* 44, 79–98.

Nicol, C.J. and Pope, S.J. (1996) The maternal feeding display of domestic hens is sensitive to perceived chick error. *Animal Behaviour* 52, 767e774.

Nicol, C.J., Lindberg, A.C., Philips, A.J., Pope, S.J., Wilkins, L.J. and Green, L.E. (2001) Influence of prior exposure to wood shavings on feather pecking, dustbathing and foraging in adult laying hens. *Applied Animal Behaviour Science* 73, 141–155.

Nielsen, B.L., Erhard, H.W., Friggens, N.C. and McLeod, J.E. (2008) Ultradian activity rhythms in large groups of newly hatched chicks (*Gallus gallus domesticus*). *Behavioural Processes* 78, 408–415.

Okuliarova, M., Skrobanek, P. and Zeman, M. (2007) Effect of increasing yolk testosterone levels on early behaviour in Japanese quail hatchlings. *Acta Veterinaria Brno* 76, 325–331.

Okuliarova, M., Sarnikova, B., Rettenbacher, S., Skrobanek, P. and Zeman, M. (2010) Yolk testosterone and corticosterone in hierarchical follicles and laid eggs of Japanese quail exposed to long-term restraint stress. *General and Comparative Endocrinology* 165, 91–96.

Okuliarova, M., Groothuis, T.G.G., Skrobanek, P. and Zeman, M. (2011) Experimental evidence for genetic heritability of maternal hormone transfer to offspring. *American Naturalist* 177, 824–834.

Panksepp, J. and Normansell, L. (1990) Effects of ACTH(1–24) and ACTH/MSH(4–10) on isolation-induced distress vocalization in domestic chicks. *Peptides* 11, 915–919.

Perre, Y., Wauters, A.M. and Richard-Yris, M.A. (2002) Influence of mothering on emotional and social reactivity of domestic pullets. *Applied Animal Behaviour Science* 75, 133–146.

Pilz, K.M., Adkins-Regan, E. and Schwabl, H. (2005) No sex differences in yolk steroid concentrations of avian eggs at laying. *Biology Letters* 1, 318–321.

Porter, R.H., Roelofsen, R., Picard, M. and Arnould, C. (2005) The temporal development and sensory mediation of social discrimination in domestic chicks. *Animal Behaviour* 70, 359–364.

Quillfeldt, P., Poisbleau, M., Parenteau, C., Trouve, C., Demongin, L., van Noordwijk, H.J. and Mostl, E. (2011) Measuring corticosterone in seabird egg yolk and the presence of high yolk gestagen concentrations. *General and Comparative Endocrinology* 173, 11–14.

Rashid N. and Andrew, R.J. (1989) Right-hemisphere advantage for topographical orientation in the domestic chick. *Neuropsychologia* 27, 937–948.

Regolin, L., Garzotto, B., Rugani, R., Pagni, P. and Vallortigara, G. (2005) Working memory in the chick: parallel and lateralised mechanisms for encoding of object- and position-specific information. *Behavioural Brain Research* 157, 1–9.

Rettenbacher, S. Moestl, E. and Groothuis, T.G.G. (2009) Gestagens and glucocorticoids in chicken eggs. *General and Comparative Endocrinology* 164, 125–129.

Riber, A.B., Nielseon, B.L., Ritz, C. and Forkman, B. (2007a) Diurnal activity cycles and synchrony in layer hen chicks (*Gallus gallus domesticus*). *Applied Animal Behaviour Science* 108, 276–287.

Riber, A.B., Wichman, A., Braastad, B.O. and Forkman, B. (2007b) Effects of broody hens on perch use, ground pecking, feather pecking and cannibalism in domestic fowl (*Gallus gallus domesticus*). *Applied Animal Behaviour Science* 106, 39–51.

Riedstra B. and Groothuis, T.G.G. (2004) Prenatal light exposure affects early feather-pecking behaviour in the domestic chick. *Animal Behaviour* 67, 1037–1042.

Robinson-Guy, E.D. and Schulman, A.H., 1980. Auditory stimulus intensity and the neonatal approach response of domestic chicks (*Gallus gallus*). *Behavioural Processes* 5, 211–225.

Roden, C. and Wechsler, B. (1998) A comparison of the behaviour of domestic chicks reared with or without a hen in enriched pens. *Applied Animal Behaviour Science* 55, 317–326.

Rodenburg, T.B., Uitdehaag, K.A., Ellen, E.D. and Komen, J. (2009a) The effects of selection on low mortality and brooding by a mother hen on open-field response, feather pecking and cannibalism in laying hens. *Animal Welfare* 18, 427–432.

Rodenburg, T.B., Bolhuis, J.E., Koopmanschap, R.E., Ellen, E.D. and Decuypere, E. (2009b) Maternal care and selection for low mortality affect post-stress corticosterone and peripheral serotonin in laying hens. *Physiology and Behavior* 98, 519–523.

Rodgers, A.B., Morgan, C.P., Bronson, S.L., Revello, S. and Bale, T.L. (2013) Paternal stress exposure alters sperm microRNA content and reprograms offspring HPA stress axis regulation. *Journal of Neuroscience* 33, 9003–9012.

Rogers, L.J. (1996) *The Development of Brain and Behaviour in the Chicken*. CABI, Wallingford, UK.

Rogers, L.J. (1997) Early experiential effects on laterality: research on chicks has relevance to other species. *Laterality* 2, 199–219.

Rogers, L.J. and Astiningsih, K. (1991) Social hierarchies in very young chicks. *British Poultry Science* 32, 47–56.

Rogers, L.J. and Deng, C. (2005) Corticosterone treatment of the chick embryo affects light-stimulated development of the thalamofugal visual pathway. *Behavioural Brain Research* 159, 63–71.

Rogers, L.J. and Ehrlich, D. (1983) Asymmetry in the chicken forebrain during development and a possible involvement of the supraoptic decussation. *Neuroscience Letters* 37, 123–127.

Rogers, L.J. and Rajendra, S. (1993) Modulation of the development of light-initiated asymmetry in chick thalamofugal visual projections by estradiol. *Experimental Brain Research* 93, 89–94.

Rogers, L.J. and Workman, L. (1989) Light exposure during incubation affects competitive behaviour in domestic chicks. *Applied Animal Behaviour Science* 23, 187–198.

Rogers, L.J., Zucca, P. and Vallortigara, G. (2004) Advantages of having a lateralized brain. *Proceedings of the Royal Society B – Biological Sciences* 271, S420–S422.

Rogers, L.J., Andrew, R.J. and Johnston, A.N.B. (2007) Light experience and the development of behavioural lateralization in chicks: III. Learning to distinguish pebbles from grains. *Behavioural and Brain Research* 177, 61–69.

Rogers, L.J., Munro, U., Freire, R., Wiltschko, R. and Wiltschko, W. (2008) Lateralized response of chicks to magnetic cues. *Behavioural Brain Research* 186, 66–71.

Rosa-Salva, O., Daisley, J.N., Regolin, L. and Vallortigara, G. (2009) Lateralization of social learning in the domestic chick, *Gallus gallus domesticus*: learning to avoid. *Animal Behaviour* 78, 847–856.

Rosa-Salva, O., Regolin, L. and Vallortigara, G. (2010) Faces are special for newly hatched chicks: evidence for inborn domain-specific mechanisms underlying spontaneous preferences for face-like stimuli. *Developmental Science* 13, 565–577.

Roy, S., Nag, T.C., Upadhyay, A., Mathur, R. and Jain, S. (2013) Repetitive auditory stimulation at a critical prenatal period modulates the postnatal functional development of the auditory as well as the visual system in chicks (*Gallus domesticus*). *Developmental Neurobiology* 73, 688–701.

Rushen, J. (1982) Development of social behaviour in chickens – a factor analysis. *Behavioural Processes* 7, 319–333.

Rushen, J. (1983) The development of sexual relationships in the domestic chicken. *Applied Animal Ethology* 11, 55–66.

Sanotra, G.S., Vestergaard, K.S., Agger, J.F. and Lawson, L.G. (1995) The relative preferences for feathers, straw, wood-shavings and sand for dustbathing, pecking and scratching in domestic chicks. *Applied Animal Behaviour Science* 43, 263–277.

Schweitzer, C., Goldstein, M.H., Place, N.J. and Adkins-Regan, E. (2013) Long-lasting and sex-specific consequences of elevated egg yolk testosterone for social behaviour in Japanese quail. *Hormones and Behavior* 63, 80–87.

Sherry, D.F. (1977) Parental food-calling and role of young in Burmese red junglefowl (*Gallus gallus spadiceus*). *Animal Behaviour* 25, 594–601.

Sherry, D.F. (1981) Parental care and the development of thermoregulation in red junglefowl. *Behaviour* 76, 250–279.

Shimmura, T., Kamimura, E., Azuma, T., Kansaku, N., Uetake, K. and Tanaka, T. (2010) Effect of broody hens on behaviour of chicks. *Applied Animal Behaviour Science* 126, 125–133.

Singh, R., Cook, N., Cheng, K.M. and Silversides, F.G. (2009) Invasive and non-invasive measurement of stress in laying hens kept in conventional cages and in floor pens. *Poultry Science* 88, 1346–1351.

Steiner, G., Bartels, T., Stelling, A., Krautwald-Junghanns, M.E., Fuhrmann, H., Sablinskas, V. and Koch, E. (2011) Gender determination of fertilized unincubated chicken eggs by infrared spectroscopic imaging. *Analytical and Bioanalytical Chemistry* 400, 2775–2782.

Stokes, A.W. (1971) Parental and courtship feeding in red jungle fowl. *Auk* 88, 21–30.

Sufka, K.J. and Hughes, R.A. (1991) Differential effects of handling on isolation-induced vocalizations, hypalgesia and hyperthermia in domestic fowl. *Physiology and Behavior* 50, 129–133.

Sufka, K.J. and White, S.W. (2013) Identification of a treatment-resistant, ketamine-sensitive genetic line in the chick anxiety-depression model. *Pharmacology Biochemistry and Behavior* 113, 63–67.

Suge, R. and McCabe, B.J. (2004) Early stages of memory formation in filial imprinting: Fos-like immunoreactivity and behaviour in the domestic chick. *Neuroscience* 123, 847–856.

Tommasi, L. and Vallortigara, G. (2001) Encoding of geometric and landmark information in the left and right hemispheres of the avian brain. *Behavioural Neuroscience* 115, 602–613.

Tommasi, L., Gagliardo, A., Andrew, R.J. and Vallortigara, G. (2003) Separate processing mechanisms for encoding of geometric and landmark information in the avian hippocampus. *European Journal of Neuroscience* 17, 1695–1702.

Vallortigara, G. and Andrew, R.J. (1994) Differential involvement of right and left hemisphere in individual recognition in the domestic chick. *Behavioural Processes* 33, 41–57.

Vestergaard, K.S. and Baranyiova, E. (1996) Pecking and scratching in the development of dust perception in young chicks. *Acta Veterinaria Brno* 65, 133–142.

Vestergaard, K.S. and Lisborg, L. (1993) A model of feather pecking development which relates to dustbathing in the fowl. *Behaviour* 126, 291–308.

Vince, M.A. (1973) Effects of external stimulation on onset of lung ventilation and time of hatching in fowl, duck and goose. *British Poultry Science* 14, 389–401.

Wauters, A.M., Richard-Yris, M.A., Pierre, J.S., Lunel, C. and Richard, J.P. (1999) Influence of chicks and food quality on food calling in broody domestic hens. *Behaviour* 136, 919–933.

Wauters, A.M., Perre, Y., Bizeray, D., Leterrier, C. and Richard-Yris, M.A. (2002) Mothering influences the distribution of activity in young domestic chicks. *Chronobiology International* 19, 543–559.

Wichman, A. and Keeling, L.J. (2008) Hens are motivated to dustbathe in peat irrespective of being reared with or without a suitable dustbathing substrate. *Animal Behaviour* 75, 1525–1533.

Wichman, A., Heikkila, M., Valros, A., Forkman, B. and Keeling, L.J. (2007) Perching behaviour in chickens and its relation to spatial ability. *Applied Animal Behaviour Science* 105, 165–179.

Wichman, A., Rogers, L.J. and Freire, R. (2009a) Visual lateralization and development of spatial and social behaviour of chicks (*Gallus gallus domesticus*). *Behavioural Processes* 81, 14–19.

Wichman, A., Rogers, L.J. and Freire, R (2009b) Light exposure during incubation and social and vigilance behaviour of domestic chicks. *Laterality: Asymmetries of Body, Brain and Cognition* 14, 381–394.

Woodcock, M.B., Pajor, E.A. and Latour, M.A. (2004) The effects of hen vocalisations on chick feeding behavior. *Poultry Science* 83, 1940–1943.

Wood-Gush, D.G.M. (1958) The effect of experience on the mating behaviour of the domestic cock. *Animal Behaviour* 6, 68–71.

Wood-Gush, D.G.M. (1971) *The Behaviour of the Domestic Fowl. Heinemann Studies in Biology* 7. Heineman Educational Books, London.

Wood-Gush, D.G.M., Duncan, I.J.H. and Savory, C.J. (1978) Observations on social behaviour of domestic fowl in the wild. *Biology of Behaviour* 3, 193–205.

Yngvesson, J. (2002) Cannibalism in laying hens: characteristics of individual hens and effects of perches during rearing. Doctoral thesis, Swedish University of Agricultural Sciences, Acta Universitatis Agriculturae Sueciae Veterinaria.

4

Behavioural Contributions to Welfare Assessment

Ever since Ruth Harrison published her landmark book detailing the intensification of animal agriculture after the Second World War (Harrison, 1964), the methods by which chickens are kept for meat and egg production have been of great concern. Members of the public raise questions about the welfare of chickens kept in very large flocks or at high stocking densities, and many people have an instinctive feeling that the birds' natural behaviours are overly curtailed or restricted in modern commercial systems. Animal scientists have raised other concerns about how the genetic background of the birds impacts on their health, and have identified new questions about how their early life experience contributes to later resilience in commercial production systems. To assess whether the public is right to be concerned about restrictions on the natural behaviour of the birds, or whether modern strains of chickens are able to cope with the challenges of commercial production systems, it is essential to develop and use validated measures of animal welfare. We cannot rely on our own feelings and intuitions to decide how chickens experience their daily lives, or how these daily experiences accumulate and affect their overall quality of life. In the first section of this chapter, we will see how behavioural studies can be used to elucidate the priorities of chickens. In the second section, the use of behaviour as a marker of internal state will be reviewed. The chapter will finish by showing how

these two approaches can be integrated so that valid indicators can be used in the assessment of chicken welfare.

Studies of Motivation, Preference and Resilience

Establishing which resources, housing environments and activities chickens value the most is critical information in a world of limited resources. With increasing pressure to produce more food at less cost for a growing human population, it makes sense to target the limited time, energy and money that humans are willing to invest to take care of chickens in the direction that will give maximum animal welfare benefits. In this first section, we will see how a study of animal motivation can be used to decide which behaviours it is essential to allow birds to perform, and how behavioural tests can be devised to reveal what chickens want. Some regard such assessments of animal preferences to be the gold standard marker of welfare, against which all other information should be compared (Barnard, 2007; Nicol et al., 2009), but they are not always straightforward to conduct or interpret, and this will also be considered. This section will also examine methods of quantifying the relative importance of difference preferences and behaviours.

a discrete choice procedure detected more subtle preferences than a free-access method (Browne et al., 2011). When the chickens in this experiment were free to enter and leave compartments whenever they pleased, their behaviour appeared largely driven by curiosity. Only when each entry to a compartment was followed by a short period of confinement, to ensure that a choice had a consequence, did the hens' marginal avoidance of a mild predatory stimulus become apparent.

Operant preference tests have also been used, whereby chickens must learn to perform a new response before they can access their preferred resource. The amount of learning can be minimal if the required response is naturalistic, such as moving against a directional wind (Faure and Lagadic, 1994) or pushing through or against a weighted door (Fig. 4.1) or narrow gap (Cooper and Appleby, 1996, 2003; de Jong et al., 2007). However, if the response can be recorded automatically, this can add precision to a preference experiment, and so chickens are often trained to peck at keys to obtain rewards (heat: Morrison et al., 1986; food illumination level: Prescott and

Wathes, 2002; food: Bokkers et al., 2004; litter or nesting materials: Smith et al., 1990; Gunnarsson et al., 2000) or to avoid punishers such as noise or vibration (Nicol et al., 1991). However, the precise nature of the operant response is not neutral and can affect the conclusions drawn from an experiment. Dawkins and Beardsley (1986) found that hens easily learned to break a photo beam to obtain access to a litter area. However, when a key-peck response was required, the motivation to obtain litter appeared to be much lower. Sumpter et al. (1995, 1998) found that hens preferred to peck a key for a food reward rather than push through a door. The hens' preference stabilized when one door push and four to five key pecks produced the same reward. Door pushing also declined as the weight required increased, and more so when the reinforcement rate was lower than when it was higher (Sumpter et al., 1998). This shows that equal schedules of reinforcement cannot be considered equivalent unless the response type is also similar.

Preference tests can be conducted only when chickens are sufficiently calm and focused to

Fig. 4.1. Measuring the motivational strength of a hen by using a swinging push-door. The door is held securely shut using a load cell. The hen must push with sufficient force to pull the metal of the door away from the electromagnet in the surrounding frame. The force required can be adjusted manually before each trial. Doors of this type have been used successfully to measure hens' motivation to reach perches, foraging materials and nest boxes. This door was supplied by Professor Linda Keeling, Swedish University of Agricultural Sciences. (Photo courtesy of Dr Poppy Statham, University of Bristol, UK.)

take part. When assessing aversive or frightening situations, chickens may either freeze or attempt to flee, rather than make active responses to avoid further exposure (Rutter and Duncan, 1991). It is therefore generally better to use passive avoidance procedures to assess aversive events such as common handling, transport and stunning practices. With passive avoidance, a chicken learns *not* to perform an active response such as walking or pecking, which it might otherwise have made, if it wishes to avoid exposure to a particular stimulus or situation. Such techniques have demonstrated with considerable precision the relative aversiveness of different transport conditions in broilers and hens (Nicol *et al.*, 1991; Abeyesinghe *et al.*, 2001a,b) or gaseous stunning or killing methods (Sandilands *et al.*, 2011).

The importance of preference tests for the assessment of chicken welfare means that it is crucial to examine the full range of factors influencing chicken choices to ensure a sound interpretation of results. Factors discussed next are the consistency of birds' choices, their ability to plan ahead and the assessment of preferences for remembered experiences.

Consistency and rationality of chicken preferences

Unlike evolutionary biologists, animal welfare scientists cannot predict in advance what the animals *should* choose. The implicit assumption of animal welfare scientists using preference testing methods is that the animals are making rational choices between complex sets of options to maximize a subjective value called 'welfare', but it is only recently that the processes underlying choice in animal welfare experiments have been explored (Browne *et al.*, 2010). Decision theory argues that animals assign each option a fixed value (utility) on a single dimension, and that they use these absolute values to choose one over another. If this is how animals make choices in preference tests then we expect them to be consistent in choosing (e.g. environment A over B) on repeated presentations. We also expect that if they choose A over B and B over C, then they should also choose A over C when this pair is presented, i.e. they should make transitive choices. Browne *et al.* (2010) discovered that individual hens showed significantly more consistency and transitivity in their choices for three environments (varying in flooring,

nesting and perching resources) than expected by chance. This was the case even though different individual birds preferred different environments (Nicol *et al.*, 2009). In addition, the most consistent birds also made the most rapid choices (Browne *et al.*, 2010).

How are chickens able to make such rational choices? As humans, our own decision making often seems a difficult cognitive exercise, one requiring us to weigh up and compare a multitude of costs and benefits associated with each option. Yet there is increasing interest in the role of emotion in human decision making, which is often less thoughtful and more intuitive than we appreciate. When choosing between an array of inexpensive consumer goods, experiments show that humans are actually more consistent and transitive in their choices when they rely on their emotional 'gut feelings' about each option than when they rely more on the cognitive processing of full written descriptions (Lee *et al.*, 2009). When fast or complex decisions are required, humans appear to rely yet further on emotional processes. It has been proposed that prior emotional reactions to the choice options are accessed as 'somatic markers' at the time of the decision (Bechara *et al.*, 1999; Bechara and Damasio, 2005). Evidence comes from specific 'gambling' tasks where humans who react emotionally to different choice situations tend to do better than participants who are less emotional or less attuned to their own emotions (Guillaume *et al.*, 2009). Early evidence suggests this could also be the case for rats (Rivalan *et al.*, 2013). As work has started only recently on the role of emotion and arousal in chicken decision making (Davies *et al.*, 2014) (Fig. 4.2), it is too early to draw conclusions about the mechanisms used when chickens choose among environments or resources that vary in many dimensions, but this is likely to be an area for fruitful future research.

Preference tests and planning ahead

It has been questioned whether chickens are able to assess the longer-term implications of the choices they make in preference tests (Fraser and Nicol, 2011). Yet evidence that chickens can choose to wait for a larger reward rather than accepting a smaller, more immediate reward provides some evidence that they can plan ahead, if only for a matter of a few seconds. In an experiment by

Fig. 4.2. Recording arousal in a chicken using a non-invasive heart-rate monitor. In these studies, the chickens are acclimatized to wearing empty backpacks for a few days before the heart-rate monitor is placed in the pack. Two electrodes are placed on the skin under the wings and the information stored in the monitor. The system was described by Lowe *et al.* (2007). (Photo courtesy of Dr Anna Davies, University of Bristol, UK.)

Abeyesinghe *et al.* (2005), chickens learned preferentially to peck a coloured button that signalled a 22 s feeding period after a 6 s delay, rather than a differently coloured button that provided food after a 2 s delay but only for a duration of 3 s. This is not quite the same as showing that chickens possess self-control (as in the famous marshmallow tests given to children; http://en.wikipedia.org/wiki/Stanford_marshmallow_experiment): in the experiment of Abeyesinghe *et al.* (2005), the chickens did not have to inhibit pecks towards food that was already in front of them in order to gain a larger, delayed reward. However, it does show that chickens may have some capacity to consider the future, with all the costs and benefits this may bring (Mendl and Paul, 2008).

Preference for a remembered experience

Most preference tests have examined the choices of birds for external objects or resources. However, chickens may also find their own internal states more or less pleasant, and preference tests can also be used in this context. Conditioned place preference or place avoidance tests make use of the ability of chickens to form learned associations between previously neutral places and their own internal state. For example, Nasr *et al.* (2013) wanted to know whether keel fractures, which are a common problem in commercial laying hens, remained painful once they had healed. Nasr *et al.* (2013) gave chickens with and without fractures a pain killing drug and once the drug had taken effect the chickens were housed in a coloured environment. On other occasions, chickens were given a saline control and housed in an environment of a different colour. After this, the chickens, by now drug free, were given a choice between the two differently coloured environments. Nasr *et al.* (2013) found that hens with no fractures had no preference for one environment over the other, but the hens with healed fractures chose the environment where the analgesic had previously been experienced. In other words, the hens formed a conditioned place preference for a previously neutral location based on their prior subjective experience in that place. Recently, it has also been shown that chicks form conditioned place aversions for white noise (Jones *et al.*, 2012). An attempt to demonstrate conditioned place preference in broiler breeder chickens was less successful (Buckley *et al.*, 2012), possibly because these birds are chronically hungry and hunger is known to interfere with learning in chickens (Nicol and Pope, 1993; Buckley *et al.*, 2011).

Strength of motivation

One of the difficulties with interpreting the preference tests described above is that they depend entirely on the options offered by the experimenter. Thus, a chicken may be offered a choice between two types of wire floor, neither of which is truly comfortable. It is useful therefore to compare how motivated chickens are to obtain certain resources or behavioural opportunities against a more standardized 'yardstick' (Dawkins, 1983b), most practically a known degree of food deprivation (Dawkins, 1983b). The strength of preference to obtain food can be assessed using techniques derived from microeconomics with its focus on how consumer demand changes with altered income and good pricing. In the context of animal welfare, a situation can be established where a defined amount of work by a chicken (e.g. pecking a key a certain number of times) results in a fixed reward (e.g. a food pellet). The amount of work is then increased using a progressive fixed ratio (FR) schedule so that the number of pecks per reward increases steadily over a course of experimental sessions, e.g. FR of 1, 2, 4, 8 and 16. Alternatively, the FR requirement can vary either upwards or downwards across sessions without progressing in any particular pattern. The chickens' responses are plotted as the log of the number of rewards obtained versus the log of the FR values across all sessions. Hungry chickens show an inelastic demand for food, expending more effort as the cost increases, so they obtain the same amount of food when it is expensive as when it is cheap. The inelastic demand for food can then be compared with the pattern of demand shown for other resources, so that the experimenter can establish which are most important from the chickens' perspective. The decline in the amount of resource obtained as it becomes more expensive is a measure called the elasticity of demand. Resources or commodities with a very elastic demand are considered to be less important for welfare than resources with a relatively inelastic demand.

Only a few demand experiments have been conducted with chickens, and these have focused on the birds' motivation for litter substrates or nest access. Cooper and Appleby (1996) calculated the rate at which demand for a nest changed with price. Others have calculated formal slopes of the bi-logarithmic plots mentioned above. Using this method, the slope of the demand curve for peat has been reported as –0.36 (Matthews *et al.*, 1995) and for straw as –0.45 (Gunnarsson *et al.*, 2000). However, any experiment that produces a single figure summarizing demand may be somewhat misleading, as the elasticity of demand can itself change with price or consumption level; for example, a chicken may have an inelastic demand for a litter substrate up to a certain price level. After this point, its demand may become highly elastic over further price increases. For these reasons and others, it has been suggested that experiments consider other measures of resource importance, including the maximum price the animal is willing to pay and the total amount of work the animal is willing to perform (Mason *et al.*, 1998; Kirkden *et al.*, 2003). Such work has been pursued in pigs, calves and horses (Jensen and Pedersen, 2008), but with little new information on chicken demand. It is surely time for a revival.

Behavioural resilience

Another way of assessing the importance of particular behaviours is to examine their resilience in the face of large-scale changes in circumstance. Houston and McFarland (1980) conceived the term behavioural resilience to describe the extent to which a given behaviour is compressed when other behaviours occupy more of the time or energy budget. Dawkins (1983b) conducted a resilience experiment with laying hens by housing them under conditions where they were allowed to access food and a litter substrate for 8, 4 or 2 h per day. At all other times, they were housed in battery cages with water *ad libitum* but little else. When 8 h was available, the hens allocated the majority of their time to litter-related activities, although they still spent over 2 h feeding. When only 2 h was available, the hens spent most of this time feeding, clearly showing that litter-related activities were less resilient. Dawkins (1983b) regarded this experiment as more of a proof of method than a conclusive demonstration of chicken priorities. She wrote:

> All that can be said for this very preliminary result is that the postulated need for litter does not clear the very highest hurdle, that of being as strong or stronger than the need for food in hungry birds. The experiment still seems worth reporting, however, because even though it is inconclusive, it illustrates a method which could be refined.

One reason why the method has not been widely used since then may be because it is difficult to apply manipulations that affect all behaviours equally (Mason et al., 1998). If the animal's time budget is reduced due to short daylight hours in winter, this could affect time-consuming behaviours (e.g. foraging) more than rarely occurring and brief-duration behaviours (e.g. wing flapping). However, in many natural situations, both energy and time budgets may be affected by external pressures such as reduced diet quality, increased egg production or chronic illness. Observations and experiments in a variety of species suggest that some behaviours are less resilient than others. Sleep is reduced in lactating baboons with a higher energy expenditure (Dunbar and Dunbar, 1988), cows reduce their use of an automated brush under high temperature and humidity conditions or when food is located at a distance (Mandel et al., 2013), climbing and other exploratory activities in mice are reduced as illness progresses (Littin et al., 2008) and play is often reduced in the face of a wide range of pressures (Held and Spinka, 2011). Experimental or observational studies of behavioural resilience in chickens facing similar pressures would be of great value.

Meta-preferences

Chickens have preferred ways in which they seek to obtain preferred resources. They often prefer to exert a degree of control or influence over events, choosing, for example, to forage for food rather than accept the same food freely given (Duncan and Hughes, 1972). Foraging for food rather than simply eating it from a feeder provides chickens with additional information about likely future resources and provides positive feedback as the chicken's behaviour directly influences the outcome. Feral chickens and wild junglefowl have a great deal of behavioural freedom as they can select where and when to forage, which food particles to ingest, whether to stand in shade or sunlight, and how close to get to other birds or humans. In captivity, individual chickens have much less control, partly because they are housed under conditions where certain resources are absent and others (e.g. diet) are uniform, and partly because exercising control can be difficult at high stocking densities, where walls block escape or

where movement is compromised due to injuries such as keel fractures.

This lack of control will have welfare consequences. It has been known for decades that exposure to uncontrollable aversive events, compared with controllable events or no stimulation, results in 'learned helplessness', increased fear, stress and susceptibility to disease (reviewed by Mineka and Hendersen, 1985). To give just one example, rats that could perform an action to terminate a shock showed a lower incidence of stress-related conditions than rats in a 'yoked' condition. The yoked rats experienced exactly the same amount of shock but without the element of control (Mineka and Hendersen, 1985). Experimental psychologists have shown that chickens too develop passivity if exposed to uncontrollable shock (Job, 1987), as well as increased fearfulness (Rodd et al., 1997) and a bias in attention towards predatory cues (Rodd et al., 1997).

If it is difficult to provide chickens with control, it may be easier to improve the predictability of their environment. The two concepts are intertwined, and a controllable event is, by its very nature, also predictable. Knocking on the door before entry is a common practice in chicken husbandry, but the benefits of this practice have been formally documented only in monkeys (Rimpley and Buchanan-Smith, 2013). It is less clear whether signalling the arrival of reward or giving an animal the opportunity to make a response to produce its own reward results in any welfare benefits beyond those associated with the reward itself. Positive effects of control over heat/light have been reported for monkeys (Buchanan-Smith and Badihi, 2012) but not for pigs (Jones and Nicol, 1998). In a yoked-design experiment, Taylor et al. (2001) found that young laying hens pecked extensively at operant keys so that they could obtain food at times when free food was not available. Additionally, some birds continued to key-peck during afternoon periods when free food was provided for all birds, and many hens also pecked a different key to obtain additional light intensity over the litter areas of their pen. The yoked groups that received identical (free but unpredictable) rewards did not peck at the nonfunctional operant keys placed within their pen. However, the welfare benefits associated with greater environmental control were difficult to quantify. Birds with control performed less preening and resting, and had significantly lower egg production, results that could be interpreted in more than one way. Taylor *et al.* (2001) argued that the birds with control appeared to have slightly lower stress levels, but more detailed experiments would be necessary to substantiate this. In contrast, a *loss* of predictability or control is met with frustration vocalizations (Zimmerman *et al.*, 2000a) and can lead to aggression (Haskell *et al.*, 2004). Overall, it seems likely that, by using simple cues to signal the onset of more frightening events (such as the intermittent operation of loud fans), welfare could be improved. The potential to improve welfare further by providing chickens with greater control in their environments awaits further research.

Behavioural Markers of Internal State

The second major area where behavioural science has contributed to the assessment of chicken welfare comes from the identification of activities, vocalizations and other responses that are markers of otherwise 'hidden' internal states. These internal states might reflect emotions such as fear, deprivation, frustration or anticipation, or relate more directly to physiological changes associated with stress, disease or pain. Sometimes such states might be signalled by unique and characteristic vocalizations or behaviours. If so, once the association between internal state and outward behaviour is recognized, the behaviour can be used as a direct indicator of welfare. At other times, the behavioural changes that occur in association with shifting motivational or emotional states will be more subtle, and may require the identification of small shifts in the timing, sequencing and patterning of behaviour. Such shifts could be useful future indicators of welfare state.

Behavioural markers of stress

Animals mount a stress response when they perceive a real or impending threat to their homeostasis. Stress is not inherently a bad thing, although if the stress response fails to counter the perturbation and return the animal to a state of balance, the chicken may become distressed and at some point its welfare will be compromised. The complex physiological stress response mounted by a chicken

may be accompanied by observable behavioural changes, and these can be very obvious; for example, the acute stress response is not called the 'fight or flight' response for nothing.

However, where chickens have been unable to avoid a threat by fleeing and are in a state of chronic stress, the behavioural signs may be less obvious. It is early days for research on behavioural organization (Asher *et al.*, 2009), but it seems that certain patterns of behaviour are associated with stress levels in chickens. Chickens are very busy and active animals, changing behaviours frequently. In one study, 70% of behaviours lasted less than 2 min before the chickens changed to another activity (Mishra *et al.*, 2005). However, transitions between activities appear to become even more frequent when chickens are housed in more barren (Pohle and Cheng, 2009) or less preferred environments where indicators of physiological stress are higher (Nicol *et al.*, 2009). In this context, the resultant shorter bout lengths give a distinct impression of agitation.

It has also been proposed that more structured sequences of behaviour occur under conditions of stress. An extreme example of a highly structured sequence would be the appearance of invariant stereotypic behaviour. Sequence variability can be quantified using a type of fractal analysis called detrended fluctuation analysis, and differences in sequence variability have been detected in chickens exposed to a variety of mild stressors, although not always in the predicted direction (Maria *et al.*, 2004) and sometimes not at all (Hocking *et al.*, 2007). Another approach to examining structural variability is to examine the relative timing of different activities in the behavioural repertoire. Feed restriction of broiler breeders produces a situation where behaviour becomes more structured, with activities such as stepping and preening becoming more predictably linked in time (Merlet *et al.*, 2005). However, more work is clearly needed before we can confidently associate structural changes with welfare outcomes in chickens (Asher *et al.*, 2009).

Behavioural markers of pain and disease

Behavioural changes are often seen in diseased chickens, with reduced feed intake, huddling and postural changes occurring, as the birds attempt to conserve energy and fight infection (Hart, 1988). However, behavioural changes can also be more subtle signs of subclinical infection, the first early-warning sign of a problem or associated with chronic pain. Toscano *et al.* (2010) found that chickens with a subclinical *Salmonella* infection showed reduced intake of their normal pelleted feed and reduced foraging among the litter but increased intake of waxworms, which were on offer for a 5 min period each day, possibly reflecting a greater need for easily processed energy. Broiler chickens with leg weakness, skeletal abnormalities or joint infections show clear behavioural changes, the most obvious of which is a reduced motivation to walk and an uneven or abnormally slow gait. The extent of lameness within a flock can therefore be assessed by people trained in standardized gait-scoring (http://www.plosone.org/article/info%3Adoi%2F10.1371%2Fjournal.pone.0001545#pone.0001545.s006; Knowles *et al.*, 2008).

Compared with sound birds, lame broiler chickens also show alterations in other patterns of behaviour, making fewer visits to feeders (Weeks *et al.*, 2000). Standardized tests of the reduced walking and standing ability of broilers have been developed and validated as good measures of leg health (Weeks *et al.*, 2002; Caplen *et al.*, 2014). The birds' latency to lie in a shallow water bath, for example, is highly correlated with its lameness state, even when other factors such as age and weight are controlled for (Caplen *et al.*, 2014). In addition, lame birds tend to show greater skin sensitivity to heat or pressure stimulation (Hothersall *et al.*, 2014). The fact that the experimental administration of analgesic drugs partially restores normal behaviour (Caplen *et al.*, 2013) suggests that these behavioural changes are good indicators of bird pain.

On a larger scale, the overall activity levels of many individual chickens can be quantified by automatic image analysis systems (Aydin *et al.*, 2010; Roberts *et al.*, 2012). Small changes in individual bird behaviour, such as a reduction in walking bout length, are detected at the macroscopic scale as changes in the overall pattern of flock movement and can provide a useful early predictor of lameness or other flock health problems (Dawkins *et al.*, 2013). Automated monitoring of broiler chicken vocalizations as an indicator of stress or disease and the use of unmanned aerial drones are both being developed at the Georgia Tech Research Institute as new methods of assessing the condition of very large flocks of commercial birds (http://www.atrp.gatech.edu/PT25_2/25_2p2.html).

More subtle behavioural changes may also be detected in laying hens with keel fractures. These birds are slower to fly down from a raised perch (Nasr *et al.*, 2012, 2013) and less likely to use the outdoor range area in free-range systems (Richards *et al.*, 2012). However, at present, these tests lack the specificity and sensitivity needed to reliably distinguish fractured from non-fractured birds on the farm.

Behavioural markers of emotional state

Validating behavioural markers of emotional state in chickens is a very recent enterprise. Indeed, all work on bird emotions is in its infancy. However, based on common features shared with humans, it seems justified to say that chickens can experience some primary emotions such as fear, frustration or attachment. Human emotions vary in intensity and valence (pleasant or unpleasant characteristics) and are often described or self-reported in terms of feelings. Chicken emotions also vary in intensity (as shown by measures of physiological arousal) and valence (as shown by approach or avoidance of situations that affect emotion) and are organized into functional and neutrally coordinated systems. However, the nature of any subjective feelings experienced by a chicken cannot be known using any current scientific methods (although, as we saw in the preceding chapter, there is also no good reason based on brain structure to *exclude* the possibility of conscious experience in these birds).

Discrete neural circuits identified using electrical brain stimulation and other methods reveal (in mammals at least) that there are at least seven primary emotional systems, described by Panksepp (2005) as Seeking, Fear, Rage, Play, Lust, Care and Panic/Grief. But which of these are possessed by chickens? Certainly they have a well-developed seeking system and develop states of frustration (when resources are difficult or impossible to access) and deprivation (when resources are absent). Anticipation is another emotional state related to the seeking system that arises when the arrival of a reward or punishment is expected. Anticipation of reward, closely related to the seeking system, would be expected to induce a state of pleasurable excitement, and recent experiments have started to investigate this possibility (Moe *et al.*, 2009). Chickens also have a well-developed fear system, characterized by vigilance and flight behaviour in response to actual or perceived threat. Without a doubt, the fear system has been studied in more detail than any other emotion in chickens (Jones, 1996). Anticipation of a punishment or a threat would be expected to induce a state of anxiety, closely related to the fear system.

Their capacity for other emotions is less clear and awaits further research. Rage may be limited to brief and momentary attack during the establishment of a dominance hierarchy. Chickens do not show the type of impulsive, violent aggression characterized by high levels of arousal that is sometimes observed in mammals such as cats and rodents. However, parts of the neural circuitry concerned with aggression and attack may be integrated with those of the seeking system (Mellor, 2012) to produce highly focused predatory attack behaviour. Anyone who has observed the chilling and effective despatch of invertebrate prey items by free-range chickens would appreciate that there may be some truth in this assertion. As reviewed in Chapter 5, the chicken's capacity for play is very restricted. Female chickens also show little evidence of lust, often seeming more to succumb to the mating demands of a cockerel than to seek out this experience. Pizzari and Birkhead (2000) stated, for example, that most matings in domestic fowl are coerced. There is little or no evidence of sexual motivation in females (van Kampen, 1994). In quail, a female bird is a reinforcer for males but the presentation of a male is not reinforcing to females (Domjan and Hall, 1986). Broody hens show great concern for their offspring, caring for them with a suite of specialized behaviours and showing evidence of emotional arousal if the chicks become distressed (Edgar *et al.*, 2011). Within the hen–chick context where social bonding is of paramount importance, the anticipation or experience of social isolation may result in anxiety disorder or even, however unlikely it may seem when discussing chickens, depression (Salmeto *et al.*, 2011). However, social bonding appears to have little relevance in modern commercial production. The broiler and layer breeding and laying stock have lost the capacity to show brooding behaviour such that their eggs are incubated

artificially, while broiler chickens themselves are slaughtered before reproductive age. As adult chickens may not develop other types of social attachment (Abeyesinghe *et al.*, 2013), emotional systems relating to care (including panic or grief if separated from a bonded companion) would seem vestigial.

Given the question marks that remain over many of the emotional systems of chickens, the following subsections will therefore review the behavioural markers, likely valence and welfare consequences of emotional states associated with the seeking (deprivation, frustration and anticipation) and fear systems. If changes in behaviour or demeanour, identified during situation-specific experiments, are sufficiently specific and characteristic markers of emotional state, it might be possible to use the same indicators on farms to infer something about the welfare of chickens in commercial flocks.

Deprivation

Deprivation implies the total absence of certain external cues or resources (perhaps sunlight, or litter substrates) that a chicken might value. The extent to which resource deprivation is a welfare problem can be examined by comparing the health and overall welfare of birds housed with and without these resources. For example, we may find that chickens housed with perches have stronger leg bones and better plumage condition than conspecifics without perches. It is also possible for chickens to act as if nothing was missing at all by performing 'vacuum' (or sham) behaviour. For example, chickens kept on wire floors will occasionally perform many of the elements of dust-bathing behaviour. Although no improvement of plumage condition is obtained in the absence of litter, simply going through the motions is (partially) satisfying for the birds (Lindberg and Nicol, 1997). However, a deeper question is whether chickens 'miss' (are aware of the absence of) these resources – in other words whether out of sight (or out of perceptual range) is also out of mind. The complexity of this question is highlighted by the possibility that rebound responses observed after periods of spatial confinement and the absence of litter (Vestergaard, 1982; Nicol, 1987b) might be responses to the novelty of stimulus representation, and not evidence after all that chickens 'missed' the resources during the deprivation period.

Normal preference and demand tests cannot resolve this issue as they always allow the test animal to perceive the options available, thereby reminding a deprived animal of something it might otherwise have forgotten. If a chicken performs an operant response, or remembers and returns to a location linked with a previous experience, this provides evidence of a degree of memory associated with particular resources. However, it could still be argued that the presence of conditioned stimuli linked with the primary resources or experiences in question (operant keys in demand experiments, or coloured environments in conditioned place experiments) provide the reminders or aide-mémoires.

One way to detect whether a chicken perceives something is missing from its environment is to examine its responses *during* the period of deprivation, to find out whether it carries on as normal or whether it is perturbed. When food, water, foraging or dustbathing materials are removed, chickens produce more gakel vocalizations (Zimmerman *et al.*, 2000a) and increase their searching and exploratory behaviour (Nicol and Guilford, 1991; Lindqvist *et al.*, 2009). In one experiment (Nicol and Guilford, 1991), litter-deprived laying hens spent 33% of their time exploring a circular tunnel compared with just 13% for non-deprived birds (both groups were tested without immediate access to litter). As the tunnel provided no litter-related external cues that could have raised motivation, this experiment suggests that chickens perceive that something is missing and take steps (literally in this case) to try to improve their situation. Cooper and Appleby (1995) similarly recorded more exploratory behaviour in hens deprived of a nest box than in hens with a nest box. These studies certainly suggest that chickens do not forget entirely about absent resources that they have experienced previously and that awareness of a missing resource (possibly represented as a mental image) may be negatively valenced and may activate the emotional seeking system.

A deeper question is whether a chicken could 'miss' a dustbath or a nest if it had been reared without any prior experience of such a resource. Cooper and Appleby's (1995) work found that the amount of exploratory behaviour did not differ between hens that had previously experienced nest boxes and hens that had never encountered this resource, suggesting that chickens may

possess innate concepts of some resources. Despite these pioneering attempts, it may be difficult or impossible to establish the consequences of depriving chickens of certain resources. I am not sure how one could assess whether indoor-reared chickens pined for the appearance of sunlight. In addition, the rather general increase in activity seen when resources are absent would be hard to detect on a commercial farm.

Frustration

If a physical or social restriction limits access or utilization of a potentially available resource, the situation is termed thwarting. When feeding is prevented by placing a transparent but impenetrable cover over a food bowl, chickens may perform short, seemingly irrelevant bouts of behaviours such as ground pecking or preening (Duncan and Wood-Gush, 1972a,b). These indicators of mild frustration are sometimes called displacement behaviours (Duncan and Wood-Gush, 1972b). Repetitive pacing is a sign of more intense and aversive frustration (Duncan and Wood-Gush, 1972a). In discrimination learning tests, chickens that made more mistakes redirected pecks towards the experimenter (Kuhne *et al.*, 2011). Chickens also give characteristic gakel calls when they are thwarted from accessing highly valued resources, the rate of calling increasing with the level of frustration and in the presence of similarly frustrated companions (Zimmerman *et al.*, 2000b, 2003). Under natural conditions, the gakel call is given by female hens during the search phase of nesting behaviour, when she is sometimes accompanied by a male bird (McBride *et al.*, 1969). This indicates one specific function of the call, but overall evidence suggests that it is a rather general signal of a frustrated internal state (Zimmerman *et al.*, 2000b).

Anticipation

Chickens can form expectations and anticipate events in the near future. When simple learned associations are formed between a neutral conditioned stimulus and a biologically meaningful unconditioned stimulus, chickens will respond to the conditioned stimulus with biological responses appropriate to the unconditioned stimulus. For example, if a sound signals the arrival of food, then chickens may pace and head bob near the feeder in anticipation. If the interval between sound and food arrival is very short, the chicken may start pecking the feeder as soon as it hears the sound. However, with longer delays, chickens show more generic anticipatory responses such as increased activity. If the unconditioned stimulus is aversive (e.g. handling or shock), then chickens can show a phenomenon called conditioned suppression where they will cease performing ongoing unrelated activities when they hear the predictive signal (Spevack et al., 1974). However, regardless of whether the expected event is positive (mealworm) or negative (water spray), chickens increase the frequency and intensity of their head movements during a signalled anticipatory period (Zimmerman et al., 2011; Moe et al., 2013; Davies et al., 2014). Other signs of anticipation include increased heart rate (Davies et al., 2014) and reduced peripheral temperature (Moe et al., 2012), although whether these physiological changes are specific to food reward or more general signs of arousal is not yet clear. One change that does seem to be specific to anticipation of a positive reward is an increase in comfort behaviour (Zimmerman et al., 2011). In these experiments, the anticipatory behaviours of the chickens had no direct influence on the arrival of the reward or aversive stimulus.

A procedure called trace conditioning shows that chickens do not simply respond to the presence of the conditioned stimulus as if it is a perfect substitute for the unconditioned stimulus. An example of trace conditioning would be a situation where a sound signal was presented but removed again before a food reward arrived. Anticipatory responses observed during the interval when neither signal nor reward is present must be based on memory of previous events leading to expectations about future ones. Chickens show an alert demeanour with open eyes, slow steps and their gaze directed appropriately towards the expected reward during trace conditioning intervals that exceed 20 s (Moe et al., 2009).

Anticipation of a reward is, in humans, often accompanied by a sensation of pleasure. Indeed, a reduced capacity to show reward anticipation is often associated with anhedonia, and occurs in long-term negative situations such as ageing (rats: Maasberg et al., 2011) and depression (humans: Smoski et al., 2009). The appearance of anticipatory behaviour may therefore be a sign that the baseline welfare of chickens is reasonably good. However, a chicken with good welfare, living in a complex environment, may show *less* of an anticipatory response to reward than a chicken housed in a more barren environment for whom the occasional reward will carry far more significance. Thus, a moderate degree of sensitivity to reward may be the best indication of good welfare, but much work will need to be done to define moderate in this context. In addition, the degree of overlap between the responses observed when chickens anticipate rewarding and aversive events makes it difficult to use indicators such as head movements or gaze direction as signs of positive emotion or affect unless the nature of the anticipated event is also known. These responses are probably more indicative of arousal than of anything more specific. Distinguishing pleasure from mild fear or frustration (because a reward has not yet materialized) is not straightforward.

If it can be shown in the future that chickens do experience anticipatory pleasure (not just anticipatory arousal), it might be possible to improve their welfare by providing more signalled anticipatory periods before reward provision. This type of strategy would work only if chickens did not come to expect the higher reward level all the time! It would also be important to check that any pleasure associated with the new signalled reward schedule did not simultaneously result in devaluation and reduced pleasure associated with the normal diet.

Fear

Fear is the reaction to a perception of danger. Fearful chickens will usually also mount a stress response (although chickens may be stressed for many reasons unrelated to fear). The underlying tendency of a chicken to react to a threat or a perceived threat, its baseline fearfulness, can be influenced greatly by its genetic background (Keer-Keer et al., 1996), epigenetic factors and its rearing environment, as reviewed in earlier chapters. However, all chickens show a similar range of behavioural responses when presented with threats such as novel objects (particularly those that appear suddenly), when placed in a novel arena or environment, when exposed to sudden noises or when potential predatory stimuli are perceived (Jones, 1996). Husbandry practices that replicate many or all of these threats simultaneously, such as catching and transportation, are particularly potent in inducing fear (Mills and Nicol, 1990). Indicators of fear in chickens include both immobility and flight responses. In some circumstances, the best response is to avoid detection by crouching or freezing. If the worst comes to the worst and the chicken is caught by a potential predator, then its best option may be to 'play dead' and enter a state of complete tonic immobility (Jones, 1986). In other circumstances, a frightened chicken will try to flee by running or flying.

Standardized procedures and tests have been developed to measure and quantify the fear reactions of chickens. One of the most common is to place individual birds in a novel arena and measure activity – the general interpretation is that the lower the activity level and the more hesitant the steps, the greater the fear level of the bird. A caveat to this interpretation is that birds that move less in the arena can sometimes be birds that have a low desire to regain contact with their companions (Forkman et al., 2007). Novel object tests can be conducted for caged or free-living birds with the general premise that fear will increase the latency with which a bird will inspect a novel object. The duration of tonic immobility following restraint by a human experimenter, under defined conditions, is also a reliable measure of fear in chickens (Jones, 1986).

All of these responses are good indicators of fear as they are relatively specific, they correlate with physiological indicators in a meaningful way and they tend to correlate with each other (Jones, 1996; Forkman et al., 2007). Measures of fear and stress are also often (Jones, 1996) but not always correlated. Certainly the terms should not be used interchangeably. For example, in one experiment, laying hens that showed a fearful, slow approach to a novel object also had lower indicators of stress (heterophil:lymphocyte ratio, glucose and acute phase protein) than hens that approached rapidly (Nicol et al., 2011b). The neural circuitry of the fear system is starting to be elucidated in chickens and their close relatives. The arcopallium and posterior pallial amygdala appear partially homologous with the mammalian amygdala and play a central role in the emotional fear system of birds (Saint-Dizier et al., 2009; Kops et al., 2013).

Anxiety

The anticipation of punishment, threat or another aversive experience may induce a chronic emotional state of anxiety. We saw in Chapter 3 that a brief period of isolation in chicks resulted in increased distress calling, which could be ameliorated to some extent by the administration of anti-anxiety drugs. However, whether the chicks enter a state of fear or a state of anxiety during social isolation depends to a great extent on which definitions are used. The distress calling and other emotional responses of the chick during social isolation are part of a normal, adaptive response to threat. The chicks are not anticipating social isolation but experiencing it directly and might be considered by many to be in a state of fear. However, their increased vulnerability may mean that they (consciously or not) have a heightened anticipation of other threats such as predation. Even so, this short-term adaptive response bears only a limited relationship to human anxiety, which is characterized by a disproportionate and chronic reaction to perceived and anticipated threats, often to the extent that this interferes with normal daily function.

However, despite the many differences between the adaptive short-term fear/anxiety state of the isolated chick and the multi-faceted anxiety states of humans, there is a hint that both may be accompanied by analogous changes in perception. Anxiety in humans and other mammals is correlated with changes in cognitive appraisal of environmental stimuli (Burman et al., 2009) and resultant biases in attention (Paul et al., 2005). One specific example of a cognitive bias is the tendency of anxious individuals to be more likely to interpret ambiguous stimuli as aversive than healthy individuals. It is therefore significant to note that young chicks isolated for just 5 min before testing are slower to approach an ambiguous visual silhouette with intermediate owl and chick characteristics than non-isolated chicks, demonstrating a short-term bias in their perception of the environment (Salmeto et al., 2011).

Depression

If isolated for more than about an hour, chicks reduce their rate of distress calling and fall into a more quiescent state. Because anti-depressant drugs, and other treatments such as environmental enrichment, can slow or reduce the onset of this state (as outlined in Chapter 3), this state has been labelled 'depression', despite the fact that its development, timing and long-term consequences share only minimal features with human depression.

However, similar to the situation with anxiety, there may be shared perceptual changes that are characteristic of both depression in humans and quiescence in the socially isolated chick. In humans, states of depression and anxiety both result in a greater tendency to judge potentially aversive but ambiguous stimuli as aversive, but depression also reduces the tendency to judge potentially appetitive stimuli as rewarding – depressed individuals can be portrayed as both more pessimistic and less optimistic than people in good mental health. When chicks that had been isolated for an hour and had fallen into a state of quiescence were given cognitive bias tests, they were slower to approach an intermediate chick/owl silhouette and also slower to approach cues indicating a food reward than non-isolated chicks (Salmeto et al., 2011). The drug imipramine, which affects many neurotransmitter systems involved in anxiety and depression, abolished these cognitive bias effects in these chicks (Hymel and Sufka, 2012).

In contrast, when adult birds were housed in an enriched versus a barren pen for some weeks, this did not induce any changes in cognitive bias, although many other social and motivational traits influenced behaviour in the test situation employed (Wichman et al., 2012). As this is a very new area of work, it is worth considering the possible reasons for a negative result of this kind. There may be technical difficulties in designing the right tests of cognitive bias for adult chickens that will be overcome with relative ease in the future. However, at a time when the value of cognitive bias tests is still being assessed, it is crucial to have independent measures of the emotional state of the subjects. In this case, as Wichman et al. (2012) pointed out, the hens had previously spent many months in cages, so in both housing treatments they may have been (relatively) elated rather than depressed. It is also possible that states of anxiety and depression will occur only in young chicks during that period of their life where they are imprinted and socially bonded to the hen, i.e. when the 'bonding/panic' system (Mellor, 2012) has the potential to be activated. Adult birds (with the possible exception of broody hens) do not appear to form social bonds and therefore show no panic on separation from any particular individual, although they may seek the general protection of the flock. For all of these reasons, establishing the significance of the prior manipulations using 'gold standard' methods is important.

The Best Possible Welfare

Behavioural resilience is a concept of considerable importance in debates about chicken welfare as it can help to identify not only situations of poor welfare but also situations of very good welfare, as explained next. Animal welfare science is occasionally portrayed as a field devoted to animals' negative experiences of pain, distress, frustration and discomfort, and there are calls for greater consideration to be given to the animals' capacity for pleasure. Sometimes, this is just a semantic issue, as wanting an animal to be free from discomfort is essentially the same as wanting it to experience comfort. I prefer the term 'better welfare' to the term 'positive welfare', as it is impossible to identify a point where negative might change to positive (unless we restrict the definition of positive welfare to situations that an animal approaches rather than avoids). However, putting definitions to one side, it is valid to seek to improve the welfare of farm animals to the greatest degree possible.

One approach to doing so is to ask experts what they believe should be provided to achieve the best possible welfare for laying hens (Edgar et al., 2013). The intention is admirable, and many of the suggestions made seem eminently sensible, but there is a risk of returning to the starting point where we learned that humans are not always good judges of animal needs (Hughes and Black, 1973). This is why indicators of better welfare are needed. In mammals, it might be possible to look for the expression of positive emotions such as play, social care or lust (Panksepp, 2005), but, as outlined above, these emotions are either not highly developed in adult chickens or have been lost in commercial breeds. We will see in future chapters that chickens do not engage in social play or have strong social affiliations. This is why behavioural resilience is such a valuable tool. If low-resilience behaviours (e.g. dustbathing, long bouts of preening and exploration) form a significant part of the chickens' daily time budget, this is a good indication that all is well. Low-resilience behaviours are sometimes described as 'luxuries', but this is misleading. In natural conditions, these behaviours are important for long-term fitness – animals that look after their plumage and gather information for an uncertain future are more likely to pass on their genes than animals that attend only to their proximate needs.

Integrated Assessments

In the real world, it is not possible or sensible to try and conduct preference tests to assess the welfare of chickens on farms. It would take too long and be far too difficult. Instead, welfare is measured using a range of physiological, behavioural or clinical indicators such as those listed in the Welfare Quality® protocol for laying hens (http://www.welfarequality.net). Auditors may visit a farm and assess whether the birds have injuries or whether their claws are overgrown. However, the basis for including some indicators in welfare assessments is obscure. Barnard (2007) argued that information about animal choice could provide a gold standard to make sense of welfare indicator data. The fact that these two approaches to animal welfare assessment – animal decisions and welfare indicators – had been pursued almost entirely in isolation for over 30 years prompted a programme of work to bring the two approaches together in a systematic way (Nicol *et al.*, 2009).

We housed hens in sequential pairs of environments and measured a broad range of putative welfare indicators. We then allowed the hens to make choices between the two environments most recently experienced and repeated the sequence of testing over many months. We then repeated the entire experiment using different hens and different environments (Nicol *et al.*, 2011a). Although different individual hens preferred different environments, we were able to link certain indicators with general environmental preference or avoidance. Regardless of the nature of the environment selected by the hens, lower levels of blood glucose, heterophil:lymphocyte ratio, corticosterone, and head shaking and higher levels (particularly durations) of preening were associated with preference. New indicators were identified (a lower moisture content in faeces was linked with preference), while some traditional measures showed no apparent associations. This work is only a start, but it demonstrates that behavioural methods have an important contribution to make in the validation of new measures and indicators. It may not be possible to conduct time-consuming preference tests on commercial farms, but those indicators that have been validated against the birds' preferences in the laboratory can be incorporated into practical and realistic on-farm assessment and auditing protocols.

References

Abeyesinghe, S.M., Nicol, C.J., Wathes, C.M. and Randall, J.M. (2001a) Development of a raceway method to assess aversion of domestic fowl to concurrent stressors. *Behavioural Processes* 56, 175–194.

Abeyesinghe, S.M., Wathes, C.M., Nicol, C.J. and Randall, J.M. (2001b) The aversion of broiler chickens to concurrent vibrational and thermal stressors. *Applied Animal Behaviour Science* 73, 199–215.

Abeyesinghe, S.M., Nicol, C.J., Hartnell, S.J. and Wathes, C.M. (2005) Can domestic fowl, *Gallus gallus domesticus*, show self-control? *Animal Behaviour* 70, 1–11.

Abeyesinghe, S.M., Drewe, J.A., Asher, L., Wathes, C.M. and Collins, L.M. (2013) Do hens have friends? *Applied Animal Behaviour Science* 143, 61–66.

Albentosa, M.J. and Cooper, J.J. (2005) Testing resource value in group-housed animals: an investigation of cage height preference in laying hens. *Behavioural Processes* 70, 113–121.

Asher, L., Collins, L.M., Ortix-Pelaez, A., Drewe, J.A., Nicol, C.J. and Pfeiffer, D.U. (2009) Recent advances in the analysis of behavioural organization and interpretation as indicators of animal welfare. *Journal of the Royal Society Interface* 6, 1103–1119.

Aydin, A., Cangar, O., Ozcan, S.E., Bahr, C. and Berckmans, D. (2010) Application of a fully automatic analysis tool to assess the activity of broiler chickens with different gait scores. *Computers and Electronics in Agriculture* 73, 194–199.

Barnard, C. (2007) Ethical regulation and animal science: why animal behaviour is special. *Animal Behaviour* 74, 5–13.

Bechara, A. and Damasio, A.R. (2005) The somatic marker hypothesis: a neural theory of economic decision. *Games and Economic Behaviour* 52, 336–372.

Bechara, A., Damasio, H., Damasio, A.R. and Lee, G.P. (1999) Different contributions of the human amygdala and ventromedial prefrontal cortex to decision-making. *Journal of Neuroscience* 19, 5473–5481.

Bokkers, E.A.M., Koene, P., Rodenburg, T.B., Zimmerman, P.H. and Spruijt, B.M. (2004) Working for food under conditions of varying motivation in broilers. *Animal Behaviour* 68, 105–113.

Browne, W.J., Caplen, G., Edgar, J., Wilson, L.R. and Nicol, C.J. (2010) Consistency, transitivity and inter-relationships between measures of choice in environmental preference tests with chickens. *Behavioural Processes* 83, 72–78.

Browne, W.J., Caplen, G., Statham, P. and Nicol, C.J (2011) Mild environmental aversion is detected by a discrete-choice preference testing method but not by a free-access method. *Applied Animal Behaviour Science* 134, 152–163.

Buchanan-Smith, H.M. and Badihi, I. (2012) The psychology of control: effects of control over supplementary light on welfare of marmosets. *Applied Animal Behaviour Science* 137, 166–174.

Buckley, L.A., McMillan, L.M., Sandilands, V., Tolkamp, B.J., Hocking, P.M. and D'Eath, R.B. (2011) Too hungry to learn? Hungry broiler breeders fail to learn a Y-maze food quantity discrimination task. *Animal Welfare* 20, 469–481.

Buckley, L.A., Sandilands, V., Hocking, P.M., Tolkamp, B.J. and D'Eath, R.B. (2012) The use of conditioned place preference to determine broiler preferences for quantitative and qualitative dietary restriction. *British Poultry Science* 53, 291–306.

Burman, O.H.P., Parker, R.M.A., Paul, E.S. and Mendl, M.T. (2009) Anxiety-induced cognitive bias in non-human animals. *Physiology and Behavior* 98, 345–350.

Caplen, G., Colborne, G.R., Hothersall, B., Nicol, C.J., Waterman-Pearson, A.E., Weeks, C.A. and Murrell, J.C. (2013) Lame broiler chickens respond to non-steroidal anti-inflammatory drugs with objective changes in gait function: a controlled clinical trial. *Veterinary Journal* 196, 477–482.

Caplen, G., Hothersall, B., Nicol, C.J., Parker, A.E., Waterman-Pearson, A.E., Weeks, C.A. and Murrell, J.C. (2014) Lameness is consistently better at predicting broiler chicken performance in mobility tests than other broiler characteristics. *Animal Welfare* 23, 179–187.

Cooper, J.J. and Appleby, M.C. (1995) Nesting behaviour of hens – effects of experience on motivation. *Applied Animal Behaviour Science* 42, 283–295.

Cooper, J.J. and Appleby, M.C. (1996) Demand for nestboxes in laying hens. *Behavioural Processes* 36, 171–182.

Cooper, J.J. and Appleby, M.C. (2003) The value of environmental resources to domestic hens: a comparison of the work-rate for food and for nests as a function of time. *Animal Welfare* 12, 39–52.

Davies, A.C., Radford, A.N. and Nicol, C.J. (2014) Behavioural and physiological expression of arousal during decision-making in laying hens. *Physiology and Behavior* 123, 93–99.

Dawkins, M. (1977) Do hens suffer in battery cages? Environmental preferences and welfare. *Animal Behaviour* 25, 1034–1046.

Dawkins, M. (1981) Priorities in the cage size and flooring preferences of domestic hens. *British Poultry Science* 22, 255–263.

Dawkins, M.S. (1982) Elusive concept of preferred group-size in domestic hens. *Applied Animal Ethology* 8, 365–375.

Dawkins, M.S. (1983a) Cage size and flooring preferences in litter-reared and cage-reared hens. *British Poultry Science* 24, 177–182.

Dawkins, M.S. (1983b) Battery hens name their price – consumer demand theory and the measurement of ethological needs. *Animal Behaviour* 31, 1195–1205.

Dawkins, M.S. (1990) From an animal's point of view – motivation, fitness and animal welfare. *Behavioral and Brain Sciences*, 13, 1–9.

Dawkins, M.S. and Beardsley, T. (1986) Reinforcing properties of access to litter in hens. *Applied Animal Behaviour Science* 15, 351–364.

Dawkins, M.S., Cain, R., Merelie, K. and Roberts, S.J. (2013) In search of the behavioural correlates of optical flow patterns in the automated assessment of broiler chicken welfare. *Applied Animal Behaviour Science* 145, 44–50.

de Jong, I.C., Wolthuis-Fillerup, M. and Van Reenen, C.G (2007) Strength of preference for dustbathing and foraging substrates in laying hens. *Applied Animal Behaviour Science* 104, 24–36.

Domjan M. and Hall, S. (1986) Sexual dimorphism in the social proximity behaviour of Japanese quail (*Coturnix coturnix japonica*). *Journal of Comparative Psychology* 100, 68–71.

Dunbar, R.I.M. and Dunbar, P. (1988) Maternal time budgets of Gelada baboons. *Animal Behaviour* 36, 970–980.

Duncan, I.J.H. (1978) The interpretation of preference tests in animal behaviour. *Applied Animal Ethology* 4, 197–200.

Duncan, I.J.H. and Hughes, B.O. (1972) Free and operant feeding in domestic fowl. *Animal Behaviour* 20, 775–777.

Duncan, I.J.H. and Wood-Gush, D.G.M. (1972a) Thwarting of feeding behaviour in domestic fowl. *Animal Behaviour* 20, 444–451.

Duncan, I.J.H. and Wood-Gush, D.G.M. (1972b) Analysis of displacement preening in domestic fowl. *Animal Behaviour* 20, 68–72.

Edgar, J.L., Lowe, J.C., Paul, E.S. and Nicol, C.J. (2011) Avian maternal response to chick distress. *Proceedings of the Royal Society B – Biological Sciences* 278, 3129–3134.

Edgar, J.L., Mullan, S.M., Pritchard, J.C., McFarlane, U.J.C. and Main, D.C.J. (2013) Towards a 'good life' for farm animals: development of a resource tier framework to achieve positive welfare for laying hens. *Animals* 3, 584–605.

Faure, J.M. and Lagadic, H. (1994) Elasticity of demand for food and sand in laying hens subjected to variable wind-speed. *Applied Animal Behaviour Science* 42, 49–59.

Forkman, B., Boissy, A., Meunier-Salauen, M.C., Canali, E. and Jones, R.B. (2007) A critical review of fear tests used on cattle, pigs, sheep, poultry and horses. *Physiology and Behavior* 92, 340–374.

Fraser, D. and Nicol, C.J. (2011) Preference and motivation research. In: Appleby, M.C., Mench, J.A., Olsson, I.A.S. and Hughes, B.O. (eds) *Animal Welfare*, 2nd edn. CABI, Wallingford, UK.

Guillaume, S., Jollant, F., Jaussent, I., Lawrence, N., Malafosse, A. and Courtet, P. (2009) Somatic markers and explicit knowledge are both involved in decision-making. *Neuropsychologia* 47, 2120–2124.

Gunnarsson, S., Matthews, L.R., Foster, T.M. and Temple, W. (2000) The demand for straw and feathers as litter substrates by laying hens. *Applied Animal Behaviour Science* 65, 321–330.

Harrison, R. (1964) *Animal Machines – the New Factory Farming Industry*. Vincent Stuart, London.

Hart, B.L. (1988) Biological basis of the behaviour of sick animals. *Neuroscience and Biobehavioural Reviews* 12, 123–137.

Haskell, M.J., Coerse, N.C.A., Taylor, P.A.E. and McCorquodale, C. (2004) The effect of previous experience over control of access to food and light on the level of frustration-induced aggression in the domestic hen. *Ethology* 110, 501–513.

Held, S.D.E. and Spinka, M. (2011) Animal play and animal welfare. *Animal Behaviour* 81, 891–899.

Hocking, P.M., Rutherford, K.M.D. and Picard, M. (2007) Comparison of time-based frequencies, fractal analysis and T-patterns for assessing behavioural changes in broiler breeders fed on two diets at two levels of feed restriction: a case study. *Applied Animal Behaviour Science* 104, 37–48.

Hothersall, B., Caplen, G., Parker, R.M.A., Nicol, C.J., Waterman-Pearson, A.E., Weeks, C.A. and Murrell, J.C. (2014) Thermal nociceptive threshold testing detects altered sensory processing in broiler chickens with spontaneous lameness. *PLoS One* 9, e97883.

Houston, A. and McFarland, D.J. (1980) Behavioural resilience and its relation to demand functions. In: Staddon, J.E.R. (ed.) *Limits to Action: The Allocation of Individual Behaviour*. Academic Press, New York.

Hughes, B.O. (1975) Spatial preference in domestic hen. *British Veterinary Journal* 131, 560–564.

Hughes, B.O. (1977) Selection of group-size by individual laying hens. *British Poultry Science* 18, 9–18.

Hughes, B.O. (1983) Floor space allowance for laying hens. *Veterinary Record*, 113, 23.

Hughes, B.O. and Black, A.J. (1973) Preference of domestic hens for different types of battery cage floor. *British Poultry Science* 14, 615–620.

Hughes, B.O. and Duncan, I.J.H. (1988a) Discrimination by hens in favour of litter rather than wire floors. *Applied Animal Behaviour Science* 21, 363–364.

Hughes, B.O. and Duncan, I.J.H. (1988b) Behavioral needs – can they be explained in terms of motivational models? *Applied Animal Behaviour Science* 19, 352–355.

Hymel, K.A. and Sufka, K.J. (2012) Pharmacological reversal of cognitive bias in the chick anxiety-depression model. *Neuropharmacology* 62, 161–166.

Jensen, M.B. and Pedersen, L.J. (2008) Using motivation tests to assess ethological needs and preferences. *Applied Animal Behaviour Science* 113, 340–356.

Jensen, P. and Toates, F.M. (1993) Who needs behavioural needs? Motivational aspects of the needs of animals. *Applied Animal Behaviour Science* 37, 161–181.

Job, R.F.S. (1987) Learned helplessness in chickens. *Animal Learning and Behavior* 15, 247–350.

Jones, A.R., Bizo, L.A. and Foster, T.M. (2012) Domestic hen chicks' conditioned place preferences for sound. *Behavioural Processes* 89, 30–35.

Jones, R. and Nicol, C.J. (1998) A note on the effect of control of the thermal environment on the well-being of growing pigs. *Applied Animal Behaviour Science* 60, 1–9.

Jones, R.B. (1986) The tonic immobility reaction of the domestic fowl: a review. *World's Poultry Science Journal* 42, 82–96.

Jones, R.B. (1996) Fear and adaptability in poultry: insights, implications and imperatives. *World's Poultry Science Journal*, 52, 131–174.

Keer-Keer, S., Hughes, B.O., Hocking, P.M. and Jones, R.B. (1996) Behavioural comparison of layer and broiler fowl: measuring fear responses. *Applied Animal Behaviour Science* 49, 321–333.

Kirkden, R.D. and Pajor, E.A. (2005) Using preference, motivation and aversion tests to ask scientific questions about animals' feelings. *Applied Animal Behaviour Science* 100, 29–47.

Kirkden, R.D., Edwards, J.S.S. and Broom, D.M. (2003) A theoretical comparison of the consumer surplus and the elasticities of demand as measures of motivational strength. *Animal Behaviour* 65, 157–178.

Knowles, T.G., Kestin, S.C., Haslam, S.M., Brown, S.N., Green, L.E., Butterworth, A., Pope, S.J., Pfeiffer, D. and Nicol, C.J. (2008) Leg disorders in broiler chickens: prevalence, risk factors and prevention. *PLoS One* 3, e1545.

Kops, M.S., de Haas, E.N., Rodenburg, T.B., Ellen, E.D., Korte-Bouws, G.A.H., Olivier, B., Gunturkun, O., Korte, S.M. and Bolhuis, J.E. (2013) Selection for low mortality in laying hens affects catecholamine levels in the arcopallium, a brain area involved in fear and motor regulation. *Behavioural Brain Research* 257, 54–61.

Kuhne, F., Adler, S. and Sauerbrey, A.F.C. (2011) Redirected behaviour in learning tasks: the commercial laying hen (*Gallus gallus domesticus*) as model. *Poultry Science* 90, 1859–1866.

Lee, L., Amir, O. and Ariely, D. (2009) In search of *Homo economicus*: cognitive noise and the role of emotion in preference consistency. *Journal of Consumer Research* 36, 173–187.

Lindberg, A.C. and Nicol, C.J. (1996a) Space and density effects on group size preferences in laying hens. *British Poultry Science* 37, 709–721.

Lindberg, A.C. and Nicol, C.J. (1996b) Effects of social and environmental familiarity on group preferences and spacing behaviour in laying hens. *Applied Animal Behaviour Science* 49, 109–123.

Lindberg, A.C. and Nicol, C.J. (1997) Dustbathing in modified battery cages: is sham dustbathing an adequate substitute? *Applied Animal Behaviour Science* 55, 113–128.

Lindqvist, C., Lind, J. and Jensen, P. (2009) Effects of domestication on food deprivation-induced behaviour in red junglefowl, *Gallus gallus*, and White Leghorn layers. *Animal Behaviour* 77, 893–899.

Littin, K., Acevedo, A., Browne, W., Edgar, J., Mendl, M., Owen, D., Sherwin, C., Wurbel, H. and Nicol, C.J. (2008) Towards humane end points: behavioural changes precede clinical signs of disease in a Huntington's disease model. *Proceedings of the Royal Society B – Biological Sciences* 275, 1865–1874.

Lowe, J.C., Abeyesinghe, S.M., Demmers, T.G.M., Wathes, C.M. and McKeegan, D.E.F. (2007) A novel telemetric logging system for recording physiological signals in unrestrained animals. *Computers and Electronics in Agriculture* 54, 74–79.

Maasberg, D.W., Shelly, L.E., Gracian, E.I. and Gilbert, P.E. (2011) Age-related differences in the anticipation of future rewards. *Behavioural Brain Research* 223, 371–375.

Mandel, R., Whay, H.R., Nicol, C.J. and Klement, E. (2013) The effect of food location, heat load, and intrusive medical procedures on brushing activity in dairy cows. *Journal of Dairy Science*, 96 6506–6513.

Maria, G.A., Escos, J. and Alados, C.L. (2004) Complexity of behavioural sequences and their relation to stress conditions in chickens (*Gallus gallus domesticus*): a non-invasive technique to evaluate animal welfare. *Applied Animal Behaviour Science* 86, 93–104.

Mason, G., McFarland, D. and Garner, J. (1998) A demanding task: using economic techniques to assess animal priorities. *Animal Behaviour* 55, 1071–1075.

Matthews, L.R., Temple, W., Foster, T.M., Walker, J. and McAdie, T.M. (1995) Comparison of the demand for dustbathing substrates by layer hens. In: Rutter, S.M., Rushen, J., Randle, H.D. and Eddison, J.C. (eds) *Proceedings of the 29th International Congress of the International Society for Applied Ethology*, Exeter, UK. Universities Federation for Animal Welfare, Potters Bar, UK, pp. 11–12.

McBride, G., Parer, I.P. and Foenander, F. (1969) The social organisation and behaviour of the feral domestic fowl. *Animal Behaviour Monographs* 2, 127–181.

McFarland, D.J. and Sibly, R.M. (1975) Behavioral final common path. *Philosophical Transactions of the Royal Society of London Series B – Biological Sciences* 270, 265–293.

Mellor, D.J. (2012) Animal emotions, behaviour and the promotion of positive welfare states. *New Zealand Veterinary Journal* 60, 1–8.

Mendl, M. and Paul, E.S. (2008) Do animals live in the present? Current evidence and implications for welfare. *Applied Animal Behaviour Science* 113, 357–382.

Merlet, F., Puterflam, J., Faure, J.M., Hocking, P.M., Magnusson, M.S. and Picard, M. (2005) Detection and comparison of time patterns of behaviours of two broiler breeder genotypes fed ad libitum and two levels of feed restriction. *Applied Animal Behaviour Science* 94, 255–271.

Mills, D.S. and Nicol, C.J. (1990) Tonic immobility in spent hens after catching and transport. *Veterinary Record* 126, 210–212.

Mineka, S. and Hendersen, R.W. (1985) Controllability and predictability in acquired motivation. *Annual Review of Psychology* 36, 495–529.

Mishra, A., Koene, P., Schouten, W., Spruijt, B., van Beek, P. and Metz, J.H.M. (2005) Temporal and sequential structure of behaviour and facility usage of laying hens in an enriched environment. *Poultry Science* 84, 979–991.

Moe, R.O., Nordgreen, J., Janczak, A.M., Spruijt, B.M., Zanella, A.J. and Bakken, M. (2009) Trace classical conditioning as an approach to the study of reward-related behaviour in laying hens: a methodological study. *Applied Animal Behaviour Science* 121, 171–178.

Moe, R.O., Stubsjoen, S.M., Bohlin, J., Flo, A. and Bakken, M. (2012) Peripheral temperature drop in response to anticipation and consumption of a signalled palatable reward in laying hens (*Gallus domesticus*). *Physiology and Behavior* 106, 527–533.

Moe, R.O., Nordgreen, J., Janczak, A.M., Spruijt, B.M. and Bakken, M. (2013) Effects of signalled reward type, food status and a mu-opioid receptor antagonist on cue-induced anticipatory behaviour in laying hens (*Gallus domesticus*). *Applied Animal Behaviour Science* 148, 46–53.

Morrison, W.D., McMillan, I., Bate, L.A., Otten, L. and Pei, D.C.T. (1986) Beahvioral observations and operant procedures using microwaves as a heat-source for young chicks. *Poultry Science* 65, 1516–1521.

Nasr, M.A.F., Murrell, J.C., Wilkins, L.J. and Nicol, C.J. (2012) The effect of keel fractures on egg-production parameters, mobility and behaviour in individual laying hens. *Animal Welfare* 21, 127–135.

Nasr, M.A.F., Browne, W.J., Caplen, G., Hothersall, B., Murrell, J.C. and Nicol, C.J. (2013) Positive affective state induced by opioid analgesia in laying hens with bone fractures. *Applied Animal Behaviour Science* 147, 127–131.

Nicol, C.J. (1986) Non-exclusive spatial preference in the laying hen. *Applied Animal Behaviour Science* 15, 337–350.

Nicol, C.J. (1987a) Effect of cage height and area on the behaviour of hens housed in battery cages. *British Poultry Science* 28, 327–335.

Nicol, C.J. (1987b) Behavioural responses of laying hens following a period of spatial restriction. *Animal Behaviour* 35, 1709–1719.

Nicol, C.J. and Guilford, T. (1991) Exploratory activity as a measure of motivation in deprived hens. *Animal Behaviour* 41, 333–341.

Nicol, C.J. and Pope, S.J. (1993) Food-deprivation during observation reduces social learning in hens. *Animal Behaviour* 45, 193–196.

Nicol, C.J., Blakeborough, A. and Scott, G.B. (1991) Aversiveness of motion and noise to broiler chickens. *British Poultry Science* 32, 249–260.

Nicol, C.J., Caplen, G., Edgar, J. and Browne, W.J. (2009) Associations between welfare indicators and environmental choice in laying hens. *Animal Behaviour* 78, 413–424.

Nicol, C.J., Caplen, G., Statham, P. and Browne, W.J. (2011a) Decisions about foraging and risk trade-offs in chickens are associated with individual somatic response profiles. *Animal Behaviour* 82, 255–262.

Nicol, C.J., Caplen, G., Edgar, J., Richards, G. and Browne, W.J. (2011b) Relationships between multiple welfare indicators measured in individual chickens across different time periods and environments. *Animal Welfare* 20, 133–143.

Panksepp J. (2005) Affective consciousness: core emotional feelings in animals and humans. *Consciousness and Cognition* 14, 30–80.

Paul, E.S., Harding, E.J. and Mendl, M. (2005) Measuring emotional processes in animals: the utility of a cognitive approach. *Neuroscience and Biobehavioural Reviews* 29, 469–491.

Pizzari, T. and Birkhead, T.R. (2000) Female feral fowl eject sperm of subdominant males. *Nature* 405, 787–789.

Pohle, K. and Cheng, H.-W. (2009) Furnished cage system and hen well-being: comparative effects of furnished cages and battery cages on behavioural exhibitions in White Leghorn chickens. *Poultry Science* 88, 1559–1564.

Prescott, N.B. and Wathes, C.M. (2002) Preference and motivation of laying hens to eat under different illuminances and the effect of illuminance on eating behaviour. *British Poultry Science* 43, 190–195.

Richards, G.J., Wilkins, L.J., Knowles, T.G., Booth, F., Toscano, M.J., Nicol, C.J. and Brown, S.N. (2012) Pop hole use by hens with different keel fracture status monitored throughout the laying period. *Veterinary Record* 170, 494–498.

Rimpley, K. and Buchanan-Smith, H.M. (2013) Reliably signalling a startling husbandry event improves welfare of zoo-housed capuchins (*Sapajus apella*). *Applied Animal Behaviour Science* 147, 205–213.

Rivalan, M., Valton V., Seriès P., Marchand A.R. and Dellu-Hagedorn F. (2013) Elucidating poor decision-making in a rat gambling task. *PLoS One* 8, e82052.

Roberts, S.J., Cain, R. and Dawkins, M.S. (2012) Predictions of welfare outcomes for broiler chickens using Bayesian regression on continuous optical flow data. *Journal of the Royal Society Interface* 9, 3436–3443.

Rodd, Z.A., Rosellini, R.A., Stock, H.S. and Gallup, G.G. (1997) Learning helplessness in chickens (*Gallus gallus*): evidence for attentional bias. *Learning and Motivation* 28, 43–55.

Rutter, S.M. and Duncan, I.J.H. (1991) Shuttle and one-way avoidance as measures of aversion in the domestic fowl. *Applied Animal Behaviour Science* 30, 117–124.

Saint-Dizier, H., Constantin, P., Davies, D.C., Leterrier, C., Levy, F. and Richard, S. (2009) Subdivisions of the arcopallium/posterior pallial amygdala complex are differentially involved in the control of fear behaviour in the Japanese quail. *Brain Research Bulletin* 79, 288–295.

Salmeto, A.L., Hymel, K.A., Carpenter, E.C., Brilot, B.O., Bateson, M. and Sufka, K.J. (2011) Cognitive bias in the chick anxiety–depression model. *Brain Research* 1373, 124–130.

Sandilands, V., Raj, A.B.M., Baker, L. and Sparks, N.H.C. (2011) Aversion of chickens to various lethal gas mixtures. *Animal Welfare* 20, 253–262.

Scholtz, B., Urselmans, S., Kjaer, J.B. and Schrader, L. (2010) Food, wood or plastic as substrates for dustbathing and foraging in laying hens: a preference test. *Poultry Science*. 89, 1584–1589.

Shields, S.J., Garner, J.P. and Mench, J.A. (2004) Dustbathing by broiler chickens: a comparison of preference for four different substrates. *Applied Animal Behaviour Science* 87, 69–82.

Smith, S.F., Appleby, M.C. and Hughes, B.O. (1990) Problem-solving by domestic hens – opening doors to reach nest sites. *Applied Animal Behaviour Science* 28, 287–292.

Smoski, M.J., Felder, J., Bizzell, J., Green, S.R., Ernst, M., Lynch, T.R. and Dichter, G.S. (2009) fMRI of alternations in reward selection, anticipation and feedback in major depressive disorder. *Journal of Affective Disorders* 118, 69–78.

Spevack, A.A., Schulman, A.H. and Cotton, B. (1974) Unconditioned and conditioned suppression in two closely related genetic lines of chickens. *Poultry Science* 53, 2020–2025.

Sumpter, C.E., Foster, T.M. and Temple, W. (1995) Predicting and scaling hens' preferences for topographically different responses. *Journal of the Experimental Analysis of Behavior* 63, 151–163.

Sumpter, C.E., Temple, W. and Foster, T.M. (1998) Response form, force, and number: effects on concurrent-schedule performance. *Journal of the Experimental Analysis of Behavior* 70, 45–68.

Taylor, P.E., Coerse, N.C.A. and Haskell, M. (2001) The effects of operant control over food and light on the behaviour of domestic hens. *Applied Animal Behaviour Science* 71, 319–333.

Toscano, M.J., Sait, L., Jorgensen, F., Nicol, C.J., Powers, C., Smith, A.L., Bailey, M. and Humphrey, T.J. (2010) Sub-clinical infection with Salmonella in chickens differentially affects behaviour and welfare in three inbred strains. *British Poultry Science* 51, 703–713.

van Kampen, H.S. (1994) Courtship food-calling in Burmese red junglefowl: I. the causation of female approach. *Behaviour* 131, 261–275.

Vestergaard, K. (1982) Dust-bathing in the domestic fowl: diurnal rhythm and dust deprivation. *Applied Animal Ethology* 8, 487–495.

Weeks, C.A., Danbury, T.D., Davies, H.C., Hunt, P. and Kestin, S.C. (2000) The behaviour of broiler chickens and its modification by lameness. *Applied Animal Behaviour Science* 67, 111–125.

Weeks, C.A., Knowles, T.G., Gordon, R.G., Kerr, A.E., Peyton, S.T. and Tilbrook, N.T. (2002) New method for objectively assessing lameness in broiler chickens. *Veterinary Record* 151, 762–764.

Wichman, A., Keeling, L.J. and Forkman, B. (2012) Cognitive bias and anticipatory behaviour of laying hens housed in basic and enriched pens. *Applied Animal Behaviour Science* 140, 62–69.

Zimmerman, P.H., Koene, P. and van Hoof, J.A.R.A.M. (2000a) Thwarting of behaviour in different contexts and the gakel-call in the laying hen. *Applied Animal Behaviour Science* 69, 255–264.

Zimmerman, P.H., Koene, P. and van Hoof, J.A.R.A.M. (2000b) The vocal expression of feeding motivation and frustration in the domestic laying hen, *Gallus gallus domesticus*. *Applied Animal Behaviour Science* 69, 265–273.

Zimmerman, P.H., Lundberg, A., Keeling, L.J. and Koene, P. (2003) The effect of an audience on the gakel-call and other frustration behaviours in the laying hen (*Gallus gallus domesticus*). *Animal Welfare* 12, 315–326.

Zimmerman, P.H., Buijs, S.A.F., Bolhuis, J.E. and Keeling, L.J. (2011) Behaviour of domestic fowl in anticipation of positive and negative stimuli. *Animal Behaviour* 81, 569–577.

5

Behavioural Needs, Priorities and Preferences

Previous chapters have reviewed genetic, sensory and developmental influences on chicken behaviour. It is now time to describe in more detail the range of behaviours that chickens perform, and the factors that influence their expression. A complete list of the behaviours that chickens exhibit under natural conditions is termed an ethogram. Sometimes behaviours are defined by their structure (e.g. raising feathers) and sometimes by their purported function (e.g. comfort activity). The benefits and drawbacks of these approaches, together with much further information on the practicalities of measuring behaviour, have been outlined by Martin and Bateson (2007). In this chapter, we will not consider the precise structural elements or neurological mechanisms that underpin behaviour but will focus primarily on the causation and function of broader behavioural patterns that are most relevant for chicken husbandry. Behaviours that form part of the chicken ethogram are usually defined so that they are mutually exclusive (a chicken should not be able to do two behaviours at the same time), although sometimes different levels of description are given. For example, a chicken could be in a certain posture (standing or sitting) and simultaneously able to engage in one of a variety of activities (e.g. preening, resting, feather pecking).

From the perspective of animal welfare, the ethogram can also be regarded as a list of candidate behaviours that might need to be performed if a chicken is going to have acceptable welfare in a captive or commercial environment. However, it is not the case that chickens have to do everything. The reality is both more subtle and more interesting, and the methods described in Chapter 4 can be used to distinguish 'behaviours that matter' (Nicol, 2011) in a given captive environment. This chapter will review the factors that influence the expression of behaviour in the chicken, with an emphasis on the relative importance of performing these behaviours for birds housed in commercial farming systems.

Feeding and Foraging

Feeding and foraging are activities performed by chickens during daylight hours. Although food is constantly available during the night in commercial systems, it is rarely touched at this time. Peaks in feeding occur early and late in the light period, with evening feeding especially pronounced in laying birds (Savory, 1980). Adult chickens provided with commercial feed rations can consume sufficient feed to maintain body weight by feeding for just 1 or 2 h day^{-1}, with short-term intake governed by the capacity of the crop. The amount of pecking activity directed towards feeders is therefore not a good indicator of food intake, as much of it is exploratory (van Rooijen, 1991). When foraging in a natural environment, the chicken's pecks to the ground are interspersed with rapid scratching movements of the feet, both activities designed to uncover and detect high-quality food items that might be hidden in the environment. Ingested particles are selected momentarily ahead of pecking, which is subsequently performed with the eyes closed. If unpalatable food is ingested, then chickens show characteristic head shaking and side-to-side bill-wiping movements (Sherwin et al., 2002). Bill wiping is also used to remove sticky food deposits from the beak. Chickens are adept at detecting small moving objects that simulate invertebrate prey items (Jones and Carmichael, 1999; Clara et al., 2009). Once any initial fear of such moving prey has been overcome (Hogan, 1965), chickens can become efficient and ruthless hunters.

Foraging behaviour tends to dominate the time budget of most chickens, and is performed for approximately 60% of daylight hours by captive junglefowl (Dawkins, 1989). In domestic fowl, both genetic (Schutz and Jensen, 2001) and environmental influences tend to reduce the time birds spend foraging compared with the semi-wild junglefowl. Often around 40% of daytime activity is devoted to foraging activities (Buier, 1996a,b; Klein et al., 2000), although the opportunity to forage during a significant portion of the day is, as we shall see in Chapter 9, an important way of reducing the risk of feather pecking. The time devoted to foraging is influenced only slightly by breed (Klein et al., 2000) but greatly by environment. Chickens forage for longer when in enriched environments where suitable substrates are available (Klein et al., 2000; Nicol et al., 2001). When housed in cages, chickens sift through the available food and simultaneously scratch at the wire floor. Conversely, foraging behaviour directed towards a litter substrate persists in non-cage systems, even when food particles are rarely or never encountered within the substrate. The birds appear to forage on the 'off chance' that they may uncover a high-quality food or prey item, or to obtain information for future use. The finding that hens exhibit a phenomenon called contrafreeloading, where they will expend effort foraging for food, even when the same food is freely available (Duncan and Hughes, 1972), supports the idea that foraging is an important way of obtaining information to secure future food supplies (Inglis et al., 1997). The tendency to contrafreeload is, however, reduced somewhat in modern breeds in comparison with red junglefowl (Lindqvist et al., 2002).

Hens readily forage in peat (Petherick and Duncan, 1989), mixed peat and sand (reviewed by Weeks and Nicol, 2006), sawdust (Vestergaard and Hogan, 1992), straw (Sanotra et al., 1995) and wood-chips (Bubier, 1996a). A wide range of materials is accepted and used for foraging, although some studies found a preference for foraging in sand and straw over feathers and wood shavings (Sanotra et al., 1995). Formal demand experiments (see methodology described in Chapter 4) have examined the desire of hens to obtain pecking and scratching substrates and researchers concluded that straw was a highly valued resource (Gunnarsson et al., 2000a). A value of −0.48 was obtained for the slope of the demand function, which the authors described as not dissimilar to the values obtained for dustbathing substrates or food. Another demand experiment examined the desire of hens to obtain sand, shavings or peat for foraging activities, by requiring hens to push doors weighted in steps from 150 to 1250 g. At the lower door weights, a high level of access was maintained for each of these substrates with more time (and greater intensity) of foraging occurring in sand. However, beyond a door weight of 500 g, the demand for foraging materials reduced, with no apparent differences in the slope of the demand curve for the different substrates (de Jong et al., 2007).

Social influences on feeding and foraging are many and varied, with social facilitation playing an important role (Tolman, 1964; Hughes, 1971; Collins and Sumpter, 2007). Indeed, the mere presence of a conspecific increases foraging activity, even when that conspecific is separated by a transparent wall reducing direct competition effects (Ogura and Matsushima, 2011). Adult domestic hens feed gregariously, and high rates of pecking and scratching attract other birds. Food running is an eye-catching activity triggered when a bird finds and picks up a large food or prey item and then runs rapidly. It seems likely that the motivation of the bird making the original discovery is to avoid conspecifics and keep the food for itself, but alternatively it has been suggested that this behaviour might attract flock mates to assist in breaking the item into smaller pieces (Wood-Gush, 1971). As we will see in Chapters 6 and 7, chickens signal and receive much information about food quality and availability, and they can learn how to obtain food via social learning.

Given the extensive literature on the importance of balancing the need to forage against dangers such as predation or exposure (e.g. Lima and Dill, 1990), it is surprising that few studies on this aspect of foraging have been conducted with chickens. Behavioural ecologists often invoke states such as fear, stress or reward in explanations of foraging behaviour, but supporting physiological evidence is rarely provided in studies of wild birds. From the limited physiological evidence available, it appears that chickens are highly aroused by the sight of cues indicating foraging rewards, exhibiting increased heart rate and head movements during a viewing period, prior to moving towards the goal (Davies *et al.*, 2014). However, over the longer term, an environment that provides extensive foraging opportunities is one that seems to reduce baseline levels of the stress hormone corticosterone (Nicol *et al.*, 2011). The importance of foraging to chickens is also shown by the role it plays in reducing behavioural and welfare problems (Chapter 9), and the positive choices that chickens make for foraging materials and environments (Bubier, 1996a,b; Nicol *et al.*, 2011).

In contrast to feeding, birds drink with their eyes open, using a scooping movement that develops over the first few days of life. If water troughs are available, they will insert the beak approximately 10 mm while drinking (Ross and Hurnik, 1983). The exact sequences involved in drinking are flexible, and chickens can reorganize the movement patterns of their jaws, tongue, larynx and head to cope with drinking small drops or with changes in beak shape (due to trimming practices, reviewed in Chapter 9) (Heidweiller *et al.*, 1992). The behavioural flexibility of drinking means that commercially kept chickens can use water troughs or nipple drinkers. Wherever possible, drinkers are sited on slatted areas to keep litter areas as dry as possible, and nipple drinkers are generally used for the same, sound reason, as dry litter is an important condition for good bird health. However, given the choice, chickens prefer to drink from troughs or bell drinkers (Houldcroft *et al.*, 2008). If, in the wider interests of overall welfare, chickens are provided with nipple drinkers, then they prefer these to be at relatively low heights (Houldcroft *et al.*, 2008).

Drinking behaviour is obviously essential to life, but once the physiological demand for water is satisfied, chickens do not appear to have any additional behavioural need to conduct drinking-related behaviours.

Nesting and Pre-laying

There is a known internal component to the motivation of nesting behaviour in chickens. Approximately 1 h before egg-laying, domestic hens exhibit behavioural changes including increased activity and restlessness. These changes are stimulated by hormonal changes preceding oviposition. In response to light stimulation directly to the brain, the hypothalamus releases gonadotrophin-releasing hormones that stimulate the release of two types of gonadotrophin hormones from the anterior pituitary, luteinizing hormone (LH) and follicle-stimulating hormone (FSH). A surge of LH occurs some 4–7 h before ovulation. Nesting behaviour is then triggered by ovulation. Oestrogen and progesterone released from the post-ovulatory follicle initiate nest site selection and nest building some 24 h later (Appleby, 1986).

The restlessness that precedes nesting behaviour indicates the start of a search phase where the hen seeks out suitable nesting sites and resources for nest building. If potential nesting sites are discovered, hens will closely examine and inspect a number of sites before choosing one to lay in. Within the chosen site, the hen will start nesting behaviour proper, alternating between sitting and building activities such as turning, floor scratching and collecting nearby pieces of straw or similar material. These materials are placed around the body or tossed over or on to her back until a rudimentary nest (Fig. 5.1) has been constructed.

Even in furnished cages, where straw-like materials are absent, hens can be observed pecking at small dust or faecal particles, which are placed

Fig. 5.1. Nest constructed by a commercial laying hen on her first exposure to hay. Prior to this, the hen had experience only of nest boxes with wooden floors and rubber mats. (Photo courtesy of Lorna Wilson, University of Nottingham, UK.)

under the breast or around the body, and grasping or tugging at the wire floor in a form of redirected litter-gathering behaviour (Sherwin and Nicol, 1993a). This pre-laying phase of behaviour can start as early as 100 min before oviposition (Sherwin and Nicol, 1993a), with the majority of birds sitting for some 25–40 min before oviposition (Sherwin and Nicol, 1993a; Cronin et al., 2005). In some studies, all birds were found to be sitting 5 min before an egg was laid (Sherwin and Nicol, 1993a.)

Preferred nest sites

Although hens will accept a variety of different types of nesting site, they also show marked preferences for some types of nest over others. The most highly preferred sites have a degree of enclosure and contain some form of manipulable nesting material. Straw, for example, is preferred over wood shavings and peat for nesting behaviours (Clausen and Riber, 2012). Early experimental work examining increasing degrees of nest site enclosure found that hens avoided the most 'open' nesting sites (even though these most resembled the natural nest scraping areas used by junglefowl) and chose instead nest boxes with walls (Appleby and McRae, 1986). However, the same hens did not seem to mind whether the nest box walls were open, constructed of mesh or solid wood. In contrast, in furnished cages, hens preferred solid partitions between nests to wire partitions or no partitions (Reed and Nicol, 1992), and the reduced calcium deposits found on the shells of eggs laid in enclosed nests suggested lower stress levels (Walker and Hughes, 1998). In many commercial systems, nests have thick and opaque plastic strips hanging over the nest entrance, providing an even greater degree of seclusion. The presence of such strips neither encourages nor discourages egg-laying (Struelens et al., 2005) but does seem to result in more settled nesting behaviour (Struelens et al., 2008a). A closed front curtain provides the greatest degree of seclusion but limits important nest inspections, which are made possible by curtain strips (Staempfli et al., 2012).

Secluded nests are likely to be darker nests and, perhaps because of this relationship, early tests suggested that hens might prefer dark nesting sites. Experiments have revealed great consistency in the nest illumination preferences of individual birds over time but a wide variation between the preferences of different individual hens for dark or illuminated nests (Appleby et al., 1984a). The colour of a nest may have a small influence too. Huber-Eicher (2004) found that hens preferred yellow nests if they had been exposed to red-, blue- or green-coloured pens as pullets. Exposure to a red pen under high light intensity increased the later preference for yellow nests (Zupan et al., 2007). The slope of the nest box floor (varying between 12 and 18°) has a small influence, with more hens sitting in boxes with a shallower slope (Staempfli et al., 2011). A more consistent influence on nest site attractiveness is the presence of suitable nesting material. Padded nest sides and artificial pecking strips increased the attractiveness of nests in prototype furnished cages (Reed and Nicol, 1992), as did lining the bases of nests with artificial grass in a much larger trial of commercial furnished cages (Wall et al., 2002). Hughes (1993) found that young hens strongly preferred to lay on artificial grass compared with a wire floor, although older hens showed little preference as they had already developed a location preference. Huber et al. (1985) examined the influence of various nesting materials on nest site selection and found that loose substrates were preferred over artificial turf or wire. The mere sight of litter material, even when no direct access is possible, is sufficient to exert a positive effect on nest site selection (Hughes et al., 1995). Birds sit for longer in nests containing litter than in boxes with no litter, although they also sit for longer in bare boxes than on bare wire floors (Freire et al., 1996). More recent studies have strongly backed these findings. Struelens et al. (2008a) discovered a whole suite of behaviours that was influenced by the provision of peat or artificial grass in nest sites when compared with coated wire. Birds showed more characteristic and settled nesting behaviour when peat or artificial grass was present, with longer nest visits, more sitting and more object pecking.

One reason why the presence of manipulable materials within a nest site is so important is because hens have a need to perform nest-building behaviour (Hughes et al., 1989). Unlike gerbils, which are sometimes satisfied if they find a burrow that someone else has constructed (Wiedenmayer, 1997), hens do not simply want to find and access a perfectly formed nest. Indeed, if a perfectly formed nest built by an individual hen on the previous day is re-presented to the same hen, rather than accepting it gratefully she will destroy it and start building again (Hughes *et al.*, 1989). The ability to mould the nest using body and feet movements seems to be of particular importance (Duncan and Kite, 1989).

A further characteristic that influences attractiveness is the memorability of the nest site. In the wild, female junglefowl lay a clutch of eggs over a number of days in the same site prior to brooding and hatching. Laying eggs in different places each day would be extraordinarily futile for the hen's reproductive success. Commercial strains of hen mostly ignore their eggs after oviposition, but these same commercial birds seem to have retained an instinct to lay consecutive eggs in the same place. Hens in cages (Sherwin and Nicol, 1992) or small pens (Zupan *et al.*, 2008) generally return to the same nest site every day. Cronin *et al.* (2012) reported a study which found that two-thirds of hens were consistent 'one-site' layers from their 11th egg onwards. In large commercial systems where hundreds of identical nest boxes are present, it is probably very difficult for a hen to remember where her previous egg was deposited. In such commercial systems, the number of floor eggs is not affected by whether nest boxes are situated centrally or against one wall (Lentfer *et al.*, 2011). However, in both types of aviary system, many hens show a preference for laying in boxes in corners or at the ends of rows. It may be that these are the most memorable positions. Alternatively, this preference for corner boxes might derive from a perception that they offer a reduced risk of predation or disturbance (Riber and Nielsen, 2013).

Although hens may have an ideal conception of a nest, and strongly prefer some types of nest over others, they are also, to an extent, adaptable birds. In relatively barren environments, the most rudimentary features of a nest will be recognized as such. Early prototype designs of furnished cages used individual plastic roll-away nest hollows, which, despite their basic nature, were accepted by over 90% of hens (Sherwin and Nicol, 1992). In modern enriched colony cages (see Chapter 9), the nest area is no more than a section of floor covered with artificial turf and screened off from the rest of the cage by solid partitions or plastic strips, and yet a high proportion of eggs is laid here.

Social factors influencing nesting

Some hens tend towards gregarious nesting and are attracted to nests where eggs or other hens are already present (Appleby *et al.*, 1984b), sometimes leading to problems with accidental injury or smothering (Clausen and Riber, 2012). The tendency to nest with others is enhanced by the threat of predation (Riber, 2012) and the occurrence of widely shared preferences for the same nest-box features such as substrate, colour and position (Clausen and Riber, 2012). Hens that have chosen a nest box in a corner or at the end of a row to avoid other hens may find they have chosen unwisely – as do people who head to the same 'secluded' beaches on a bank holiday. Gregarious nesting was a feature of early designs of furnished cages where separate plastic nest hollows were provided (Appleby, 1990; Sherwin and Nicol, 1992). Sherwin and Nicol (1993a) found approximately equal numbers of hens nesting in a solitary or a gregarious manner in such furnished cages. In modern free-range, barn and colony cage systems, large group nests are the norm, with trends to increase nest size further, despite hens preferring smaller nests (Ringgenberg *et al.*, 2014). These developments favour the needs of gregarious nesters over hens with more solitary preferences.

Dominant birds can exclude subordinates from nest sites (Freire *et al.*, 1998), and the greater number of aggressive pecks a hen receives, the shorter the time she will stay in the nest box (Lundberg and Keeling, 1999). For these reasons, a relatively high ratio of nests to hens is usually provided. The RSPCA Freedom Food (RSPCA, 2013) standards recommend one nest per five birds, or 1 m² of nesting substrate per 120 hens for group nesting. However, even when it is unlikely that birds are being competitively excluded, some hens lay their eggs on wire or litter floors indoors (often in corners), or on the outdoor range, at considerable economic cost to farmers. Although the majority of hens will accept even a few rudimentary pieces of plastic flooring screened by a curtain as a nest, a variable minority seem to have formed a different conception of a nest. In one study, the presence of cockerels in experimental flocks led to a reduction in floor laying (Rietveld-Piepers *et al.*, 1985), possibly because vigilant cockerels may reduce the hen's perception of threat when choosing an enclosed nest box. It has also been suggested that the provision of heterogenous nest sites that are more easily distinguished by hens could (partially) solve the problem of floor eggs (Appleby and McRae, 1986; Zupan et al., 2008), but this remains to be tested on a commercial scale.

Not all birds use nests

The birds' view of what constitutes a suitable nest site seems to develop prior to the onset of sexual maturity (Appleby and McRae, 1983; Rietveld-Piepers et al., 1985; Sherwin and Nicol, 1992), but as few birds are exposed to potential nest sites during the rearing phase, it is not surprising that some hens develop idiosyncratic notions. Early exposure to nest boxes can be beneficial, as shown by Sherwin and Nicol (1993b), who placed pullets in furnished cages with nests at 14 or 16 weeks (before lay) or at 18 or 21 weeks of age (after onset of lay). The younger the birds were at placement, the fewer floor eggs were laid. Pullets reared on litter were less inclined to accept the plastic nests than pullets reared on wire, suggesting that they had developed a different conception of a suitable nest site. Once formed, nest site preferences can be long lasting. Hens that have learned to lay on wire floors will not necessarily revise their preferences when more attractive artificial turf nests are provided (Hughes, 1993). So, however inconvenient for the farmer, are these floor-laying birds simply expressing a different preference or have they failed to find any oviposition site that meets their needs?

This question can be addressed by seeing whether the birds' nesting behaviour is settled or unsettled. Unsettled behaviour, with high levels of locomotion, nest inspections and restlessness during the pre-lay period, is a sign that birds have not found a suitable nest site. This is what is typically observed in conventional battery cages where no nest sites are available at all. Hens in conventional cages or bare wire-floor pens show disturbed nesting and oviposition behaviour, with exaggerated or stereotyped pacing prior to lay, and many elements of nesting behaviour such as straw tossing, turning and pre-lay sitting reduced or absent (Wood-Gush, 1975; Meijsser and Hughes, 1989). Similar patterns of behaviour are often, although not always, observed in hens that have chosen to

lay on the floor in preference to a nest box (Cooper and Appleby, 1996; Zupan et al., 2008). When the only alternative to a nest box is a wire floor or a shallow coating of shavings, unsettled pre-lay behaviour does seem to indicate frustration (Mills and Wood-Gush, 1985; Meijsser and Hughes, 1989). This interpretation is supported by a study by Cooper and Appleby (1997), who found that both nest layers and floor layers would pay a cost (squeezing through a narrow gap) to move from a home pen to an exploration area where they could look for possible nesting sites. When traditional nest boxes were placed in the home pen, the nest layers stopped exploring. In contrast, neither a solid floor covered in 20 mm of shavings (the floor laying site) nor the traditional nest boxes met the needs of the floor-laying birds, who continued to squeeze through the narrowest gaps to explore further.

Where more attractive alternatives to traditional nest boxes are available, altered pre-lay behaviour may not always indicate frustration. Zupan et al. (2008) found relatively unsettled and exploratory behaviour in a minority of birds that had a preference for nesting in an open litter tray (with a 100 mm depth of shavings provided) but argued that this reflected an anti-predation strategy rather than indicating frustration. This interpretation was supported by Kruschwitz et al. (2008) who found that the hens that preferred to lay in an open litter area were just as strongly motivated to reach their open litter tray as the nest layers were to reach an enclosed nest box. Further support comes from findings that increased pre-lay activity is not associated with elevated stress (Cronin et al., 2012). As these authors found a positive correlation between interrupted (more bouts, shorter duration) sitting behaviour and increased plasma corticosterone, they suggested that sitting-phase behaviour might be a more relevant indicator of welfare. Future studies of floor-laying birds should therefore focus on this phase of the nesting sequence.

Strength of nesting motivation

The above account shows that once a hen has developed a conception of a suitable nest, she is highly motivated to access this resource to perform nesting behaviour. Hens will work to gain access to nests during the search phase (Duncan and Kite, 1987). In one study, where doors designed to close off a litter area were installed, hens learned to open the doors to gain access to the litter area at nesting time. Different techniques were used, including levering the side or the base of the door with the beak, or wedging the head through a slot in the door and pulling backwards (Smith *et al.*, 1990). Thus, hens are motivated to search for nests, but they are not always prepared to pay the highest costs until the next phase of the nesting sequence. As the sitting phase of nesting approaches, hens will pay increasingly higher costs to access the nest site, so recording the timing of motivation tests relative to the time of oviposition is critical. During the sitting phase, hens are prepared to pay high costs (see Chapter 4) to reach a nest, including walking past dominant or unfamiliar birds in narrow corridors (Freire *et al.*, 1997), squeezing through narrow gaps (Cooper and Appleby, 1996) and pushing through weighted doors (Cooper and Appleby, 2003).

Cooper and Appleby (2003) demonstrated how the motivation to access a nest increases as the time of oviposition draws near. Hens were trained to continue to push at a door that was sealed with an electromagnet. The birds' work rate (a combined measure of pushing force (N) and time spent pushing) increased as the time of oviposition approached. Birds tested at 20 min prior to oviposition had a work rate approximately three times higher than when they were tested at 40 min or more prior to oviposition. This work rate was also significantly higher than that shown for access to feed after a 4 h deprivation period.

The strong evidence that nesting is a behavioural need would logically lead to a prediction that preventing hens from accessing a suitable nest site would result in a state of frustration, typified by an enhanced physiological stress response. Indirect evidence comes from Sherwin *et al.* (2010), who detected lower stress-related calcium deposits on the shells of hens in furnished cages with nests compared with eggs from hens in conventional battery cages. However, in a more direct investigation, Yue and Duncan (2003) found no apparent increase in stress-related calcium deposits on eggshells after access to the hens' habitual nest box was blocked (although signs of rapid pacing were observed). Similarly, no consistent increases in egg albumen corticosterone level were detected when access to a nest box was prevented (Cronin

et al., 2012). However, both these indicators depend on stress levels being elevated for some time to allow deposition of corticosterone in albumen, or to provoke the retention of the egg for longer than normal. It seems plausible from the behavioural evidence that any frustration arising from a denial of nest access may be severe but rather short lived. Measures of stress would need to be taken 20 min or less prior to oviposition to investigate this.

Perching and Roosting

Under natural or semi-natural conditions, chickens make use of elevated structures in the environment (most often tree branches) to perch during the day (Wood-Gush *et al.*, 1978) and to rest or sleep at night (Blokhuis, 1984). During the day, perching in elevated locations gives chickens a vantage point from which they can monitor their environment and, day or night, perching in an elevated location provides protection from most ground predators. Red junglefowl perch more frequently than domestic chickens, and in a more synchronous manner (Eklund and Jensen, 2011), but domesticated chickens also readily learn to perch and appear to feel more secure in environments where perches are present. The need for daytime vigilance may have decreased during domestication, but it certainly has not been eliminated. An early study reported more daytime perching by white leghorn birds than by brown strains (Faure and Jones 1982), but a more recent study found that brown hybrids used a roosting area more than white hens (Schrader and Muller, 2009).

Feral domestic hens seek out branches for perching in response to falling light levels during the late afternoon or early evening, when they are more photosensitive. This ensures that the majority of birds are safely elevated when night falls. The anti-predator function of accessing an elevated structure is supported by the finding that perch usage rates are higher in small groups where individuals have to be more vigilant (Newberry *et al.*, 2001). Birds may seek elevation to avoid poor litter quality, although this has not been examined experimentally. Chickens may also use perches or elevated structures to escape from other birds, an idea supported by the results of Cordiner and

Savory (2001). These authors sequentially provided small groups of 20 adult hens with no perches, or combinations of perches in low (17.5 cm above ground), medium (highest perch 35 cm above ground) or high (highest perch 70 cm above ground) configurations. The frequency of aggressive interactions was significantly lower with the high perch than when no perch was provided. In two of the four groups, perches of all heights were used more by subordinate birds than by dominant birds, suggesting that the perches may have provided a refuge.

Development of perching

For the first days of life, young chicks are brooded by their mothers on the ground. This ground roosting behaviour is triggered by falling afternoon light levels in the same way as elevated perching (Kent et al., 1997). Chicks start to perch when they are around 2 weeks of age (earlier if they are brooded by a hen; Riber et al., 2007), and the proportion of daytime spent perching increases steadily until the chicks are around 6 weeks of age. At this time, they also begin to show settled night-time roosting behaviour (Heikkila et al., 2006). From around 6 weeks of age, under feral conditions chicks follow their mother to night-time roosts and roost with her at progressively higher elevations as the chicks gain strength (McBride et al., 1969). In natural environments, groups of hens return to the same roosting sites at night (McBride et al., 1969; Wood-Gush et al., 1978). In commercial systems, birds reared without perches are less able to use perches in adult life due to a combination of reduced muscle tone, spatial awareness and balance.

Utilization of perches by laying hens

In commercial systems where perches are provided, there is considerable variation in the extent to which laying hens make use of the perches during the day. In furnished cages, perch utilization during the day varies between about 25 and 40% (Braastad, 1990; Appleby et al., 1992, 1993; Duncan et al., 1992). In experimental group pens, utilization rates are more variable (from as little as 3% to over 40%) and depend on factors such as perch

design (Lambe and Scott, 1998), perch height (60 cm perches are preferred over 20 cm perches; Cordiner and Savory, 2001; Newberry et al., 2001) and group size (Newberry et al., 2001).

In contrast, most studies report a large majority of, if not all, hens perching at night (Braastad, 1990). For example, in furnished cages, 80–95% of hens roost on perches at night (Appleby et al., 1992, 1993). Higher proportions are observed where sufficient perch length (suggested to be 140 mm per hen) is available (Appleby, 1995). Again, individual variation is observed, with some hens roosting consistently on the cage floor at night (Appleby et al., 1992).

Perching and roosting motivation and preferences in hens

The high percentage of perch usage at night is, in itself, some indication that roosting is a highly motivated behaviour. Additionally, when roosting is prevented, hens become agitated (Olsson and Keeling, 2000). The high motivation for night-time roosting is demonstrated by studies showing that hens will use a push-door to access a perch at night. Olsson and Keeling (2002) increased the force required for a hen to open the push-door from 25% of an individual's capacity to 100%, the maximum possible for that individual bird (calculated from the force overcome to reach food after 24 h of deprivation). The median resistance overcome to access the perch was 75% of capacity, compared with 0% for a sham perch. However, these studies do not reveal whether the primary motivation of the birds is to grasp a perch (wrap their feet around a suitably sized structure, as shown in Fig. 5.2) or to obtain an elevated position. Chickens possess touch receptors in the pads of their feet, which assist in grasping and balancing movements. However, their feet are not particularly specialized for grasping in comparison with other bird species (Sustaita et al., 2013). Chickens also use their feet for locomotion and for scratching for food, so they may not have a specific grasping motivation.

The motivation to achieve an elevated position appears to take precedence over a motivation to grasp. These two components were examined separately by offering hens different combinations of high (60 cm) or low (15 cm) perches or high or

Fig. 5.2. A free-range hen grasps a perch. Chickens seek elevated structures for refuge and for night-time safety. However, the extent to which they are motivated to perform this grasping action is not yet clear. (Photo courtesy of Lorna Wilson, University of Nottingham, UK.)

low flat plastic grids for night-time roosting (Schrader and Muller, 2009). Hens preferred high structures to low, and perches to grids, but when offered a choice between high grids or low perches, the height was by far the most important component in the hens' preference. Newberry *et al.* (2001) also found that birds preferred 60 cm perches to 20 cm perches in the daytime. Generally, in non-cage systems, greater utilization rates have also been reported for high perches at night (Oden *et al.*, 2002; Wichman *et al.*, 2007). However, when hens are housed in cages or other small-group systems, a preference for high perches appears to be influenced by the overall cage structure and height, such that birds are reluctant to perch too close to the cage roof. Provided there is a minimum distance of approximately 20 cm between a top perch and the cage roof, then hens prefer to perch on the highest perches available over a range from 6 to 36 cm (Struelens *et al.*, 2008b). However, if the cage height is less than 55 cm, the preference for the highest perches vanishes and overall perch use and comfort behaviour performance is reduced (Struelens *et al.*, 2008b). Others have even found that non-elevated perches are preferred in cage environments (e.g. Rönchen *et al.*, 2010). Chen *et al.* (2014) found that hens in 65 cm height cages preferred perches at heights of 10 or 20 cm more than perches at heights of 30 or 40 cm.

Preferences for perch widths and shapes appear even more variable than preferences for perch height. Given the choice, laying hens tend to avoid perching on narrow perches of just 1.5 cm diameter (Struelens *et al.*, 2009), and on perches of less than 4.5 cm (the most frequently used commercial perch diameter), hens find it more difficult to balance (Pickel *et al.*, 2010). In some studies, hens prefer 3 cm perches to 5 cm perches and rectangular perches to round (Chen *et al.*, 2014), although other studies found no strong preferences among perches of widths varying between approximately 3 and 11 cm. All of these perch diameters were well utilized during daytime perching and night-time roosting (Struelens *et al.*, 2009), perhaps reflecting an evolved natural tolerance for variably sized branches. Perch material has an influence on bird welfare, with rubber perches facilitating good grip and steel perches facilitating heat loss (Pickel *et al.*, 2010).

Although there is evidence that laying hens are frustrated if they are unable to access perches previously experienced, evidence of a deprivation effect is equivocal. No one has examined whether chickens 'miss' perches using methods used to examine deprivation of other resources (see Chapter 4). The extent to which the actual presence of higher structures stimulates the birds' preference for elevated perching (i.e. provides a cue that the birds would not miss in its absence) remains to be investigated. It is also possible that if other types of shelter or protection were provided, the motivation to seek an elevated structure might be reduced. For example, behaviours that increase vulnerability, such as resting or preening, are performed preferentially on perches but are also performed preferentially under cover (Newberry and Shackleton, 1997).

The extent to which chickens need perches in cage systems can be debated, as when pullets or laying hens are housed in conventional or furnished cages either with or without the provision of a perch, no differences in stress response have been observed (Barnett *et al.*, 2009; Yan *et al.*, 2013), despite the fact that the perches are well used when provided. In small experimental pens, no effects of perch provision on fear level were detected, although hens with perches did have lower heterophil:lymphocyte ratios after 15 weeks, an indicator of lower stress (Campo *et al.*, 2005). The perches in cages do not provide much in the way of elevation, but they do allow the birds' feet to be placed in a grasping formation. However, as mentioned briefly above, little is known about whether chickens have a grasping motivation. In addition, chickens in cages on the upper tiers of large commercial houses already have considerable elevation from the ground and they may be less likely to seek additional elevation of a few centimetres.

In commercial non-cage systems, the effects of perches are more pronounced and seemingly more beneficial. Direct welfare benefits relating to the satisfaction of a perching motivation are complemented by a range of indirect welfare benefits that arise when hens are able to make use of vertical height in a large house to rest or to avoid aggressive birds or those with a tendency to feather peck. A direct comparison of free-range hens in flocks of 7000 or 8000 with or without access to aerial perches found that birds in flocks with perches were less aggressive towards each other and less fearful of humans (Donaldson and O'Connell, 2012). The protective effects of perches in reducing injurious pecking are increasingly recognized. Access to perches during rear or lay can decrease feather pecking (Huber-Eicher and Audigé, 1999) and cloacal cannibalism (Gunnarsson et al., 1999), but access during rear is particularly important to ensure full development of spatial skills. Although no differences in the physical ability to reach a 40 cm height grid were noted between young hens reared with and without perches, hens reared without perches performed relatively poorly in a more complex food-finding task that required navigation between grids placed at various heights (Gunnarsson et al., 2000b).

There is no apparent relationship between chicks' ability to solve two-dimensional detour tasks and their later perching prowess. However, Wichman et al. (2007) detected a positive relationship between the ability of individual hens to find mealworms in a radial arm maze test and the extent of their initial use of perches in a laying environment shortly afterwards.

In practice, the provision of perches within commercial systems brings with it a complex mixture of welfare advantages and disadvantages relating to foot condition, keel bone shape, overall skeletal health and the risk of bone fractures via collisions.

Perching behaviour in broilers

Broiler chickens use perches far less than laying hens. As perching behaviour does not fully develop until the age of 6 weeks (an age when most broiler chickens will have been slaughtered), this is not totally surprising. However, we would expect to see increasing amounts of perching from 2 weeks of age onwards. The developmental pattern observed shows an increase in broiler perching between weeks 2 and 4, followed by a decrease thereafter, particularly in the heaviest birds (LeVan et al., 2000). Overall daytime perching is far lower than seen in layer-strain chicks, with just 1–3% of broilers using conventional perches at any one time (Su et al., 2000; Pettit-Riley and Estevez, 2001). Even when perching is actively encouraged by using low-height perches at low stocking densities, utilization rates seem to peak at around 25% at 4 weeks of age, falling back to just 10% by 6 weeks (Ventura et al., 2012), or averaging 10% across the full growing period (Bizeray et al., 2002a). Night-time perching in broilers is also far lower than seen in laying hens, with only between 8 and 19% of broilers seen perching at midnight in a study that provided low-height perches (Nielsen, 2004).

These studies suggest that broilers are inhibited or prevented from perching due to their extremely rapid growth and heavy body mass. However, low-level barriers (at a height of 8–15 cm) can be provided in broiler pens to encourage increased activity. If the barriers are placed strategically between food and water, then broilers have to walk further between these resources and they can also use the barriers for perching (Bizeray et al., 2002a; Ventura et al., 2012). Perches can also be designed to slope upwards from the floor at a shallow angle to improve access (LeVan et al., 2000). Such well-designed broiler perches and barriers have sometimes reaped benefits in reduced disturbance and aggression (Ventura et al., 2012), but not always (Pettit-Riley et al., 2002). They can also improve foot or leg health (Birgul et al., 2012) but not in all studies (Su et al., 2000; Bizeray et al., 2002b; Ventura et al., 2010). With perch utilization rates so much lower than those observed in laying hens, it is not surprising that the potential benefits of perching are often not realized. When ambient temperature is high, the use of conventional perches by broiler chickens is particularly low, and there are indications that broilers can be more stressed in the presence of perches than in their absence (Heckert et al., 2002). A possible exception to this conclusion is the way in which broilers prefer (Estevez et al., 2002) and utilize water-cooled perches in hot environments. In contrast to an age-related decline in the usage of normal perches, cool perches are used at increasing frequencies as broilers age (Zhao *et al.*, 2013), with associated improvements in leg health and production (Zhao *et al.*, 2012). These promising experimental results have yet to be fully evaluated or adopted under commercial conditions.

Locomotion

Energetic movements are often triggered by external causal factors such as a threat that prompts a rapid escape response or the discovery of a sufficiently exciting food item that prompts food running. The desire to perch or roost is also, as we have just seen, influenced to a large extent by the presence and availability of suitable elevated structures, with perhaps a small internally motivated component. In addition to walking and running, chickens can, to varying extents, jump and fly, but formal studies of the motivation of these activities have not been conducted.

Energetic locomotor behaviours are observed far more frequently in laying hens than in broilers, where factors related to selection for rapid growth, including high bodyweight and increased susceptibility to leg deformities and joint infections, greatly limit walking and running (Weeks *et al.*, 1994, 2000). Fast-growing strains of broiler are less active than slower-growing strains from the age of 4 weeks (Nielsen *et al.*, 2004). However, there is no indication that broilers have lost the motivation or desire to move. Rather, it appears that certain movements have become too difficult or costly to perform (Bokkers and Koene, 2004) and by 6 weeks of age, broilers are essentially flightless birds. Pain may also be a limiting factor, with studies showing that treatment with non-steroidal anti-inflammatory drugs (NSAIDs) can improve broilers' walking ability assessed using a simple subjective 'gait score' (McGeown *et al.* 1999). Recently, more objective methods of assessing gait and distinguishing different types of locomotor difficulty in chickens have been used with broiler chickens, including kinematic analysis using three-dimensional motion capture systems (Caplen *et al.*, 2012). To assess locomotion in this way, chickens have to be fitted with bright reflective markers that can be detected by the cameras, as shown in Fig. 5.3. Using this technique, Caplen *et al.* (2013) found that NSAIDs increased the

Fig. 5.3. Broiler chicken fitted with fluorescent kinematic markers for motion detection studies. (Photo courtesy of Dr Gina Caplen, University of Bristol, UK.)

velocity and stride length of broilers with locomotor difficulties.

Unlike broiler chickens, commercial strains of laying hens flap their wings at a similar rate to junglefowl and have not lost the instinct to jump (often with accompanying wing movements) or to fly (Provine *et al.*, 1984). Most hens can jump across a gap of at least 50 or 60 cm (Scott and Parker, 1994; Moinard *et al.*, 2004a), but only a minority voluntarily jump a wider 80 cm gap, and jumping downwards appears to be more difficult than jumping up (Scott *et al.*, 1997; Moinard *et al.*, 2004a). Hens with healed fractures of the keel bone are more reluctant to attempt to jump from a perch to the floor than uninjured hens, their reluctance increasing with perch height (Nasr *et al.*, 2012a). Such hens placed on high perches will often pace along the perch and make intention movements as if about to jump before seemingly changing their minds. This indecisiveness is somewhat ameliorated by the experimental administration of some types of NSAID or opioid analgesic drug (Nasr *et al.*, 2012b). Collisions and mistakes in landing occur when hens attempt to jump or fly downwards over gaps of 80 cm or more (Moinard *et al.* 2004b). Overall, these results suggest that the hesitation of injured hens reflects a cautious approach to avoid incurring further damage.

Preening

Preening comprises the cleaning and aligning of feathers with the beak and the distribution of oil on to the feathers from the uropygial gland and is an activity largely influenced by internal causal factors. Wood-Gush (1971) noted that preening took place as birds became sleepy when 'many stimuli in the immediate environment lose effectiveness while tactile stimuli gain in relative strength'. Delius (1988) added his own observations on the motivational control of preening arguing that 'By and large, as a preventive body-surface maintenance activity, [preening] does not have to be done at any particular time as long as it is done sometime and at intervals that are not too far apart.' From a motivational perspective, therefore, preening behaviour takes place when more immediate needs have been satisfied. It could be described as a low-resilience behaviour, one that is temporarily foregone under conditions of hardship or difficulty. Thus, much time spent in preening behaviour can be considered an indicator that much else is well with the world. Delius (1988) noted that aviary-kept seagulls nearly doubled the time they spent preening compared with their more hard-pressed wild conspecifics. More preening is also observed under conditions of fasting, when time normally allocated to feeding behaviour is made available (Webster, 1995; Hocking et al., 2007).

Preening with the beak develops in chickens at just 1 day or so of age (Kruijt, 1964; Williams and Strungis, 1977), but oiling behaviour is not observed until birds are approximately 1 month old (Williams and Strungis, 1977). Feather lipid concentrations increase with bird age, partly due to oiling behaviour but also as accumulations due to sebaceous secretions (Sandilands et al. 2004a,b). However, preening behaviour is not greatly influenced by the concentration of lipids on the feathers. This is not to say that external cues play no role – the presence of skin parasites, plumage misalignment or rain can stimulate bouts of preening – but these appear as adjustments to an underlying rhythm governed by diurnal cues and the need to prioritize other activities.

In contrast to the settled bouts of preening that are used to maintain plumage condition, very short, frequent bouts of preening are sometimes observed. Experimentally, adrenocorticotropic hormone administration shortens preening bout length in pigeons (Delius, 1988). Frequent, short preening bouts occur during periods of recovery from mild stress or (as described in Chapter 4) during periods of frustration. These bouts of preening can be so short that they are not functional, with no feather alignment or cleaning actually taking place.

There is a need for more information, but a tentative summary of available evidence points to short, frequent bouts of preening indicating raised levels of arousal in chickens, often in a negative context but occasionally in a positive situation. Unfortunately, many previous studies have used only instantaneous or scan sampling to record preening, and so the bout structure is not clear and results have to be interpreted cautiously. Scan sampling studies, which may well detect short bouts, show increases in preening in response to a lack of operant control (Taylor et al., 2001) or associated with a high stocking density and aggression (Zimmerman et al., 2006). Because frustration can sometimes be a stimulus to action, it is interesting to note a recent finding whereby hens that showed more displacement preening learned an extinction task more quickly (Kuhne et al., 2013). The frequency of preening also seems to increase during a period of reward anticipation (Zimmerman et al., 2011), supporting the view that it is an indicator of arousal rather than valence (whether the arousal is perceived positively or negatively by the bird).

In seemingly direct contrast, preening is also described as an indicator of post-consummatory relaxation (Seehuus et al., 2013) and the number of preening bouts can be inversely related to corticosterone concentration (Jones and Harvey, 1987). This paradox may be resolved if we appreciate the importance of separately examining preening frequency, preening duration and preening bout length. High frequencies of preening (probably short bout) have been associated with aversive responses in chickens (Nicol et al., 2011). However, a greater overall time spent preening is associated with environmental preference (Nicol et al., 2009, 2011). Longer preening durations are noted when hens observe content chicks than when they observe mildly distressed chicks (Edgar et al., 2011). Future work on preening needs to specify the form, structure and bout length of the behaviour to enable progress.

Dustbathing

Dustbathing is a behaviour that removes stale lipids (van Liere, 1992) and ectoparasites (Martin and Mullens, 2012) from the chickens' feathers.

Hens that used an available dustbath reduced their mite and lice infestations by 80–100% within a week in comparison with non-users, with sulphur a particularly effective substrate in this regard (Martin and Mullens, 2012). Given the opportunity, hens will dustbathe for many minutes on most days, and in commercial housing systems this often results in the excavation of hollows that are used exclusively for this behaviour (Weeks and Nicol, 2006). Substrate preferences are more pronounced than for foraging. Preferences depend partly on previous experience (Olsson et al., 2002a) but can be adjusted by exposure to new substrates in adult birds (Nicol et al., 2001; Wichman and Keeling, 2009), and fine-grained substrates are readily accepted by inexperienced adults (Wichman and Keeling, 2008). Substrates with fine particles, such as peat, sand or lignocellulose, are preferred over shavings, straw or food particles, probably because the fine particles are better at penetrating and cleaning the plumage (Petherick and Duncan, 1989; van Liere et al., 1990; de Jong et al., 2005; Scholz et al., 2010). Wood shavings that have been used previously and are therefore degraded in structure are accepted more readily for dustbathing than new wood shavings (Moesta et al., 2008). Interestingly, the lipid content of the available substrate has an influence, with hens preferring to dustbathe in substrates with lower lipid content (Scholz et al., 2011). This finding is relevant to the use of furnished cages where birds are expected to dustbathe on Astro-Turf mats sprinkled with relatively high-lipid food particles, a substrate that is incapable of acting to remove feather lipids (Scholz et al., 2014). Formal demand experiments confirm the findings from simpler choice tests, with a shallow demand curve, greater maximum price paid and greater total effort expended via a push-door to access peat for dustbathing compared with sand or wood shavings (de Jong et al., 2007). Fewer substrates have been tested with broiler chickens, but they too appear to form a strong preference for sand over rice hulls, wood shavings or paper (Shields et al., 2004).

A full sequence of dustbathing comprises many elements. First, the chicken scratches and rakes through the substrate with its beak, as if investigating the quality of the material. If satisfied with the substrate, it lies down and turns slightly on its side, with raised feathers. Dustbathing proper then commences, with episodes of vertical wing shaking and head rubbing interspersed with further bill raking and scratching with the available leg. After this vigorous phase has been completed, the chicken may remain lying for some time performing more occasional rubbing or scratching movements, before standing up and giving a thorough body shake to remove loose substrate (van Liere, 1991, cited by Olsson and Keeling, 2005). Long bouts of dustbathing that incorporate all of these possible elements are a good indicator of positive welfare. The performance of a complete bout of dustbathing demonstrates a lack of frustration and the chicken's acceptance of the substrate available. Generally, such complete bouts are accompanied by a return to baseline levels of motivation (van Liere, 1992). If a fine particulate substrate is not available, then chickens will select mildly abrasive substrates such as string-coated wire or AstroTurf in preference to wire, rubber or slats. However, their dustbathing behaviour on these less-preferred substrates is generally more unsettled, with shorter bouts indicating a degree of frustration (Merrill and Nicol, 2005; Merrill et al., 2006; Wichman and Keeling, 2009; Alvino et al., 2013).

In housing systems where a good substrate is not available, chickens may perform partial or incomplete sequences of dustbathing ('sham dustbathing') behaviour on wire floors, often scattering feed as a substitute for a proper substrate (Lindberg and Nicol, 1997). Sham dustbathing has been reviewed by Olsson and Keeling (2005). It is not as satisfying as dustbathing on a particulate substrate, and it does not return motivation to baseline levels (Vestergaard, 1982; Vestergaard et al., 1999; Olsson et al., 2002a).

Dustbathing appears to be a low-resilience behaviour that is forfeited when other needs are more pressing. It has been suggested that chickens allocate time to this behaviour as and when the opportunity arises (Widowski and Duncan, 2000). In quail, a rebound in some elements of dustbathing behaviour (dust tossing and head rubbing) occurs after a period of litter restriction (Borchelt et al., 1973), but this is not always seen in hens (Guesdon and Faure, 2008). This may explain why experimental studies show it to be a lesser priority than other behaviours (Petherick et al., 1993; Keeling, 1994). The opportunistic timing of dustbathing may also explain why many birds are often seen to dustbathe simultaneously in groups. An alternative explanation for synchronous dustbathing might be that the behaviour is socially facilitated, but experimental studies, such as using video images or careful exposure to the behaviour of other birds, provide little support for social facilitation (Olsson *et al.*, 2002b; Lundberg and Keeling, 2003).

Comfort Behaviours

The term 'comfort behaviour' was used by Kruijt (1964) to describe the movements of stretching, yawning, preening, dustbathing, head shaking and tail shaking in the junglefowl. One of the first studies of comfort behaviour in the domestic fowl was by Black and Hughes (1974), who compared the behaviour of hens housed in cages or pens. However, the term has been used in different ways by different authors, and rarely with any independent evidence relating to bird comfort. Activities such as body shaking (Fig. 5.4) and tail wagging are included by some authors but not others. A summary is given in Table 5.1.

Other authors have used terms such as 'self-maintenance behaviours' to describe similar groupings of activities (Albentosa *et al.*, 2007). Authors also vary in whether they subsequently analyse the component activities separately or as a group, and whether they separate one particular activity (e.g. dustbathing or preening) for separate analysis.

As preening and dustbathing have already been discussed in this chapter, the following discussion considers the influences on the shorter stretching and grooming components of comfort behaviour.

Spatial allowance has a profound effect on the ability of a chicken to perform comfort behaviour. Broiler chickens do more wing flapping in floor pens than in cages (Fortomaris *et al.*, 2007), but the majority of studies have been conducted with laying hens in relation to debates about the welfare consequences of conventional battery cages. The space provided for a laying hen in a conventional battery cage prior to legislative changes imposed in Europe from 1999 was 450 cm^2 per hen, after which it was increased to 550 cm^2 per hen. In the USA, United Egg Producers, a certification body that covers approximately 80% of the industry, has developed industry guidelines and auditing procedures designed to promote a gradual increase in space allowance per bird from the 2002 standard of 360 cm^2 per white hybrid bird or 400 cm^2 per brown hybrid bird to a slightly more generous 430 cm^2 per white hybrid and 490 cm^2 per brown hybrid bird. However, even these new allowances will impose very severe restrictions on the behaviour of the laying hens. Dawkins and Hardie (1989) filmed birds and showed, for example, that unconstrained wing stretching occupied a minimum of 653 cm^2, wing flapping a minimum of 860 cm^2 and preening a minimum of 814 cm^2. Thus, if birds are to perform these

Fig. 5.4. A free-range hen shakes her body, a movement that is always accompanied by raised feathers and which serves to re-align the plumage. Body shaking is often described as a comfort behaviour. (Photo courtesy of Lorna Wilson, University of Nottingham, UK.)

Table 5.1. Definitions of activities considered to be 'comfort behaviour' by different authors.

Reference(s)	Dustbathe	Preen (sometimes specific body areas)	Body shake/ feather ruffle	Feather raise	Wing stretch, wing flap, wing raise or wing/ leg stretch	Tail wag/tail shake	Head rub	Bill wipe	Stretch	Head or body scratch	Yawn	Lie, sit, sleep
Black and Hughes (1974)		X	X	X	X					X		
Tebbe et al. (1986)		X	X	X	X	X						
Nicol (1989)		X	X		X	X				X		
Appleby et al. (2002)			X		X	X		X	X	X		
Carmichael et al. (1999)			X		X	X						
Albentosa et al. (2004, 2007)			X	X	X	X					X	
Shimmura et al. (2006, 2007)	X	X	X		X	X	X	X				
Balazova and Baranyiova (2010)		X										X
Zimmerman et al. (2011)		X	X		X	X				X		X

behaviours at all in conventional cages, they will be squeezed against other birds or the cage walls. It is therefore not surprising that birds reduce their performance of these behaviours in conventional cages (Tebbe et al., 1986; Nicol, 1987a). This matters because these behaviours matter to the hens. Comfort behaviours may be performed relatively rarely, often just once an hour or less, but birds are highly motivated to do these behaviours. Nicol (1987b) reported that wing/leg stretching, wing flapping, tail wagging and preening increased dramatically in rate after prevention by spatial restriction, indicating a strong internal component to their motivation.

Even the 550 cm^2 allowance implemented in the EU in 1999 constrains and restricts comfort behaviour in laying hens. Evidence to this effect was crucial in support of a ban of the conventional cage system and its replacement with larger, furnished cages. In the lead-up to the 2012 ban on conventional cages, studies showed that many comfort behaviours were performed more readily at around 750 cm^2 than at lesser allowances (Appleby et al., 2002; Shimmura et al., 2006, 2007; Pohle and Cheng, 2009). At still greater space allowances, comfort behaviours increase further (Albentosa and Cooper, 2004), although the birds' preference and demand for additional space tapers once the birds have 700–800 cm^2. Stocking density in aviaries or free-range systems does not appear to have a limiting effect on comfort behaviour in laying hens (Carmichael et al., 1999).

Space is not the only influence on comfort behaviour in laying hens. Social factors also have an influence. In one study, tail wagging, body shaking and preening increased when conspecifics were visible or in close proximity compared with when they were distant or behind a screen (Nicol, 1989). Keeling and Duncan (1991) similarly found a positive association between flock cohesion and proximity and the performance of preening behaviour. Negative correlations between aggression and comfort behaviour have also been noted (Nicol, 1989; Bradshaw, 1992), and wing flapping increases in anticipation of a positive event (Zimmerman et al., 2011).

The one exception to the general picture that these movements are helpful in plumage and musculoskeletal condition and comfort is the occurrence of head shaking. Head shaking in cockerels can occur in response to the crowing of other males. In hens, head shaking has been described as an 'alerting response' as it similarly increases in response to the appearance of disturbing stimuli (Hughes, 1983). These observations are in line with other findings that head shaking occurs more often during fasting (Webster, 1995) and during anticipation of a negative event (Zimmerman et al., 2011). Most strikingly, head shaking is an indicator that hens are in a less-preferred environment, and high rates of head shaking may therefore be a valid indicator of poor welfare (Nicol et al., 2009, 2011).

Rest and Sleep

We have already reviewed how brooding, early learning and lighting influence the sleep of young chicks (see Chapter 3). Once broiler chickens reach 5 or 6 weeks of age, the vast majority of their time is spent lying, and they appear to doze or sleep for at least a quarter of this lying time and to be otherwise inactive for the rest of the lying time (Weeks et al., 2000). Blokhuis (1983) found that junglefowl did more sleeping than commercial hybrids, but both types of chicken also showed dozing behaviour, usually in a sitting posture. When a chicken starts to doze, it withdraws its neck and makes small head movements with open eyes and the tail slightly down. As dozing progresses, the head is lowered and the eyes close, while the feathers are slightly raised and the tail and wings begin to droop. Broiler chickens may sleep in a dozing position, but adult hens often tuck their heads into their feathers. Uninterrupted resting or dozing behaviour is important, and broiler chickens will preferentially choose to rest against walls, particularly at higher stocking densities, to avoid disturbance from other birds (Buijs et al., 2010). Layer chicks also benefit from the reduced disturbance of rest provided by the use of dark brooders (reviewed in Chapter 3).

True sleep can be distinguished from waking by EEG recordings and, as in mammals, both high-amplitude slow-wave and low-voltage fast-wave (paradoxical) EEG sleep patterns are seen in chickens. The chicken EEG changes are essentially similar to those of mammals, although the sleep spindles observed in the mammalian neocortex are lacking in the chicken cerebrum (Ookawa, 2004). Sleep comprises only a small proportion of the time budget of adult chickens, with fast-wave sleep occupying less than 1% of time (reviewed by Blokhuis, 1983). As discussed in Chapter 3, chickens (like other birds) can sleep with one cerebral hemisphere, while the other hemisphere remains awake.

Play

Play is a low-resilience behaviour, one that is expressed when more pressing needs have been met and one that is lost from the behavioural repertoire when an animal is subjected to other pressures (Held and Spinka, 2011). It therefore has the potential to be a valid (although complex) indicator of welfare state (Held and Spinka, 2011). However, if chickens play at all, it is to a limited extent during the first couple of weeks of life. Guhl (1958, cited by Wood-Gush, 1971) noted that at about 1 week of age, chicks performed frolicking activity, comprising spontaneous running with the wings raised, and by 2 weeks of age, frolicking was accompanied by sparring in which chicks jumped up and down in the manner of adult birds engaged in a fight but without the exchange of aggressive pecks. These activities do possess some of the characteristics of play defined by Burghardt (2005; cited by Held and Spinka, 2011) being incomplete forms of adult flying and fighting behaviours that occur during the formation of a social hierarchy. However, they are rather invariant activities, with no apparent innovation or invention, and they are soon lost from the ethogram. It would be difficult to argue that they provide an example of social play.

Social play, where individuals respond to each other in a vigorous, varied and reciprocal manner, is widespread among mammals but has been documented in only a few avian species such as parrots and corvids. Members of these avian species engage in play fights, chases and social object play (Diamond and Bond, 2003), producing complex behavioural sequences not generally observed in chickens.

Exploration

Exploration is a behaviour that facilitates the acquisition of information about new or altered features of the environment. Chickens have a strong and highly developed urge to explore novel objects or features of their environment, provided that these objects are not too threatening, but does this mean that chickens can feel bored if novelty is absent? Another question is whether animals will search for novelty in the absence of any reminder cues, a possible internal motivation that has been investigated in pigs (Wood-Gush and Vestergaard, 1991; Rushen, 1993) but not in chickens.

In the past, specialists in animal exploration drew a distinction between extrinsic exploration where investigatory behaviour is targeted towards a particular resource or goal (food, nesting material), and intrinsic exploration where an animal shows investigatory behaviour out of pure curiosity or an interest in novel stimuli for their own sake (Hughes, 1997). In a test of the exploratory tendencies of chickens, Newberry (1999) housed broilers in a home pen, to which a supplementary pen was attached and intermittently available. For some groups of chickens, the supplementary pen was empty, or simply contained the same essential resources as the home pen (food, water and heat). A third treatment provided resources in the supplementary pen that were expected to stimulate functional behaviours such as foraging or dustbathing. For the final treatment, the supplementary pen was populated with changing sets of novel objects of low biological relevance including 'a wooden chair, a beach ball, a tin of marbles, a mirror, a rubber boot, a bamboo stalk and a tray of sand'. During the first 5 min of daily access, the chickens from this novel object group were the most likely to enter the supplementary pen, often running in as soon as a gate was opened. Across the experiment as a whole, the time spent with the novel objects was significantly higher than time spent in the empty control, and approached that spent with the functional resources. Newberry (1999) suggested that these results provided evidence of intrinsic motivation in that the chickens that rushed into the novel object pen were not seeking to meet current biological needs. However, she also admitted that some of the novel objects may have had minor functional benefits and it seems difficult to draw an absolute distinction here. Chickens may well investigate novel objects on the off-chance that they could prove useful in the future. Pecking at objects such as string or polystyrene may look functionless but may be an expression of foraging behaviour in an artificial environment.

However, other evidence does suggest that chickens are interested in novel stimuli for their own sake. When artificial and biologically neutral moving video images (e.g. of cartoon fish) were presented to chicks, they showed an initial fear response, but they showed a positive attraction to the images after three daily exposures whether housed individually (Jones *et al.*, 1996) or in small groups (Jones *et al.*, 1998). Similar results were obtained with singly housed adult hens (Clarke and Jones, 2000).

An understanding of the chicken's motivation to seek novelty for its own sake is highly relevant to the issue of environmental enrichment. A review of the chickens' most important behavioural priorities suggests that efforts to enrich and improve chicken housing should first and foremost be directed towards the provision of resources that meet their most important foraging, nesting and comfort behaviour needs. However, from the limited evidence available, there may also be a role for non-threatening toys or other stimulation to keep chickens gainfully occupied and to prevent them from directing pecking behaviours towards each other. These aspects will be considered in more detail in the later chapters of this book.

References

Albentosa, M.J. and Cooper, J.J. (2004) Effects of cage height and stocking density on the frequency of comfort behaviours performed by laying hens housed in furnished cages. *Animal Welfare* 13, 419–424.

Albentosa, M.J., Cooper, J.J., Luddem, T., Redgate, S.E., Elson, H.A. and Walker, A.W. (2007) Evaluation of the effects of cage height and stocking density on the behaviour of laying hens in furnished cages. *British Poultry Science* 48, 1–11.

Alvino, G.M., Tucker, C.B., Archer, G.S. and Mench, J.A. (2013) Astroturf as a dustbathing substrate for laying hens. *Applied Animal Behaviour Science* 146, 88–95.

Appleby, M.C. (1986) Hormones and husbandry: control of nesting behaviour in poultry production. *Poultry Science* 65, 2352–2354.

Appleby, M.C. (1990) Behaviour of laying hens in cages with nest sites. *British Poultry Science* 31, 71–80.

Appleby, M.C. (1995) Perch length in cages for medium hybrid laying hens. *British Poultry Science* 36, 23–31.

Appleby, M.C. and McRae, H.E. (1983) Floor-laying by domestic hens. *Applied Animal Behaviour Science* 11, 202.

Appleby, M.C. and McRae, H.E. (1986) The individual nest box as a super-stimulus for domestic hens. *Applied Animal Behaviour Science* 15, 169–176.

Appleby, M.C., McRae, H.E. and Peitz, B.E. (1984a) The effect of light on the choice of nests by domestic hens. *Applied Animal Ethology* 11, 249–254.

Appleby, M.C., McRae, H.E., Duncan, I.J.H. and Bisazza, A. (1984b) Choice of social conditions by laying hens. *British Poultry Science* 25, 111–117.

Appleby, M.C., Smith, S.F. and Hughes, B.O. (1992) Individual perching behaviour of laying hens and its effects in cages. *British Poultry Science* 33, 227–238.

Appleby, M.C., Smith, S.F. and Hughes, B.O. (1993) Nesting, dust-bathing and perching by laying hens in cages: effects of design on behaviour and welfare. *British Poultry Science* 34, 835–847.

Appleby, M.C., Walker, A.W., Nicol, C.J., Lindberg, A.C., Freire, R., Hughes, B.O. and Elson, H.A. (2002) Development of furnished cages for laying hens. *British Poultry Science* 43, 489–500.

Balazova, L. and Baranyiova, E. (2010) Broiler response to open field test in early ontogeny. *Acta Veterinaria Brno* 79, 19–26.

Barnett, J.L., Tauson, R., Downing, J.A., Janardhana, V., Lowenthal, J.W., Butler, K.L. and Cronin, G.M. (2009) The effects of a perch, dust bath, and nest box, either alone or in combination as used in furnished cages, on the welfare of laying hens. *Poultry Science* 88, 456–470.

Birgul, O.B., Mutaf, S. and Alkan, S. (2012) Effects of different angled perches on leg disorders in broilers. *Archiv fur Geflugelkunde* 76, 44–48.

Bizeray, D., Estevez, I., Letterier, C. and Faure, J.M. (2002a) Effects of increasing environmental complexity on the physical activity of broiler chickens. *Applied Animal Behaviour Science* 79, 27–41.

Bizeray, D., Estevez, I., Letterier, C. and Faure, J.M. (2002b) Influence of increased environmental complexity on leg condition, performance and level of fearfulness in broilers. *Poultry Science* 81, 767–773.

Black A.J. and Hughes, B.O. (1974) Patterns of comfort behaviour and activity in domestic fowls – a comparison between cages and pens. *British Veterinary Journal* 130, 23–33.

Blokhuis, H.J. (1983) The relevance of sleep in poultry. *World's Poultry Science Journal* 39, 33–37.

Blokhuis, H.J. (1984) Rest in poultry. *Applied Animal Behaviour Science* 12, 289–303.

Bokkers, E.A.M. and Koene, P. (2004) Motivation and ability to walk for a food reward in fast- and slow-growing broilers to 12 weeks of age. *Behavioural Processes* 67, 121–130.

Borchelt, P.L., Eyer, J. and McHenry, D.S. (1973) Dustbathing in Bobwhite quail (*Colinus virginianus*) as a function of dust deprivation. *Behavioural Biology* 8, 109–114.

Braastad, B.O. (1990) Effects on behaviour and plumage of a key-stimuli floor and a perch in triple cages for laying hens. *Applied Animal Behaviour Science* 27, 127–139.

Bradshaw, R.H. (1992) Effects of social status on the performance of noninteractive behaviours in small groups of laying hens. *Applied Animal Behaviour Science* 33, 77–81.

Bubier, N.E. (1996a) The behavioural priorities of laying hens: the effect of cost/no cost multi-choice tests on time budgets. *Behavioural Processes* 37, 225–238.

Bubier, N.E. (1996b) The behavioural priorities of laying hens: the effects of two methods of environment enrichment on time budgets. *Behavioural Processes* 37, 239–249.

Buijs, S., Keeling, L.J., Vangestel, C., Baert, J., Vangeyte, J. and Tuyttens, F.A.M. (2010) Resting or hiding? Why broiler chickens stay near walls and how density affects this. *Applied Animal Behaviour Science* 124, 97–103.

Campo, J.L., Gil, M.G., Davila, S.G. and Munoz, I. (2005) Influence of perches and footpad dermatitis on tonic immobility and heterophil:lymphocyte ratio of chickens. *Poultry Science* 84, 1004–1009.

Caplen, G., Hothersall, B., Murrell, J., Nicol, C.J., Waterman-Pearson, A., Weeks, C.A. and Colborne, B. (2012) Kinematic analysis quantifies gait abnormalities associated with lameness in broiler chickens and identifies evolutionary gait differences. *PLoS One* 7, e40800.

Caplen, G., Colborne, G.R., Hothersall, B., Nicol, C.J., Waterman-Pearson, A.E., Weeks, C.A. and Murrell, J.C. (2013) Lame broiler chickens respond to non-steroidal anti-inflammatory drugs with objective changes in gait function: a controlled clinical trial. *Veterinary Journal* 196, 477–482.

Carmichael, N.L., Walker, A.W. and Hughes, B.O. (1999) Laying hens in large flocks in a perchery system: influence of stocking density of location, use of resources and behaviour. *British Poultry Science* 40, 165–176.

Chen, D.H., Bao, J., Meng, F.Y. and Wei, C.B. (2014) Choice of perch characteristics by laying hens in cages with different group size and perching behaviours. *Applied Animal Behaviour Science* 150, 37–43.

Clara, E., Regolin, L., Vallortigara, G. and Rogers, L.J. (2009) Chicks prefer to peck at insect-like elongated stimuli moving in a direction orthogonal to their longer axis. *Animal Cognition* 12, 755–765.

Clarke, C.H. and Jones, R.B. (2000) Responses of adult laying hens to abstract video images presented repeatedly outside the home cage. *Applied Animal Behaviour Science* 67, 97–110.

Clausen, T. and Riber, A.B. (2012) Effect of heterogeneity of nest boxes on occurrence of gregarious nesting in laying hens. *Applied Animal Behaviour Science* 142, 168–175.

Collins L.M. and Sumpter, D.J.T. (2007) The feeding dynamics of broiler chickens. *Journal of the Royal Society Interface* 4, 65–72.

Cooper, J.J. and Appleby, M.C. (1996) Demand for nestboxes in laying hens. *Behavioural Processes* 36, 171–182.

Cooper, J.J. and Appleby, M.C. (1997) Motivational aspects of individual variation in response to nestboxes by laying hens. *Animal Behaviour* 54, 1245–1253.

Cooper, J.J. and Appleby, M.C. (2003) The value of environmental resources to domestic hens: a comparison of the work-rate for food and for nests as a function of time. *Animal Welfare* 12, 39–52.

Cordiner, L.S. and Savory, C.J. (2001) Use of perches and nestboxes by laying hens in relation to social status, based on examination of consistency of ranking orders and frequency of interaction. *Applied Animal Behaviour Science* 71, 305–317.

Cronin, G.M., Butler, K.L., Desnoyers, M.A. and Barnett, J.L. (2005) The use of nest boxes by hens in cages: what does it mean for welfare? *Animal Science Papers and Reports* 23, 121–128.

Cronin, G.M., Barnett, J.L. and Hemsworth, P.H. (2012) The importance of pre-laying behaviour and nest boxes for laying hen welfare: a review. *Animal Production Science* 52, 398–403.

Davies, A.C., Radford, A.N. and Nicol, C.J. (2014) Behavioural and physiological expression of arousal during decision-making in laying hens. *Physiology and Behavior* 123, 93–99.

Dawkins, M.S. (1989) Time budgets in red junglefowl as a baseline for the assessment of welfare in domestic fowl. *Applied Animal Behaviour Science* 24, 77–80.

Dawkins, M.S. and Hardie, S. (1989) Space needs of laying hens. *British Poultry Science* 30, 413–416.

de Jong, I.C., Fillerup, M. and van Reenen, K. (2005) Substrate preferences in laying hens. *Animal Science Papers and Reports* 23 (Suppl. 1), 143–152.

de Jong, I.C., Wolthuis-Fillerup, M. and van Reenen, C.G. (2007) Strength of preference for dustbathing and foraging substrates in laying hens. *Applied Animal Behaviour Science* 104, 24–36.

Delius, J.D. (1988) Preening and associated comfort behaviour in birds. *Annals of the New York Academy of Sciences* 525, 40–55.

Diamond, J. and Bond, A.B. (2003) A comparative analysis of social play in birds. *Behaviour* 140, 1091–1115.

Donaldson, C.J. and O'Connell, N.E. (2012) The influence of access to aerial perches on fearfulness, social behaviour and production parameters in free-range laying hens. *Applied Animal Behaviour Science* 142, 51–60.

Duncan, E.T., Appleby, M.C. and Hughes, B.O. (1992) Effect of perches in laying cages on welfare and production of hens. *British Poultry Science* 33, 25–35.

Duncan, I.J.H. and Hughes, B.O. (1972) Free and operant feeding in domestic fowl. *Animal Behaviour* 20, 775–777.

Duncan, I.J.H. and Kite, V.G. (1987) Some investigations into motivation in the domestic fowl. *Applied Animal Behaviour Science* 18, 387–388.

Duncan, I.J.H. and Kite, V.G. (1989) Nest site selection and nest-building behaviour in domestic fowl. *Animal Behaviour* 37, 215–231.

Edgar, J.L., Lowe, J.C., Paul, E.S. and Nicol, C.J. (2011) Avian maternal response to chick distress. *Proceedings of the Royal Society B – Biological Sciences* 278, 3129–3134.

Eklund, B. and Jensen, P. (2011) Domestication effects on behavioural synchronization and individual distances in chickens (*Gallus gallus*). *Behavioural Processes* 86, 250–256.

Estevez, I., Tablante, N., Pettit-Riley, R.L. and Carr, L. (2002) Use of cool perches by broiler chickens. *Poultry Science* 81, 62–69.

Faure, J.M. and Jones, R.B. (1982) Effects of sex, strain and type of perch on perching behaviour in the domestic fowl. *Applied Animal Ethology* 8, 281–293.

Fortomaris, P., Arsenos, G., Tserveni-Gousi, A. and Yannakopoulis, A. (2007) Performance and behaviour of broiler chickens as affected by the housing system. *Archiv fur Geflugelkunde* 71, 97–104.

Freire, R., Appleby, M.C. and Hughes, B.O. (1996) Effects of nest quality and other cues for exploration on pre-laying behaviour. *Applied Animal Behaviour Science* 48, 37–46.

Freire, R., Appleby, M.C. and Hughes, B.O. (1997) Assessment of pre-laying motivation in the domestic hen using social interaction. *Animal Behaviour* 54, 313–319.

Freire, R., Appleby, M.C. and Hughes, B.O. (1998) Effects of social interactions on pre-laying behaviour in hens. *Applied Animal Behavour Science* 56, 47–57.

Guesdon, V. and Faure, J.M. (2008) A lack of dust-bathing substrate may not frustrate laying hens. *Archiv fur Geflugelkunde* 72, 241–249.

Gunnarsson, S., Keeling, L.J. and Svedburg, J. (1999) Effect of rearing factors on the prevalence of floor eggs, cloacal cannibalism and feather pecking in commercial flocks of loose housed laying hens. *British Poultry Science* 40, 12–18.

Gunnarsson, S., Matthews, L.R., Foster, T.M. and Temple, W. (2000a) The demand for straw and feathers as litter substrates by laying hens. *Applied Animal Behaviour Science* 65, 321–330.

Gunnarsson, S., Yngvesson, J., Keeling, L.J. and Forkman, B. (2000b) Rearing without early access of perches impairs the spatial skills of laying hens. *Applied Animal Behaviour Science* 67, 217–228.

Heckert, R.A., Estevez, I., Russek-Cohen, E. and Pettit-Riley, R. (2002) Effects of density and perch availability on the immune status of broilers. *Poultry Science* 81, 451–457.

Heidweiller, J., Vanloon, J.A. and Zweers, G.A. (1992) Flexibility of the drinking mechanism in adult chickens (*Gallus gallus*). *Zoomorphology* 111, 141–159.

Heikkila, M., Wichman, A., Gunnarsson, S. and Valros, A. (2006) Development of perching behaviour in chicks reared in enriched environment. *Applied Animal Behaviour Science* 99, 145–156.

Held, S.D.E. and Spinka, M. (2011) Animal play and animal welfare. *Animal Behaviour* 81, 891–899.

Hocking, P.M., Rutherford, K.M.D. and Picard, M. (2007) Comparison of time-based frequencies, fractal analysis and T-patterns for assessing behavioural changes in broiler breeders fed on two diets at two levels of feed restriction: a case study. *Applied Animal Behaviour Science* 104, 37–48.

Hogan, J.A. (1965) An experimental study of conflict and fear – an analysis of behaviour of young chicks toward a mealworm. 1. Behavior of chicks which do not eat the mealworm. *Behaviour* 25, 45–49.

Houldcroft, E., Smith, C., Mrowicki, R., Headland, L., Grieveson, S., Jones, T.A. and Dawkins, M.S. (2008) Welfare implications of nipple drinkers for broiler chickens. *Animal Welfare* 17, 1–10.

Huber, H.V., Folsch, D.W. and Stahli, U. (1985) Influence of various nesting materials on nest site selection of the domestic hen. *British Poultry Science* 26, 367–373.

Huber-Eicher, B. (2004) The effect of early colour preference and of a colour exposing procedure on the choice of nest colours in laying hens. *Applied Animal Behaviour Science* 86, 63–76.

Huber-Eicher, B. and Audigé, L. (1999) Analysis of risk factors for the occurrence of feather pecking in laying hen growers. *British Poultry Science* 40, 599–604.

Hughes, B.O. (1971) Allelomimetic feeding in domestic fowl. *British Poultry Science* 12, 359–363.

Hughes, B.O. (1983) Headshaking in fowls: the effect of environmental stimuli. *Applied Animal Ethology* 11, 45–53.

Hughes, B.O. (1993) Choice between artificial turf and wire floor as nest sites in individually caged laying hens. *Applied Animal Behaviour Science* 36, 327–335.

Hughes, B.O., Petherick, J.C., Brown, M.F. and Waddington, D. (1989) The performance of nest building by domestic hens – is it more important than the construction of a nest? *Animal Behaviour* 37, 210–214.

Hughes, B.O., Petherick, J.C., Brown, M.F. and Waddington, D. (1995) Visual recognition of key nest-site stimuli by laying hens in cages. *Applied Animal Behaviour Science* 42, 271–281.

Hughes, R.N. (1997) Intrinsic exploration in animals: motives and measurement. *Behavioural Processes* 41, 213–226.

Inglis, I.R., Forkman, B. and Lazarus, J. (1997) Free food or earned food? A review and fuzzy model of contrafree-loading. *Animal Behaviour* 53, 1171–1191.

Jones R.B. and Carmichael, N.L. (1999) Responses of domestic chicks to selected pecking devices presented for varying durations. *Applied Animal Behaviour Science* 64, 125–140.

Jones, R.B. and Harvey, S. (1987) Behavioral and adrenocortical responses of domestic chicks to systematic reductions in group size and to sequential disturbance of companions by the experimenter. *Behavioural Processes* 14, 291–303.

Jones, R.B., Larkins, C. and Hughes, B.O. (1996) Approach/avoidance responses of domestic chicks to familiar and unfamiliar video images of biologically neutral stimuli. *Applied Animal Behaviour Science* 48, 81–98.

Jones, R.B., Carmichael, N. and Williams, C. (1998) Social housing and domestic chicks' responses to symbolic video images. *Applied Animal Behaviour Science* 56, 231–243.

Keeling, L.J. (1994) Inter-bird distances and behavioural priorities in laying hens: the effect of spatial restriction. *Applied Animal Behaviour Science* 39, 131–140.

Keeling, L.J. and Duncan, I.J.H. (1991) Social spacing in domestic fowl under seminatural conditions – the effect of behavioural activity and activity transitions. *Applied Animal Behaviour Science* 32, 205–217.

Kent, J.P., McElligott, A.G. and Budgey, H.V. (1997) Ground-roosting in domestic fowl (*Gallus gallus domesticus*) in the Gambia: the anticipation of night. *Behavioural Processes* 39, 271–278.

Klein, T., Zeltner, E. and Huber-Eicher, B. (2000) Are genetic differences in foraging behaviour of laying hen chicks paralleled by hybrid-specific differences in feather pecking? *Applied Animal Behaviour Science* 70, 143–155.

Kruijt, J.P. (1964) *Ontogeny of Social Behaviour in Burmese Red Junglefowl (Gallus gallus spadiceus)*. Brill, Leiden.

Kruschwitz, A., Zupan, M., Buchwalder, T. and Huber-Eicher, B. (2008) Nest preference of laying hens (*Gallus gallus domesticus*) and their motivation to exert themselves to gain nest access. *Applied Animal Behaviour Science* 112, 321–330.

Kuhne, F., Sauerbrey, A.F.C. and Adler, S. (2013) The discrimination-learning task determines the kind of frustration-related behaviours in laying hens (*Gallus gallus domesticus*). *Applied Animal Behaviour Science* 148, 192–200.

Lambe, N.R. and Scott, G.B. (1998) Perching behaviour and preferences for different perch designs among laying hens. *Animal Welfare* 7, 203–216.

Lentfer, T.L., Gebhardt-Henrich, S.G., Frohlich, E.K.F. and van Borell, E. (2011) Influence of nest site on the behaviour of laying hens. *Applied Animal Behaviour Science* 135, 70–77.

LeVan, N.F., Estevez, I. and Stricklin, W.R. (2000) Use of horizontal and angled perches by broiler chickens. *Applied Animal Behaviour Science* 65, 349–365.

Lima, S.L. and Dill, L.M. (1990) Behavioral decisions made under the risk of predation – a review and prospectus. *Canadian Journal of Zoology* 68, 619–640.

Lindberg, A.C. and Nicol, C.J. (1997) Dustbathing in modified battery cages: is sham dustbathing an adequate substitute? *Applied Animal Behaviour Science* 55, 113–128.

Lindqvist, C.E.S., Schutz, K.E. and Jensen, P. (2002) Red junglefowl have more contra free-loading than white leghorn layers: effects of food deprivation and consequences for information gain. *Behaviour* 139, 1195–1209.

Lundberg, A. and Keeling, L.J. (1999) The impact of social factors on nesting in laying hens (*Gallus gallus domesticus*) *Applied Animal Behaviour Science* 64, 57–69.

Lundberg, A.S. and Keeling, L.J. (2003) Social effects on dustbathing behaviour in laying hens: using video images to investigate effects of rank. *Applied Animal Behaviour Science* 81, 43–57.

Martin, C.D. and Mullens, B.A. (2012) Housing and dustbathing effects on northern fowl mites (*Ornithonyssus sylviarum*) and chicken body lice (*Menacanthus stramineus*) on hens. *Medical and Veterinary Entomology* 26, 323–333.

Martin, P. and Bateson, P.P.G. (2007) *Measuring Behaviour, an Introductory Guide*, 3rd edn. Cambridge University Press, Cambridge, UK.

McBride, G., Parer, I.P. and Foenander, F. (1969) The social organisation and behaviour of the feral domestic fowl. *Animal Behaviour Monographs* 2, 125–181.

McGeown, D., Danbury, T.C., Waterman-Pearson, A.E. and Kestin, S.C. (1999) Effect of carprofen on lameness in broiler chickens. *Veterinary Record* 144, 668–671.

Meijsser, F.M. and Hughes, B.O. (1989) Comparative analysis of pre-laying behaviour in battery cages and in three alternative systems. *British Poultry Science* 30, 747–760.

Merrill, R.J.N. and Nicol, C.J. (2005) The effects of novel flooring on dustbathing, pecking and scratching behaviour of caged hens. *Animal Welfare* 14, 179–186.

Merrill, R.J.N., Cooper, J.J., Albentosa, M.J. and Nicol, C.J. (2006) The preferences of laying hens for perforated Astroturf over conventional wire as a dustbathing substrate in furnished cages. *Animal Welfare* 15, 173–178.

Mills, A.D. and Wood-Gush, D.G.M. (1985) Pre-laying behaviour in battery cages. *British Poultry Science* 26, 247–252.

Moesta, A., Knierim, U., Briese, A. and Hartung, J. (2008) The effect of litter condition and depth on the suitability of wood shavings for dustbathing behaviour. *Applied Animal Behaviour Science* 115, 160–170.

Moinard, C., Statham, P., Haskell, M.J., McCorquodale, C., Jones, R.B. and Green, P.R. (2004a) Accuracy of laying hens in jumping upwards and downwards between perches in different light environments. *Applied Animal Behaviour Science* 85, 77–92.

Moinard, C., Statham, P. and Green, P.R. (2004b) Control of landing flight by laying hens: implications for the design of extensive housing systems. *British Poultry Science* 45, 578–584.

Nasr, M.A.F., Murrell, J., Wilkins, L.J. and Nicol, C.J. (2012a) The effect of keel fractures on egg-production parameters, mobility and behaviour in individual laying hens. *Animal Welfare* 21, 127–135.

Nasr, M.A.F., Nicol, C.J. and Murrell, J.C. (2012b) Do laying hens with keel bone fractures experience pain? *PLoS One* 7, e42420.

Newberry, R.C. (1999) Exploratory behaviour of young domestic fowl. *Applied Animal Behaviour Science* 63, 311–321.

Newberry, R.C. and Shackleton, D.M. (1997) Use of visual cover by domestic fowl: a Venetian blind effect? *Animal Behaviour* 54, 387–395.

Newberry, R.C., Estevez, I. and Keeling, L.J. (2001) Group size and perching behaviour in young domestic fowl. *Applied Animal Behaviour Science* 73, 117–129.

Nicol, C.J. (1987a) Effect of cage height and area on the behaviour of hens housed in battery cages. *British Poultry Science* 28, 327–335.

Nicol, C.J. (1987b) Behavioural responses of laying hens following a period of spatial restriction. *Animal Behaviour* 35, 1709–1719.

Nicol, C.J. (1989) Social influences on the comfort behaviour of laying hens. *Applied Animal Behaviour Science* 22, 75–81.

Nicol, C.J. (2011) Behaviour as an indicator of animal welfare. In: Webster, A.J.F. (ed.) *The UFAW Farm Handbook: Management and Welfare of Farm Animals*, 5th edn. Wiley-Blackwell, Chichester, UK, pp. 31–67.

Nicol, C.J., Lindberg, A.C., Phi/iips, A.J., Pope, S.J., Wilkins, L.J. and Green, L.E. (2001) Influences of prior exposure to wood shavings on feather pecking, dustbathing and foraging in adult laying hens. *Applied Animal Behaviour Science* 73, 141–155.

Nicol, C.J., Caplen, G., Edgar, J. and Browne, W.J. (2009) Associations between welfare indicators and environmental choice in laying hens. *Animal Behaviour* 78, 413–424.

Nicol, C.J., Caplen, G., Statham, P. and Browne, W.J. (2011) Decisions about foraging and risk trade-offs in chickens are associated with individual somatic response profiles. *Animal Behaviour* 82, 255–262.

Nielsen, B.L. (2004) Breast blisters in groups of slow-growing broilers in relation to strain and the availability and use of perches. *British Poultry Science* 45, 306–315.

Nielsen, B.L., Kjaer, J.B. and Friggens, N.C. (2004) Temporal changes in activity measured by passive infrared detection (PID) of broiler strains growing at different rates. *Archiv fur Geflugelkunde* 68, 106–110.

Oden, K., Keeling, L.J. and Algers, B. (2002) Behaviour of laying hens in two types of aviary systems on 25 commercial farms in Sweden. *British Poultry Science* 43, 169–181.

Ogura, Y. and Matsushima, T. (2011) Social facilitation revisited: increase in foraging efforts and synchronization of running in domestic chicks. *Frontiers in Neuroscience* 5, 91.

Olsson, I.A.S. and Keeling, L.J. (2000) Night-time roosting in laying hens and the effect of thwarting access to perches. *Applied Animal Behaviour Science* 68, 243–256.

Olsson, I.A.S. and Keeling, L.J. (2002) The push-door for measuring motivation in hens: laying hens are motivated to perch at night. *Animal Welfare* 11, 11–19.

Olsson, I.A.S. and Keeling, L.J. (2005) Why in earth? Dustbathing behaviour in jungle and domestic fowl reviewed from a Tinbergian and animal welfare perspective. *Applied Animal Behaviour Science* 93, 259–282.

Olsson, I.A.S., Keeling, L.J. and Duncan, I.J.H. (2002b) Why do hens sham dustbathe when they have litter? *Applied Animal Behaviour Science* 76, 53–64.

Olsson, I.A.S., Duncan, I.J.H., Keeling, L.J. and Widowski, T.M. (2002a) How important is social facilitation for dustbathing in laying hens? *Applied Animal Behaviour Science* 79, 285–297.

Ookawa, T. (2004) The electroencephalogram and sleep in the domestic chicken. *Avian and Poultry Biology Reviews* 15, 1–8.

Petherick, J.C. and Duncan, I.J.H. (1989) Behaviour of young domestic fowl directed towards different substrates. *British Poultry Science* 30, 229–238.

Petherick, J.C., Seawright, E. and Waddington, D. (1993) Influence of motivational state on choice of food or a dustbathing foraging substrate by domestic hens. *Behavioural Processes* 28, 209–220.

Pettit-Riley, R. and Estevez, I. (2001) Effects of density on perching behavior of broiler chickens. *Applied Animal Behaviour Science* 71, 127–140.

Pettit-Riley, R., Estevez, I. and Russek-Cohen, E. (2002) Effects of crowding and access to perches on aggressive behaviour in broilers. *Applied Animal Behaviour Science* 79, 11–25.

Pickel, T., Scholz, B. and Schrader, L. (2010) Perch material and diameter affects particular perching behaviours in laying hens. *Applied Animal Behaviour Science* 127 37–42.

Pohle, K. and Cheng, H.W. (2009) Furnished cage system and hen well-being: comparative effects of furnished cages and battery cages on behavioural exhibitions in White Leghorn chickens. *Poultry Science* 88, 1559–1564.

Provine, R.R., Strawbridge, C.L. and Harrison, B.J. (1984) Comparative analysis of the development of wing-flapping and flight in the fowl. *Developmental Psychobiology* 17, 1–10.

Reed H.J. and Nicol, C.J. (1992) Effects of nest lining, pecking strips and partitioning on nest use and behaviour in modified battery cages. *British Poultry Science* 33, 719–727.

Riber, A.B. (2012) Gregarious nesting – an anti-predator response in laying hens. *Applied Animal Behaviour Science* 138, 70–78.

Riber, A.B. and Nielsen, B.L. (2013) Changes in position and quality of preferred nest box: effects on nest box use by laying hens. *Applied Animal Behaviour Science* 148, 185–191.

Riber, A.B., Wichman, A., Braastad, B.O. and Forkman, B. (2007) Effects of broody hens on perch use, ground pecking, feather pecking and cannibalism in domestic fowl (*Gallus gallus domesticus*). *Applied Animal Behaviour Science* 106, 39–51.

Rietveld-Piepers, B., Blokhuis, H.J. and Wiepkema, P.R. (1985) Egg-laying behaviour and nest-site selection of domestic hens kept in small floor-pens. *Applied Animal Behaviour Science* 14, 75–88.

Ringgenberg, N., Fröhlich, E.K.F., Harlander-Matauschek, A., Würbel, H. and Roth, B.A. (2014) Does nest size matter to laying hens? *Applied Animal Behaviour Science* 155, 66–73.

Rönchen, S., Scholz, B., Hamann, H. and Distl, O. (2010) Use of functional areas, perch acceptance and selected behavioural traits in three different layer strains kept in furnished cages, small group systems and modified small group systems with elevated perches. *Archiv Fur Geflugelkunde* 74, 256–264.

Ross, P.A. and Hurnik, J.F. (1983) Drinking behaviour of broiler chicks. *Applied Animal Ethology* 11, 23–31.

RSPCA (2013) RSPCA welfare standards for laying hens. http://www.freedomfood.co.uk/media/30295/RSPCA_welfare_standards_for_laying_hens___September_2013___web.pdf.

Rushen, J. (1993) Exploration in the pig may not be endogenously motivated. *Animal Behaviour* 45, 183–184.

Sandilands, V., Savory, J. and Powell, K. (2004a) Preen gland function in layer fowls: factors affecting morphology and feather lipid levels. *Comparative Biochemistry and Physiology A –Molecular and Integrative Physiology* 137, 217–225.

Sandilands, V., Powell, K., Keeling, L.J. and Savory, C.J. (2004b) Preen gland function in layer fowls: factors affecting preen oil fatty acid composition. *British Poultry Science* 45, 109–115.

Sanotra, G.S., Vestergaard, K.S., Agger, J.F. and Lawson, L.G. (1995) The relative preference for feathers, straw, wood-shavings and sand for dustbathing, pecking and scratching in domestic chicks. *Applied Animal Behaviour Science* 43, 263–277.

Savory, C.J. (1980) Diurnal feeding patterns in domestic fowls – a review. *Applied Animal Ethology* 6, 71–82.

Scholz, B., Urselmans, S., Kjaer, J.B. and Schrader, L. (2010) Food, wood or plastic as substrates for dustbathing and foraging in laying hens: a preference test. *Poultry Science* 89, 1584–1589.

Scholz, B., Kjaer, J.B., Urselmans, S. and Schrader, L. (2011) Litter lipid content affects dustbathing behaviour in laying hens. *Poultry Science* 90, 2433–2439.

Scholz, B., Kjaer, J.B., Petow, S. and Schrader, L. (2014) Dustbathing in food particles does not remove feather lipids. *Poultry Science* 93, 1877–1882.

Schrader, L. and Muller, B. (2009) Night-time roosting in the domestic fowl – the height matters. *Applied Animal Behaviour Science* 121, 179–183.

Schutz, K.E. and Jensen, P. (2001) Effects of resource allocation on behavioural strategies: a comparison of red junglefow (*Gallus gallus*) and two domesticated breeds of poultry. *Ethology* 107, 753–765.

Scott, G.B. and Parker, C.A.L. (1994) The ability of laying hens to negotiate between horizontal perches. *Applied Animal Behaviour Science* 42, 121–127.

Scott, G.B., Lambe, N.R. and Hitchcock, D. (1997) Ability of laying hens to negotiate horizontal perches at different heights, separated by different angles. *British Poultry Science* 38, 48–54.

Seehuus, B., Mendl, M., Keeling, L.J. and Blokhuis, H. (2013) Disrupting motivational sequences in chicks: are there affective consequences. *Applied Animal Behaviour Science* 148, 85–92.

Sherwin, C.M. and Nicol, C.J. (1992) Behaviour and production of laying hens in three prototypes of cages incorporating nests. *Applied Animal Behaviour Science* 35, 41–54.

Sherwin, C.M. and Nicol, C.J. (1993a) A descriptive account of the pre-laying behaviour of hens housed individually in modified cages with nests. *Applied Animal Behaviour Science* 38, 49–60.

Sherwin, C.M. and Nicol, C.J. (1993b) Factors influencing floor-laying by hens in modified cages. *Applied Animal Behaviour Science* 36, 211–222.

Sherwin, C.M., Heyes, C.M. and Nicol, C.J. (2002) Social learning influences the preferences of domestic hens for novel food. *Animal Behaviour* 63, 933–942.

Sherwin, C.M., Richards, G. and Nicol, C.J. (2010) A comparison of the welfare of layer hens in 4 housing systems used in the UK. *British Poultry Science* 51, 488–499.

Shields, S.J., Garner, J.P. and Mench, J.A. (2004) Dustbathing by broiler chickens: a comparison of preference for four different substrates. *Applied Animal Behaviour Science* 87, 69–82.

Shimmura, T., Eguchi, Y., Lietake, K. and Tanaka, T. (2006) Behavioral changes in laying hens after introduction to battery cages, furnished cages and an aviary. *Animal Science Journal* 77, 242–249.

Shimmura, T., Eguchi, Y., Uetake, K. and Tanaka, T. (2007) Behavior, performance and physical condition of laying hens in conventional and small furnished cages. *Animal Science Journal* 78, 323–329.

Smith, S.F., Appleby, M.C. and Hughes, B.O. (1990) Problem-solving by domestic hens – opening doors to reach nest sites. *Applied Animal Behaviour Science* 28, 287–292.

Staempfli, K., Roth, B.A., Buchwalder, T. and Frohlich, E.K.F. (2011) Influence of nest-floor slope on the nest choice of laying hens. *Applied Animal Behaviour Science* 135, 286–292.

Staempfli, K., Buchwalder, T., Frohlich, E.K.F. and Roth, B.A. (2012) Influence of front curtain design on nest choice by laying hens. *British Poultry Science* 53, 553–560.

Struelens, E., Tuyttens, F.A.M., Janssen, A., Leroy, T., Audoorn, L., Vranken, E., De Baere, K., Ödberg, F., Berckmans, D., Zoons, J. and Sonck, B. (2005) Design of laying nests in furnished cages: influence of nesting material, nest box position and seclusion. *British Poultry Science* 46, 9–15.

Struelens, E., van Nuffel, A., Tuyttens, F.A.M., Audoorn, L., Vranken, E., Zoons, J., Berckmans, D., Odberg, F., van Dongen, S. and Sonck, B. (2008a) Influence of nest seclusion and nesting material on pre-laying behaviour of laying hens. *Applied Animal Behaviour Science* 112, 106–119.

Struelens, E., Tuyttens, F.A.M., Duchateau, L., Leroy, T., Cox, M., Vranken, E., Buyse, J., Zoons, J., Berckmans, D., Odberg, F. and Sonck, B. (2008b) Perching behaviour and perch height preference of laying hens in furnished cages varying in height. *Poultry Science* 49, 381–389.

Struelens, E., Tuyttens, F.A.M., Ampe, B., Odberg, F., Sonck, B. and Duchateau, L. (2009) Perch width preferences of laying hens. *British Poultry Science* 50, 418–423.

Su, G., Sorensen, P. and Kestin, S.C. (2000) A note on the effects of perches and litter substrate on leg weakness in broiler chickens. *Poultry Science* 79, 1259–1263.

Sustaita, D., Pouydebat, E., Manzano, A., Abdala, V., Hertel, F. and Herrel, A. (2013) Getting a grip on tetrapod grasping: form, function and evolution. *Biological Reviews* 88, 380–405.

Taylor, P.E., Coerse, N.C.A. and Haskell, M. (2001) The effects of operant control over food and light on the behaviour of domestic hens. *Applied Animal Behaviour Science* 71, 319–333.

Tebbe, V., Bogner, H., Kraeusslich, H., Klein, F.W. and Sprengel, D. (1986) The behaviour of laying hens in cages with different floor area per animal. *Bayerisches Landwirtschaftliches Jahrbuch* 63, 219–242.

Tolman, C.W. (1964) Social facilitation of feeding behaviour in domestic chick. *Animal Behaviour* 12, 245–250.

van Liere, D.W. (1992) Dustbathing as related to proximal and distal feather lipids in laying hens. *Behavioural Processes* 26, 177–188.

van Liere, D.W., Kooijman, J. and Wiepkema, P.R. (1990) Dustbathing behaviour of laying hens as related to quality of dustbathing material. *Applied Animal Behaviour Science* 26, 127–141.

van Rooijen, J. (1991) Feeding behaviour as an indirect measure of food intake in laying hens. *Applied Animal Behaviour Science* 30, 105–115.

Ventura, B.A., Siewerdt, F. and Estevez, I. (2010) Effects of barrier perches and density on broiler leg health, fear and performance. *Poultry Science* 89, 1574–1583.

Ventura, B.A., Siewerdt, F. and Estevez, I. (2012) Access to barrier perches improves behaviour repertoire in broilers. *PLoS One* 7, e29826.

Vestergaard, K. (1982) Dustbathing in the domestic fowl – diurnal rhythm and dust deprivation. *Applied Animal Ethology* 8, 487–495.

Vestergaard, K. and Hogan, J.A. (1992) The development of a behaviour system – dustbathing in the Burmese junglefowl: effects of experience on stimulus preference. *Behaviour* 121, 215–230.

Vestergaard, K.S., Damm, B.I., Abbott, U.K. and Bildsoe, M. (1999) Regulation of dustbathing in feathered and featherless domestic chicks: the Lorenzian model revisited. *Animal Behaviour* 58, 1017–1025.

Walker, A.W. and Hughes, B.O. (1998) Egg shell colour is affected by laying cage design. *British Poultry Science* 39, 696–699.

Wall, H., Tauson, R. and Elwinger, K. (2002) Effect of nest design, passages and hybrid on use of nest and production performance of layers in furnished cages. *Poultry Science* 81, 333–339.

Webster, A.B. (1995) Immediate and subsequent effects of a short fast on the behaviour of laying hens. *Applied Animal Behaviour Science* 45, 255–266.

Weeks, C.A. and Nicol, C.J. (2006) Behavioural needs, priorities and preferences of laying hens. *World's Poultry Science Journal* 62, 296–307.

Weeks, C.A., Nicol, C.J., Sherwin, C.M. and Kestin, S.C. (1994) Comparison of the behaviour of broiler-chickens in indoor and free-range environments. *Animal Welfare* 3, 179–192.

Weeks, C.A., Danbury, T.D., Davies, H.C., Hunt, P. and Kestin, S.C. (2000) The behaviour of broiler chickens and its modification by lameness. *Applied Animal Behaviour Science* 67, 111–125.

Wichman, A. and Keeling, L.J. (2008) Hens are motivated to dustbathe in peat irrespective of being reared with or without a suitable dustbathing substrate. *Animal Behaviour* 75, 1525–1533.

Wichman, A. and Keeling, L.J. (2009) The influence of losing or gaining access to peat on the dustbathing behaviour of laying hens. *Animal Welfare* 18, 149–157.

Wichman, A., Heikkila, M., Valros, A., Forkman, B. and Keeling, L.J. (2007) Perching behaviour in chickens and its relation to spatial ability. *Applied Animal Behaviour Science* 105, 165–179.

Widowski, T.M. and Duncan, I.J.H. (2000) Working for a dustbath: are hens increasing pleasure rather than reducing suffering. *Applied Animal Behaviour Science* 68, 39–53.

Wiedenmayer, C. (1997) Causation of the ontogenetic development of stereotypic digging in gerbils. *Animal Behaviour* 53, 461–470.

Williams, N.S. and Strungis, J.C. (1977) Development of grooming behaviour in the domestic chicken (*Gallus gallus domesticus*). *Poultry Science* 58, 469–472.

Wood-Gush, D.G.M. (1971) *The Behaviour of the Domestic Fowl*. Heinemann Educational Books, London.

Wood-Gush, D.G.M. (1975) The effect of cage floor modification on pre-laying behaviour in poultry. *Applied Animal Ethology* 1, 113–118.

Wood-Gush, D.G.M. and Vestergaard, K. (1991) The seeking of novelty and its relation to play. *Animal Behaviour* 42, 599–606.

Wood-Gush, D.G.M., Duncan, I.J.H. and Savory, C.J. (1978) Observations on the social behaviour of domestic fowl in the wild. *Biology of Behaviour* 3, 193–205.

Yan, F.F., Hester, P.Y., Enneking, S.A. and Cheng, H.W. (2013) Effects of perch access and age on physiological measures of stress in caged White Leghorn pullets. *Poultry Science* 92, 2853–2859.

Yue, S. and Duncan, I.J.H. (2003) Frustrated nesting behaviour: relation to extra-cuticular shell calcium and bone strength in White Leghorn hens. *British Poultry Science* 44, 175–181.

Zhao, J.P., Jiao, H.C., Jiang, Y.B., Song, Z.G., Wang, X.J. and Lin, H. (2012) Cool perch availability improves the performance and welfare status of broiler chickens in hot weather. *Poultry Science* 91, 1775–1784.

Zhao, J.P., Jiao, H.C., Jiang, Y.B., Song, Z.G., Wang, X.J. and Lin, H. (2013) Cool perches improve the growth performance and welfare status of broiler chickens reared at different stocking densities and high temperatures. *Poultry Science* 92, 1962–1971.

Zimmerman, P.H., Lindberg, A.C., Pope, S.J., Glen, E., Bolhuis, J.E. and Nicol, C.J. (2006) The effect of stocking density, flock size and modified management on laying hen behaviour and welfare in a non-cage system. *Applied Animal Behaviour Science* 101 111–124.

Zimmerman, P.H., Buijs, S.A.F., Bolhuis, J.E. and Keeling, L.J. (2011) Behaviour of domestic fowl in anticipation of positive and negative stimuli. *Animal Behaviour* 81, 569–577.

Zupan, M., Kruschwitz, A. and Huber-Eicher, B. (2007) The influence of light intensity during early exposure to colours on the choice of nest colours by laying hens. *Applied Animal Behaviour Science* 105, 154–164.

Zupan, M., Kruschwitz, A., Buchwalder, T., Huber-Eicher, B. and Stuhec, I. (2008) Comparison of the prelaying behavior of nest layers and litter layers. *Poultry Science* 87, 399–404.

6

Social Behaviour

Under natural conditions, chickens live in social groups where they recognize and respond differentially to other individuals. The exact nature of the group alters as chicks are hatched, reared and disperse, or as food reserves and seasons change. A typical chicken group size cannot be specified too closely, but unconstrained chickens are often seen associating in groups of between three and 30 individuals. Once group size exceeds around 100 individuals, chickens cannot distinguish between familiar and unfamiliar birds (D'Eath and Keeling, 2003). The disadvantages of group living, such as increased competition for resources, risk of disease or parasitic infection, and overall visibility to predators, are outweighed for most individuals by the benefits of increased chances of finding food, an earlier warning and diluted risk of predation, and access to mates. Behaviours that favour group cohesion include courtship and reproductive behaviours, communication about food and threats, recognition and affiliation with familiar individuals, a social dominance structure that minimizes overt aggression and a tendency towards social facilitation that can increase behavioural synchrony and spatial clustering. All of these cohesive behaviours provide a counterbalance to the risks of aggressive fighting that might otherwise be engendered by resource competition.

Courtship and Reproductive Behaviours

Most bird species are serially monogamous over successive breeding seasons, with a minority of species showing lifetime monogamy (Emery et al., 2007). Against this background, chickens are relatively unusual in having a polygynous mating system whereby the male birds each attempt to mate with more than one female. In junglefowl, several males are attached to each small group of females (Collias and Collias, 1996), remaining in close proximity and engaging in social encounters. Dominant males attain more matings (Guhl et al., 1945; Jones and Mench, 1991; Johnsen et al., 2001), and DNA fingerprinting reveals that there is usually a high correlation between mating frequency and paternity, such that dominant males father more offspring than subordinates (Jones and Mench, 1991; Wilson et al., 2008), although Bilcik et al. (2005) provides an exception. As a cockerel does not recognize his own offspring (Ligon and Zwartjes, 1995), his mating success is the best guide the male bird has as to his own paternity. Bantam cockerels with a high mating success record tend to give more alarm calls, a correlation that holds even after controlling for the influence of dominance (Wilson et al., 2008). Similarly, Wilson and Evans (2008) found that male birds with current or recent mating experience gave one-third more alarm calls than birds that had not been allowed to mate. In this latter experiment, Golden Sebright cockerels were either pair-housed with females that they could access, or pair-housed with females but separated from them by a wire mesh partition. The relationship between mating success and alarm calling makes adaptive sense, as alarm calls are costly and potentially dangerous to produce, and therefore should most fruitfully be directed towards the caller's own vulnerable offspring (Wilson and Evans, 2008). The general reproductive success of dominant birds is partly an effect of male–male competition, whether via direct agonistic encounters or indirect effects; for example, subordinate males have been found to reduce their attempts at courtship if a more-dominant male is nearby (Wilson et al., 2009). The influence of male–male competition is also suggested by the finding that cockerels with a low paternity success rate in a multi-male competitive environment can show high fertility when housed individually with females (Bilcik et al., 2005). The success of dominant birds is also governed by female preference, as we will see later.

The courtship behaviour of the cockerel, described in detail by Wood-Gush (1971), comprises a mix of behaviours designed to intimidate other males and behaviours designed to attract females. Wood-Gush (1971) described an increase in non-specific activities such as wing flapping, preening and tail-wagging, together with the appearance of more specialized displays such as tit-bitting. Tit-bitting is a ritualized form of feeding behaviour whereby the male bird picks up and drops food while emitting repetitive calls that function to attract female birds (Wood-Gush, 1971). Other displays, such as waltzing (where the male bird circles around the female with his far wing hanging slack) and vigorous wing flapping while in an erect posture, occur during both courtship and agonistic encounters and appear to be somewhat threatening to the females (Wood-Gush, 1971). Wood-Gush did not mention alarm calls as a courtship display, but these are also important. However, he did characterize two further displays specific to the courtship condition, namely, cornering (a foot-stamping display associated with tit-bitting) and rear approach: 'In the rear approach the cock approaches the hens, generally from behind, with his neck stretched and possibly his ruff raised. His deportment varies; he either holds himself high and moves with very high steps, or approaches more rapidly with less exaggeratedly high steps and the body kept lower.... It is often followed by mating.'

Sexual harassment is common in small mixed flocks of feral or free-ranging domestic chickens (McBride et al., 1969; Pizzari and Birkhead, 2001), and a proportion of matings are coerced in these flocks (Pizzari and Birkhead, 2000) and in small pens of broiler breeder birds (Bilcik and Estevez, 2005). In some strains of broiler breeder birds, forced copulations (where females struggled and attempted to escape) occurred in approximately one-third of all mating attempts (Millman et al., 2000), a much higher rate than observed in layer strain birds. Broiler breeder males have sometimes been reported to show abnormally high levels of aggression towards females (Millman et al., 2000; de Jong et al., 2009). This problem has a genetic basis, but the high levels of aggression may also be influenced by commercial rearing and housing practices that affect both male and female birds (see Chapter 8).

Females can respond directly to male harassment, but this is often costly and can lead to injury (Lovlie and Pizzari, 2007). We will see later how more subtle female tactics are adopted to redress the balance. Among other reasons, males may harass females in an attempt to reduce the hens' opportunities to mate with other males. The act of mounting by a cockerel, for example, results in a sharp if temporary reduction in the tendency or ability of a female to mate with another male, even if the mounting is not accompanied by insemination (Lovlie et al., 2005). That cockerels are sensitive to the possibility that females will mate with other males is shown by their ability to reduce the amount of sperm they produce when mating with a familiar female but instantaneously to increase the amount of sperm they produce when mating with a new female (Pizzari et al., 2003). In broiler breeders, sperm quality and quantity are just as important in securing paternity as mating frequency (Bilcik and Estevez, 2005).

At first sight, the female role in mating behaviour appears somewhat passive. The primary display made by the female is the adoption of a crouching posture, which may be an appeasement gesture as much as a signal of sexual receptivity. Females also crouch when approached by a human or in response to a mild threat, and the posture is often regarded as an index of mild fearfulness (Jones *et al.*, 1981; Hemsworth *et al.*, 1993). Dominant hens are courted less than subordinate females (Guhl *et al.*, 1945, cited by Wood-Gush, 1971) and subordinate females are more likely to crouch than dominant hens (Guhl, 1964; van Kampen, 1994). There is more than one way to interpret this information. van Kampen (1994) suggested that crouching was primarily a sign of sexual receptivity, and argued that the more sexually receptive (subordinate) females were less frightened by the cockerel and therefore more likely to crouch to facilitate mating. However, it could also be argued that dominant females are more able to avoid the unwanted attentions of male birds, showing both reduced fear and reduced crouching.

Whatever the relationship between fear and sexual behaviour in females, it is clear that hens possess a subtle set of strategies to ensure their eggs are fertilized by the male bird they most prefer. First, when given the choice, female birds will select males on the basis of their social and behavioural attributes. The relative mating success of dominant males is due not only to their ability to fight off challengers but due also to the fact that they are strongly preferred by females (Graves *et al.*, 1985; Pizzari and Birkhead, 2000; Pizzari, 2003). In the absence of information about the actual dominance status of the males, females will select males with physical attributes that correlate with dominance, testosterone and display intensity.

The most robust finding in junglefowl (reviewed by Parker and Ligon, 2003) is a general female preference for males with larger comb size and brighter comb colour (Zuk *et al.*, 1990, 1995a,b; Ligon *et al.*, 1998). This preference is shown also by domestic hens (Graves *et al.*, 1985). Comb size is not a permanent characteristic. The combs of dominant males increase in size when placed in a larger group, whereas the combs of subordinate males become smaller (Zuk and Johnsen, 2000). Females reserve the right not to make a choice at all if they do not like the look of any of the males available (Zuk *et al.*, 1990) and they are not easily fooled by experimental manipulations of male ornamentation (Zuk *et al.*, 1992). Comb size is not a perfect marker of quality: in small-scale experiments, it can be negatively related to fertility (Vonschantz *et al.*, 1995) and it is not *always* related to dominance (Johnsen *et al.*, 2001).

These are all reasons why, in commercial breeds (Millman and Duncan, 2000b) and in more natural settings, male behaviour appears to be a more salient influence on female choice.

Hens may select cockerels that display vigorously or that possess extravagant ornaments for two rather different reasons. First, their preferences for certain features possessed by males may exist simply because this is what other females also prefer. If there is a genetic component to the possession of a large comb or the tendency to produce eye-catching behavioural displays, then the sons of the female making this choice will themselves become more attractive cockerels in the future. This idea, proposed by Fisher (1930), has often been referred to subsequently as the 'sexy sons' hypothesis. Alternatively, an attribute such as comb size or display intensity may be a marker of male 'quality' (and his ability to pass on good genes). The theoretical concern with this scenario is the possibility that poor-quality males could invest their limited efforts in the features that are attractive to females, i.e. they could 'cheat'. Much work on chickens has been conducted to elucidate these different ideas about female choice and to determine how an ornament such as a comb or a behavioural display such as alarm calling could be regarded as a reliable and 'honest' signal of male quality. One possibility is that the features of greatest attractiveness to females, including comb size and display intensity, are both testosterone dependent (Gyger *et al.*, 1988) and so they may provide reliable evidence that a male is not currently devoting substantial resources to fighting infection or parasites (Hamilton and Zuk, 1982; Wilson *et al.*, 2008). As a general biological principle, there is something of a trade-off between investment in immune function and in hormonal allocation of resources (Folstad and Karter, 1992). Generally, comb size and colour do seem to be good markers of some aspects of current health and condition of the cockerel (Johnson *et al.*, 1993), including lower parasite loads (Johnsen and Zuk, 1998) and an ability to father heavier chicks (Johnson *et al.*, 1993), particularly heavier male chicks (Parker, 2003). It should, however, be recognized that the majority of work on the characteristics that influence female choice has been conducted with red junglefowl, and the extent to which the same mechanisms remain extant in domestic fowl is less known.

A meta-analysis of 90 studies encompassing research on many different invertebrate and vertebrate species provides greater support for Fisher's hypothesis than for the 'good genes' proposal (Prokop *et al.*, 2012). An additional complication is that females themselves may invest more when their offspring have been fathered by attractive males. Very few studies have been able to disentangle these effects, although an exception was a study of red junglefowl. Parker (2003) housed groups of four junglefowl hens with a vasectomized cockerel who had either a large or a small comb. The hens were artificially inseminated with sperm from independently housed male birds with large or small combs to separate the influence of male genetic contribution from any influence of maternal investment. Although there were no effects of maternal investment on chick weight, heavier hens produced larger clutch sizes when housed with large-combed males, suggesting that at least one component of maternal investment was influenced by the perceived attractiveness of the hens' mate.

Even when females are unable to exert a direct choice over which male to mate with, they can adopt many other strategies to improve their own reproductive success. For example, hens can avoid fertilization by subordinate males by reorganizing their daily routine to avoid mating when they already have sperm stored in their oviduct (Lovlie and Pizzari, 2007). If mounted by subordinate males, they can give distress calls to attract the attention of more-dominant cockerels who may respond by disrupting proceedings (Pizzari, 2001). Most surprisingly, females have developed post-copulation strategies to avoid fertilization by unwanted males. For example, hens can differentially reject sperm that has been inseminated by a subordinate male (Pizzari and Birkhead, 2000; Dean *et al.*, 2011).

The relevance of these studies for the management of commercial breeding flocks, where large flock size may interfere with the formation of dominance relationships and individual bird recognition, is not clear. On the one hand, mating opportunities may be more evenly shared between males (Craig *et al.*, 1977). On the other hand, when birds are unable to recognize each other as individuals, courtship behaviour (in broiler breeder flocks at least) may be severely curtailed (Millman *et al.*, 2000), with higher levels of aggression (Millman and Duncan, 2000a).

Communication

Communication between conspecifics is used to convey information about food availability and quality, threats and danger, and social status. Around 30 different vocalizations have been described for adult and juvenile chickens (Collias and Joos, 1953; Wennrich, 1981). Although not all have been interpreted and understood, we know that many have a specific role in conspecific communication. For example, Pizzari and Birkhead (2001) suggested that the loud repetitive cackle vocalization that is given occasionally by hens shortly after oviposition may signal a lack of sexual receptivity and reduce harassment at that time. The cockerel's crow is a signal of his social status, one to which females and subordinate males pay little attention, but which encourages other dominant males to approach and possibly challenge (Leonard and Horn, 1995).

Some chicken calls are both referential, in that they provide specific information about an event, and volitional, in that they reflect decisions (about whether to call and whether to respond), which are subtly adjusted according to circumstance. There has been huge interest in the communicative calls of chickens, as the evidence now suggests a degree of sophistication that was previously suspected to occur only in primate species, if not uniquely in humans.

The food call of the cockerel provides an introduction to this fascinating aspect of animal communication. An encounter with food, or stimuli that are strongly associated with food, can stimulate the male bird to produce rhythmic repetitive food calls, often interspersed with picking up and dropping food particles or items, in a ritualized display known as tit-bitting (Wood-Gush, 1971; Smith and Evans, 2008). This display is directed towards female birds and is an attempt by the male to attract females with whom he can mate (van Kampen, 1999). Females are generally found in closer proximity to tit-bitting males than to non-signalling males (Smith and Evans, 2009), and they take note of both the visual and audible components of the tit-bitting display (Smith and Evans, 2008, 2009). The visual aspects of the display are exaggerated and are easily distinguished from normal male feeding behaviour (Smith and Evans, 2011). Subordinate males may give a silent visual tit-bitting display to attract nearby females while avoiding unwanted detection by a dominant male (Smith et al., 2011). The role of male ornamentation during the display has received much less research attention than the food-calling aspect (Smith et al., 2009). In particular, there has been sustained focus on the manner in which male calling is modified by context, whereby the male bird takes account of the nature of his environment, available food and potential audience when formulating his vocalizations. Although cockerels will produce occasional calls in response to food arrival or discovery when no hens are present (Evans and Marler, 1994), their rate of food calling declines markedly if females remain absent for many hours (van Kampen, 1999). In contrast, the presence of hens, especially if these are or appear unfamiliar, stimulates an immediate and significant increase in the rate of food calling (Evans and Marler, 1994; van Kampen and Hogan, 2000). However, females of a different strain from that of the cockerel are a less potent stimulus for male food calling (Evans and Marler, 1992). The cockerel therefore invests most effort in impressing new females who could potentially allow new mating opportunities. Accordingly, the male tendency to give food calls decreases after a successful mating has been achieved (van Kampen and Hogan, 2000).

From a female perspective, the most impressive thing a cockerel can do is give a genuine and reliable signal of a high-quality food discovery. Female birds do not seem to approach food-calling males in order to engage in sexual activity; rather, their motivation in approaching seems related to their exploratory tendencies (van Kampen, 1994). In playback experiments, hens respond to male food calls by looking downwards as if anticipating a food item (Evans and Evans, 1999). Of course, females need to be alert to the possibility that the male bird is making a fuss about nothing, and their discernment seems to have some influence on male behaviour. Cockerels do tend to call at a higher rate and frequency when they encounter a high protein or other high-quality food item (e.g. a mealworm) than when they encounter lower-quality items (Marler et al., 1986a; Gyger and Marler, 1988; van Kampen and Hogan, 2000), and females respond accordingly with an increased likelihood of approach (Marler et al., 1986b). Male perception of food quality is affected by internal state so that the frequency and duration of food calling is enhanced when males encounter food after a period of food deprivation (van Kampen

and Hogan, 2000). Occasionally, cockerels will make food calls when no apparent food is available or in the presence of a low quality of sparse food, and this is most likely to occur in the presence of unfamiliar (Marler et al., 1986b) or distant females who are less able to check the call accuracy (Gyger and Marler, 1988).

The response of the female birds may also depend on their internal state. Although in one study food-deprived females were no more likely to approach a calling male than non-deprived females (van Kampen, 1994), other studies have found that the female's prior history of food reward affects their response. Presentation of a food call induced immediate food searching activity in bantam hens placed in a small soundproof test chamber, but this response to the call was entirely abolished if the same hens had just received a few kernels of highly preferred maize. Evans and Evans (2007) suggested that, in the latter situation, the food call provided no information additional to that already possessed by the hens. It is another example of the volitional nature of chicken communication – the hens are not in any sense programmed to respond blindly or in a fixed manner to any particular call.

Broody hens give a similar food-calling display directed towards their chicks. Broody junglefowl hens give more calls when larger amounts of food are discovered (Wauters and Richard-Yris, 2003), and they give longer, more intense calls to high-quality food items such as mealworms than to ordinary food (Moffatt and Hogan, 1992; Wauters et al., 1999a; Wauters and Richard-Yris, 2003). The identification of high-quality food is somewhat subjective, with good correlations between a hen's relative food preferences, her hunger level and the number of food calls she makes (Wauters et al., 1999b). However, food palatability is not the only influence on maternal food calling – broody hens also take the location and behaviour of their brood into account. Hens give more food calls when chicks are at a slight distance but within hearing range than when they are already in close proximity (Wauters et al., 1999a), suggesting that food calling is used partially to attract and regroup the brood (although in the absence of food, the brood is summoned by normal clucking; Collias, 1952). This interpretation is supported by evidence that maternal food calling is reinforced and persists for longer when it is followed by the return and appearance of missing chicks (Wauters and Richard-Yris, 2001). Hens produce more food calls when their chicks are not feeding (Wauters and Richard-Yris, 2002) or when the chicks are pecking at potentially harmful or low-quality food items (Nicol and Pope, 1996). The behaviour of the chicks is also influenced by a mix of innate tendency to respond to the maternal food call combined with a growing knowledge of reward contingencies. In a laboratory setting, the speed with which chicks run down a runway is increased initially by playback of a maternal 'high-quality food' call, but, after a few days of experience, running speed is increasingly determined by the chick's own knowledge of the likely food reward (Moffatt and Hogan, 1992).

Distinct aerial alarm calls in the form of a loud, high-pitched scream are given by cockerels, and very occasionally by hens, when the movement of an object is detected overhead (Collias and Joos, 1953; Wood-Gush, 1971; Gyger et al., 1987). Dominant cockerels give more alarm calls than subordinates, probably because dominant birds are more likely to have kin who will benefit from protection due to their greater mating success (Wilson and Evans, 2008; Kokolakis et al., 2010). Unlike food calls, alarm calls do not increase male attractiveness (Wilson and Evans, 2010). Sometimes 'weaker' alarm calls are given to harmless stimuli such as small flying birds, leaves or distant aeroplanes, with the strongest signals reserved for large, imminent threats (Gyger et al., 1987). Aerial predator calls are structurally quite distinct from the beat-like 'scolding' calls indicating the detection of a ground predator (Gyger et al., 1987). Distinct aerial and ground predator calls can be elicited not only by genuine predatory stimuli but also by appropriate video images, such as of hawks and racoons, respectively (Evans et al., 1993a). Relevant information is encoded in these vocalizations. On nearly 50% of occasions when female birds heard an aerial predator call, they ran for cover, whereas on hearing the ground predator call, hens adopted an erect, vigilant posture but did not move towards overhead shelter (Evans et al., 1993a).

Unlike food calling, the production of alarm calls does not increase male attractiveness to females (Wilson and Evans, 2010), but, as with food calling, social factors mediate the production of aerial predator alarm calls by cockerels. In response to overhead models of hawk silhouettes, cockerels gave more alarm calls in the presence of male, female (real or video image) or juvenile conspecifics than when they were alone or in the presence of a bird of a different species (Karakashian et al., 1988; Evans and Marler, 1991). This audience effect is most apparent for aerial alarm calls, with ground predator calls less dependent on an audience (Evans, 1997). Chickens are highly sensitive to the size and apparent speed of approach of an aerial predator, with a greater rate of alarm calling elicited by a more rapid predator approach (Evans et al., 1993b). Both male and female (broody) bantam chickens adjust their alarm calling according to the size of the aerial predator relative to the size of their own growing chicks. Alarm calls are given in response to small hawks only when chicks are very young and vulnerable (Palleroni et al., 2005).

In addition to considering his conspecific audience, the male bird must also consider the extent to which he is putting himself at risk by calling, as this could advertise his presence to the predator that he has spotted. Males are rather astute in assessing these risks, and give more alarm calls and longer-duration calls when they are less than 1 m from a safe shelter than when they are at a greater distance from a potential refuge (Kokolakis et al., 2010). One way in which this risk is managed is by limiting the production of the initial high-amplitude warning pulse. This part of the call alerts conspecifics to danger, but because it may provide information to the predator about the cockerel's location, it is rarely repeated (Bayly and Evans, 2003). Instead, the cockerel may repeat the other, less traceable parts of his alarm call until the threat has passed.

The subtlety and context specificity of the referential communication of chickens is quite astonishing. Very little is currently known about the mechanisms that chickens use to combine information from a great many sources to create such flexible communication, but this will surely be a fruitful area for future research.

Social Recognition and Preference

Both auditory and visual cues are used to distinguish familiar companions from other conspecifics. Chicks direct fewer aggressive pecks towards familiar companions than unfamiliar (Zajonc et al., 1975; Riedstra and Groothuis, 2002; Porter et al., 2006) and also prefer to be closer to familiar chicks (Marin *et al.*, 2001; Guzman and Marin, 2008). Chicks that have formed a preference for a familiar partner during the first few days after hatching will retain that preference during a separation period of some hours provided that social stimulation during the separation is minimal. If chicks that have been separated encounter a new companion during the separation period, they appear to lose their recognition or preference for their initial partner due to interference processes (Porter *et al.*, 2006). Vision plays a key role in social discrimination in chicks (Porter *et al.*, 2005), but it is not the only sense that is used. When placed behind an opaque screen where visual cues were not available, the distress call characteristics of young chicks differed according to whether companions on the other side of the screen were familiar or unfamiliar (Koshiba *et al.*, 2013). Even though the number of companions present was confounded with familiarity in this experiment, it does suggest that chicks can detect familiarity via auditory cues.

Young broiler chicks vary considerably in the speed with which they will approach familiar companions in laboratory tests. This probably reflects variation in their social reinstatement motivation, or at least in the extent to which their social reinstatement motivation overcomes their fear responses. Chicks that emerge quickly from a T-maze to rejoin their companions subsequently gain more weight, have a lower plasma corticosterone response to an acute stressor and are more sociable than their slower conspecifics (Marin *et al.*, 1999, 2003).

Adult birds also distinguish familiar from unfamiliar birds and tend, like chicks, to direct aggression towards unfamiliar individuals (D'Eath and Stone, 1999). Visual cues seem to be primarily important in distinguishing familiar from unfamiliar birds. Bradshaw (1992a) allowed subject hens placed in a central box to choose where to stand in relation to other boxes placed around the periphery. These additional boxes contained either two familiar hens, two unfamiliar hens or an empty box, and subjects showed a clear preference for the familiar birds. However, in tests that lasted 9 h, avoidance of the unfamiliar birds diminished, presumably as they became more familiar or as uncertainty (relating to relative rank) diminished. Grigor *et al.* (1995) also found that hens were more reluctant to move down a runway for a food reward if they had to pass an unfamiliar

caged bird on the way than if they had to pass a familiar bird (dominant or subordinate) or an empty cage. Similarly, Dawkins (1995) observed that hens chose preferentially to feed from a bowl with a familiar companion than from one with an unfamiliar bird. As this preference was expressed only when subjects were able to make their choice at close range (approximately 15 cm) not at a greater distance (70 or 140 cm from the companion bird), Dawkins (1995) proposed that close scrutiny of another bird's head region is essential in identifying familiar birds. A further study showed the same preference for a familiar feeding companion when the only visible features of the companion were the frontal or lateral aspects of its head and neck (Dawkins, 1996). In addition, hens released at a distance of approximately 2 m approached a familiar or unfamiliar stimulus bird in equal measure. However, once close scrutiny of the companion bird had taken place through the wooden bars used for confinement, subjects subsequently chose to be closer to familiar rather than unfamiliar birds (Dawkins, 1996). Together, these studies show that hens discriminate between familiar and unfamiliar companions and generally prefer to associate with familiar conspecifics. As with broiler chicks, there is variation in the speed with which individuals will approach familiar companions in laboratory tests. For example, hens of the Lohmann Tradition strain were quicker to leave a start box and return to familiar companions than Isa Brown birds (Ghareeb *et al.*, 2008).

A preference for familiar birds over unfamiliar ones could occur because of some group-level signal such as a general odour (see Chapter 2) or pattern of behaviour shared among all flockmates. Showing that hens recognize each other as specific individuals is much harder. Ryan (1982) found that bantam cockerels could learn to use images of individual chickens, or of individual pigeons for that matter (Ryan and Lea, 1994), as discriminative stimuli in operant experiments. This shows that chickens can detect particular features of a visual image, but it does not show that they recognize that image as a representation of reality or that they treat those images as equivalent to a real or specific individual. To assess individual recognition, the subject must treat all images of a certain individual as equivalent, i.e. as a category distinct from categories of other individuals. Bradshaw and Dawkins (1993) found no evidence of such category formation using slides of laying hens, although Ryan (1982) reported a degree of generalization among different slides of the same individual where the entire body was visible. Of course, difficulties in this early work arose as much from a lack of knowledge about how chickens might respond to two-dimensional images presented via various media (see Chapter 2) as by a lack of knowledge about their social capabilities. Bradshaw and Dawkins' (1993) firmest conclusion was that their subject chickens did not perceive the slides as representative of real birds, something borne out by later work using other visual media (Dawkins, 1996). As discussed in Chapter 2, high-definition flicker-free video seems to be a better medium for the presentation of social stimuli to chickens, something that was not available for earlier research work.

Individual discrimination has only recently been demonstrated convincingly (Abeyesinghe et al., 2009). After a period of initial training in either a Y-maze or an operant discrimination chamber, test hens were required to discriminate between two familiar companions (hens positive (P) and negative (N)) in order to obtain a food reward. Food could be obtained by only approaching hen P in the Y-maze or by pecking a key near to hen P in the operant task.

The position of hen P was varied randomly so that test hens could not use location as a cue to solve the task. As both birds were familiar, the test hen had to learn that it was the individual identity of one of her companions that provided the solution. This complex discrimination was learned by most test birds in the Y-maze experiment but by very few in the operant chamber.

Social Structure and Dominance

The ability of chickens to discriminate between familiar and unfamiliar birds, and possibly among individual animals within each category, has wide-ranging implications for the social structure of the flock. It is widely known that chickens housed in small flocks establish pecking orders (Schjelderup-Ebbe, 1922; Guhl, 1958; Wood-Gush, 1971; Rushen, 1982a). In chicks, frolicking, jumping and sparring behaviours occur before aggressive pecks are delivered (Guhl, 1958; Rushen 1982b). A first encounter between two adult chickens commonly results in aggression, the result of which will determine where each hen finds itself

in the pecking order (Dawkins, 1996; D'Eath and Stone, 1999). On subsequent meetings, the two birds remember their relative ranks, with social dominance signalled by threat and posture, in a way that avoids the need for further overt aggression. The extent of this effect was demonstrated by Craig et al. (1969), who reported agonistic behaviour within flocks subjected to weekly random regrouping of members to be between 22 and 122% higher than in undisturbed flocks. Once stable dominance relationships have been established, subordinate birds do not attempt to avoid close proximity with more-dominant individuals (Grigor et al., 1995).

Often, a network of social dominance relationships is established so that each bird in a small flock knows its position relative to every other bird. Sometimes the resulting dominance hierarchy is linear, whereby bird A is dominant over birds B, C and D, B is dominant over C and D, and C is dominant only over D. However, relationships are not always this simple; sometimes reversals or triangular relationships can exist in which bird A pecks bird B, which pecks C, which in turn pecks A (Wood-Gush, 1971). In this case, it would not be easy to ascribe dominance to any bird. It is also misleading to attempt to rank individuals in the absence of perfect information by *trying* to construct the most linear hierarchy possible. This technique risks the construction of an apparently linear hierarchy when encounters between individuals are actually random (Appleby, 1983). The most common method used to ascribe a dominance status to an individual is therefore to weight an individual's rank according to the rank of its opponent, although other methods can also be used. Indeed Bayly et al. (2006) assessed the social structure of eight groups of cockerels using eight different indices of dominance. Given a reasonable level of linearity within the flock, these different methods generally produced similar estimates of the social status of an individual (Bayly et al., 2006). However, this was not the case when linearity was weak. If reversals or intransitive relationships are detected, then the choice of method for ascribing dominance requires detailed consideration (Bayly et al., 2006).

The attributes associated with a bird gaining higher or lower social status have been of interest to scientists for nearly a century (Cloutier and Newberry, 2000). There is no evidence that birds that attain dominance perform better on simple learning and discrimination tasks than other birds (Croney *et al.*, 2007), although testing on a wider range of cognitive tasks would be needed before concluding that dominant birds have no intelligence advantage. Most studies over the years have examined the role of factors such as the relative body weight, comb size and comb shape of the participants, as well as their previous experience of social victory or defeat (reviewed by Cloutier and Newberry, 2000). In female junglefowl, physical attributes were not good predictors of social rank in newly formed flocks, but previous flock experience and aggression levels were predictive (Kim and Zuk, 2000). However, in domestic female hens, comb size varies greatly (Fig. 6.1), and possession of a larger comb can be one of (Bradshaw, 1992b) or even the most important (O'Connor *et al.*, 2011) predictor of social dominance.

Comb colour is also predictive, with birds having darker (Bradshaw, 1992b) or yellower (O'Connor *et al.*, 2011) combs attaining higher dominance scores. Using a modelling approach, Cloutier and Newberry (2000) found a combined influence of some of these factors. In two different strains, birds of higher body weight were more likely to have won recent fights and were more likely to attack others when regrouped. A larger comb was associated with winning recent fights in one group where variation in comb size was substantial but not in a group where variation was less. More generally, they argued that, when the extent of physical differences between participants was low, the most important factor predicting the attainment of dominance was recent experience of winning an encounter.

The stability of the social structure of small flocks relies on birds using the social recognition abilities reviewed above. For example, when hens are housed in groups of ten, they act aggressively towards unfamiliar but not towards familiar test birds presented in a cage (D'Eath and Keeling, 2003). Hens can even extrapolate information by observing the outcomes of interactions between other birds. For example, by observing the outcome when a familiar conspecific fights with a stranger, a hen can evaluate her own potential to defeat the stranger (Hogue *et al.*, 1996). With this ability, an individual has the huge benefit of avoiding the risk of fights that it is likely to lose. Such reasoning may occur through transitive inference, where if you know A>B and B>C, you know that A>C. The study set up three conditions for assessment of a bystander's social learning ability. The first showed individuals their prior dominant being defeated by a stranger and then introduced them to that stranger. The second showed them their prior dominant defeating a

Fig. 6.1. The combs of closely related laying hens vary greatly in size and shape, as shown by these photos of flockmates of the same age. When hens are kept in small groups, hens with larger combs tend to have a higher dominance status. In large groups, where dominance hierarchies are not formed, hens with larger combs still tend to be more aggressive. (Photos courtesy of Annabel Osborn, University of Bristol, UK.)

stranger. In the first condition, the hen should infer that because the dominant hen was defeated she would also be defeated. This was exactly the case, with none of the bystanders initiating an attack on this 'powerful' stranger. If the stranger initiated an attack, then these hens would immediately show signs of submission. In the second condition, the bystander cannot be sure whether it is able to defeat the stranger or not. It could be said that it has a 50/50 chance to win or lose. The hens were found to attack the stranger 50% of the time and win 50% of the time. These conditions both showed clearly that the bystanders were able to learn from what they had witnessed and act accordingly. The evaluative reasoning involved in this behaviour is considerable; however, further study is required to show whether transitive inference is utilized.

Increasing group size can place pressure on the ability of birds to distinguish familiar and unfamiliar birds. When hens were housed in flocks of 120 individuals, there was no difference in the behaviour shown towards supposedly familiar test birds (from the same group) or towards unfamiliar test birds that were presented in a cage (D'Eath and Keeling, 2003). This has wide-ranging implications for the social structure of most commercial flocks, as dominance relationships built on individual recognition and memory do not seem to be employed. It has been proposed that in large groups it is more beneficial for chickens to employ simple rules of thumb to signal status, avoiding any need for individual recognition (Pagel and Dawkins, 1997). D'Eath and Keeling (2003) found that, within the larger 120-bird groups, it was the heavier birds with larger combs that tended to attack test birds. Such a strategy based on direct signalling rather than recognition means that birds do not invest cognitive effort in remembering and establishing relationships with strangers (a strategy also used by humans using the tube in London).

Dominant birds generally have priority of access to food and water resources and to mates, as discussed above. In furnished cages, dominant laying hens are better able to access dustbathing areas than subordinates (Shimmura et al., 2007). However, when resources are not limited or are widely dispersed, dominant birds do not attempt defence, and subordinate birds can obtain equal access (Shimmura et al., 2008). Subordinate hens are not inhibited from dustbathing in the presence of dominant birds, although the quantity and quality of dustbathing behaviour is reduced in the presence of unfamiliar birds (Shimmura et al., 2010). Laying hens learn more about how to obtain food by watching dominant birds than by watching subordinate birds, even when there are no overt differences in the pecking behaviour or foraging success of these two types of demonstrator (Nicol and Pope, 1994, 1999). This suggests that hens simply pay more attention to dominant birds.

Agonistic Behaviour

Agonistic behaviours comprise both aggressive acts, such as pecks, threats and chases, and responses to those aggressive acts, such as avoidance or submission. In actors (and recipients), the giving (or receiving) of head pecks and threats is highly correlated (Rushen, 1984). One of the benefits of living in a group with a dominance hierarchy is that overt aggression over every resource can be replaced by more subtle and less costly signalling. As dominance relationships are formed, individuals are progressively less likely to threaten or headpeck other individuals from whom they receive frequent threats or head pecks, and are more likely to show spontaneous avoidance (Rushen, 1984). The initial aggression can be regarded as an investment designed to reap a more peaceful future existence for most birds (Pagel and Dawkins, 1997). However, in some cases initial aggression is supplemented by a level of ongoing attack towards the individual closest in rank to the aggressor, i.e. the bird most likely to continue to contest its position (Forkman and Haskell, 2004). Contrary to common belief, aggression is not triggered when one bird invades the 'personal space' of another. Rather, it tends to occur when an energetic behaviour is directed towards another and interpreted as a threat (Rodriguez-Aurrekoetxea and Estevez, 2014).

Once a hierarchy has been established, the most dominant birds are not necessarily the most aggressive. Indeed, aggression is more likely to occur repeatedly when differential social status has not been fully established, as might be the case when birds are unfamiliar or when birds cannot remember the outcome of previous encounters. Situations of social uncertainty are more likely to occur in commercial systems where chickens are kept in very large groups than in small flocks of backyard chickens or in natural groups of junglefowl. However, Pagel and Dawkins (1997) argued that aggression could be maintained at surprisingly low levels within large flocks of chickens if the birds switched from a system of individual recognition to a cue-based system as group size increased. Such a switch would be favoured, as the costs of individual recognition would become very high with increasing group size. Instead, visible bodily cues (correlated with general fighting ability but not individually ascribed) could be used to establish relative status. However, chickens would need to have experienced sufficient variation in group size during their evolutionary history for natural selection to have favoured such a switching strategy (Pagel and Dawkins, 1997). Estevez et al. (1997) presented a slightly different hypothesis, arguing that aggression should be lower in large groups simply because there becomes a point where it is no longer possible for an animal to defend limited resources from a large number of competitors. Whatever the theoretical basis, the evidence that aggression decreases with group size is quite compelling (Rodenburg and Koene, 2007). Lindberg and Nicol (1996) reported lower aggression in groups of 44 than in groups of four, and Nicol et al. (1999) noted lower aggression in groups of more than 200 than in groups of 72 housed in an experimental perchery. Hughes et al. (1997) provided a summary of previous literature showing that, while aggression seemed to increase with group size up to about ten or 12 birds, with an approximate average rate of aggression of ten per bird h[-1], it thereafter showed a linear decline with flock size such that rates were more like one per bird h[-1] in flocks of over 500 birds. Low levels of aggression have also been reported in commercial flocks of laying hens (Hughes et al., 1997; Carmichael et al., 1999). Rates of fewer than 0.5 aggressive acts per bird h[-1] were reported by Zimmerman et al. (2006) for hens in aviary flocks of between 2450 and 4200 birds.

In a more direct test of the effect of group size, Estevez et al. (2002, 2003) observed aggressive interactions in groups of 15–120 pullets and confirmed that lower rates were evident in the larger groups. However, one of the major influences on aggression was the number of birds present at a feeding patch, irrespective of the past group size in which the birds had been housed. In addition,

aggressive interactions were not distributed evenly, and certain individuals received more aggressive pecks in large groups than in small groups (Estevez et al., 2003). The few birds that receive an excessive amount of aggression (or other forms of pecking) from others in the flock are colloquially termed 'pariah' birds (Freire et al., 2003; Appleby et al., 2004). Their welfare is a cause of considerable concern. These victimized birds often scurry rapidly from one hiding place to another, their abnormal behaviour seemingly attracting further aggression as they attempt to find shelter under fittings or in nest boxes. They may rarely venture out for food and water due to the extreme risk of attack.

Modern strains of broiler and layer cockerels are less aggressive towards other males (Campo et al., 2005) and towards models of other males (Millman and Duncan, 2000a) than are older strains of bird that were bred for fighting.

Temporal Synchrony and Spatial Clustering within Groups

Many of the environmental and social pressures that favour group living, including protection from predators and an increased chance of resource discovery, will induce a degree of synchrony in the performance of certain behaviours. For example, if a new resource has been discovered, it may be beneficial to all individuals to exploit it before it runs out. Similarly, a shared threat may mean that all individuals rest or hide together to be as inconspicuous as possible, or show simultaneous flight or escape behaviour. A degree of synchrony in the performance of behaviours such as feeding, dustbathing and resting is a common occurrence in small and large flocks of chickens (Savory et al., 1978; L.M. Collins et al., 2011). The most common measure of behavioural synchrony is the proportion of individuals within a group that are performing the same behaviour at the same time, although other measures have also been used, and it is important to control for the degree of synchrony that could occur by chance in groups of different sizes (Asher and Collins, 2012).

Synchrony can arise because each individual responds independently but in the same manner to the current environment. It is not surprising that, for example, the majority of birds feed together at every run of the chain feeder, or that most laying hens fly up to the perches in an aviary system at lights out. Scientific studies of synchrony in chicken flocks show other environmental influences. For example, higher light intensities can increase synchrony in broiler flocks, with birds reared under 200 lx showing greater synchronized activity during the day (with fewer birds attempting to rest) than birds reared under 5 or 50 lx (Alvino *et al.*, 2009). If light periods are too long (23 h light), then commercial broiler chickens lose their circadian and often synchronized activity rhythms, leading to more disturbance and sleep disruption in the flock (Schwean-Lardner *et al.*, 2014). In these various studies, the synchronous behaviour of the chickens occurs largely because the environment provides circadian or other cues that influence individuals in the same way.

In his classic studies, Tolman (1968) showed that the 'mere presence' of an active non-feeding companion chick facilitated feeding behaviour in other chick subjects. In adult hens, Nicol (1989) found that the close proximity of other hens facilitated preening behaviour. These examples would not lead to an increase in group synchrony but are mentioned because such 'mere-presence' effects are termed social facilitation in some studies and there is a potential for confusion. Mere-presence effects most likely occur because the proximity of a companion reduces subject fearfulness. Certain activities that would be risky to perform if isolated are disinhibited in the presence of a companion who can keep an eye out for predators (Nicol, 1995).

However, chickens can also attend to the behaviour of their companions, and replicate or follow the actions of other birds. Such social effects will act to increase the degree of synchrony within the group. This phenomenon has variously been termed allelomimetic behaviour, contagious behaviour, response facilitation (Hoppitt *et al.*, 2007) or (again, but with a different meaning from that described above) social facilitation. All of these terms are used essentially to describe how social influences lead to the matched performance of normal species-specific behaviours (Nicol, 1995). Often such social matching of behaviour is quite instinctive, as in food pecking in chicks. However, it is also possible for chickens to learn to match the behaviour of a companion because, for example,

a companion pecking at a fast rate might become associated with the availability of high-quality or abundant food (Nicol, 1995). Some have suggested that the term response facilitation is the most useful as it makes no assumptions about whether behavioural matching is instinctive or learned (Hoppitt *et al.*, 2007).

To show a social influence on matching behaviour, the possibility of similar but socially independent responses to environmental change must be strictly controlled. When this is accomplished, there is good evidence that chickens are highly influenced by the responses of their companions. For example, in addition to the influence of the mere presence of a companion, feeding behaviour in subject chicks is strongly influenced by the feeding behaviour of a companion. If subject chicks hear or see (real or model) companion chicks increasing their pecking rate, they show a corresponding increase themselves (Tolman and Wilson, 1965; Tolman, 1967a,b). In broiler chickens feeding at commercial troughs, the presence of one bird feeding for at least 15 s almost doubles the chance of another bird joining it compared with joining an unoccupied length of trough (Collins and Sumpter, 2007). Similar results have been found for feeding behaviour in adult domestic hens (Hughes, 1971), even when video images are used in place of real companions (Keeling and Hurnik, 1993). In adult domestic hens, an increased companion drinking rate stimulates an increased rate of drinking and swallowing movements in subject birds, although no increase in actual water consumption (Forkman, 1996). This latter example suggests that one advantage of synchrony in behaviour may be that individuals are less conspicuous if they mimic the behaviours of their companions, even if this is unnecessary for their own resource consumption.

The idea that synchrony reduces conspicuousness is especially relevant when considering behaviours, such as dustbathing and nesting, which place chickens in a vulnerable position. Most laying hens nest in the morning as they respond to the same circadian cues, but the timing of their nesting is not influenced by that of their companions (Appleby, 2004). In contrast, dustbathing is an interesting example of a behaviour often performed synchronously (Wood-Gush, 1971) but where the nature of the social influence is complex. The mere presence of other birds does act to increase dustbathing in otherwise isolated birds (Duncan *et al.*, 1998), particularly if those birds are familiar (Shimmura *et al.*, 2010), but beyond this, hens appear to be responding independently to the same environmental cues rather than matching each other's behaviour. Olsson *et al.* (2002) found that hens showed some interest in the dustbathing behaviour of others, but the experience of watching other hens dustbathing did not increase either simultaneous or later dustbathing motivation of the subject birds. In response to video images, high-ranking hens showed a slight tendency to start dustbathing earlier if they observed a dustbathing hen than if they observed a standing hen (Lundberg and Keeling, 2003), but there was no effect on the duration or intensity of dustbathing. However, when considering a complex behaviour of this type, the advantages of video images in providing a consistent conspecific stimulus must be weighed against the disadvantages of a lack of reality and the possible difficulties hens have in interpreting complex images from two-dimensional sources (reviewed in Chapter 2).

Studies that reveal greater synchrony in behaviour within groups than between groups (provided these groups are all kept in the same environment) are also informative about social influences. Webster and Hurnik (1994) housed hens in pairs in adjacent conventional cages and found a significantly greater tendency to synchronize behaviour within cages than between cages. This was true for feeding, head shaking, resting, sitting and standing, and, to a lesser extent, preening, but not for drinking, walking or pecking. However, within a very confined space, it may physically be difficult for one bird to rest while another is actively walking. It is interesting, therefore, to see that in less-confined and slightly larger groups, similar results have been obtained. Hoppitt *et al.* (2007) examined the synchrony of behaviour in two (visually isolated) groups of adult hens kept in adjacent pens in the same environment. Compelling evidence was obtained that behavioural synchrony in sitting, dustbathing and, most notably, preening was stronger within groups than between groups, consistent with response facilitation acting as the primary influence on behaviour. Within groups, the proportion of visible conspecifics preening was highly significantly associated with the rate of onset of other birds preening, and this was not just a result of birds choosing the same favoured

preening locations within each pen. To control for location effects on a finer scale, Hoppitt and Laland (2008) arranged three bowls in a triangle, two containing water and one containing food or, in a related experiment, two containing food and one containing water. They then observed small groups of chickens as they interacted with the bowls. The probability that a subject bird would start to feed or drink was increased if a companion bird was already either drinking or feeding. Thus, there was a general presence and location effect of the companion that facilitated both types of ingestive behaviour. However, drinking behaviour was facilitated to a significantly greater extent than feeding if the companion bird was already drinking. This shows that the specific behaviour of the companion was an important additional influence on subject drinking. The specific effects of companion feeding were less clear in this experiment due to some interference from food competition. Overall, it seems that the presence and location of companions can facilitate behavioural performance, but for certain behaviours, especially drinking and preening, these effects are complemented by stronger response-facilitation effects that lead to increased social synchrony. Much further work will be required to examine exactly which aspects of the appearance or sound of a companion's behaviour produce response facilitation (Nicol, 1995), although the importance of vision is clearly demonstrated by the fact that genetically blind chickens show reduced behavioural synchrony (S. Collins *et al.*, 2011).

Where social factors play a role in promoting temporal synchrony, they may also promote spatial clustering. For example, Collins and Sumpter (2007) modelled the social feeding behaviour of large flocks of commercial broiler chickens and found that social attraction would lead to clustering at the feed trough, regardless of stocking density. Even when broiler chickens are not feeding, they tend to be clustered more than would be expected by chance (Febrer *et al.*, 2006). A simulation approach found that the degree of clustering observed in real flocks at approximately 35 days of age was best explained by a model in which simulated birds avoided being too far from other companions, i.e. it suggested a degree of social attraction (Febrer *et al.*, 2006). However, other studies of broiler chickens have found that a tendency to cluster when young is progressively diminished as age and stocking density increase, with birds adopting more even patterns of spacing at the highest stocking densities (Buijs *et al.*, 2011). In small groups of laying hens kept in small pens, unfamiliar birds cluster more than familiar birds (Lindberg and Nicol, 1996). When small groups of birds are familiar, then observed levels of clustering are explained by shared resource preferences more than social attraction (L.M. Collins *et al.*, 2011; Asher *et al.*, 2013), and in furnished cages, laying hens prioritize an even distribution (and maximum space between birds) over any preference for cage height (Albentosa and Cooper, 2005). Given the complexity of chicken social preferences reviewed above, it is likely that the balance between social attraction and repulsion will vary according to factors such as bird age, familiarity, fearfulness and spatial allowance. Environmental temperature and the need for thermoregulation will also play an important part.

It is essential that housing features are designed to take into account the tendencies of chickens towards temporal synchrony and spatial clustering, regardless of whether these arise because of individual responses to the same environmental cues or from social or response facilitation. For example, the length of feed troughs should be sufficient to enable all birds to feed at once (Knierim, 2000). When calculating spatial or resource requirements, it is not acceptable to assume that chickens will share their use of these resources evenly throughout the day. For example, if (hypothetically) each individual chicken in a group of six needed to perch for 4 out of 24 h, then just one perch would be sufficient if the birds were indifferent as to their actual perching times. However, shared preferences and social facilitation effects will almost certainly mean that one perch is insufficient. These arguments were expanded by Appleby (2004) in a consideration of housing design principles for hens kept in furnished cages, and by Bokkers *et al.* (2011) when considering the space needs of broiler chickens.

In very large flocks, it is possible that certain individuals have a preference for a certain location within the house, or to be with specific birds, processes that could lead to the formation of subgroups. Oden *et al.* (2000) found some evidence for location preferences, but other studies (D'Eath and Keeling, 2003) including studies with tagged birds (Freire *et al.*, 2003) do not suggest strong segregation of birds into subgroups but rather a continuous distribution of preference for different areas of the house.

Affiliation

In some bird species, behaviours such as allopreening, bill twining, food sharing and play can help to maintain affiliative bonds (Emery *et al.*, 2007) and act as a counterbalance to the forces of competition and group dispersal. Although we have seen that chickens do engage in food sharing in certain courtship and parental contexts, they do not seem to do so when kept in single-sex groups, and neither do they show obvious signs of other affiliative behaviours. There have been occasional reports of allopreening between unrelated adult or juvenile chickens although allopreening is not mentioned in most behavioural studies, but sometimes it is explicitly noted that no allopreening has been observed (e.g. Savory and Mann, 1997, in a study of the behavioural development of pullets of different strains). Allopecking (rather than preening) is sometimes mentioned, but we have to remember that behavioural definitions can change over time. Some of the gentle nibbling movements directed towards other birds (Leonard *et al.*, 1995) might today be regarded as a form of gentle feather pecking (reviewed in Chapter 9). This is particularly likely in papers published before the distinction between gentle and severe feather pecking was appreciated. In other cases, gentle pecks by one bird to another may be an attempt to access food particles from the plumage or beak of a conspecific (Rushen, 1982b; Zimmerman *et al.* 2006), rather than being a social affiliative behaviour. There is rather better evidence that some forms of gentle peck take place during social exploration, with gentle pecks directed towards the feet, beak, wattles or combs of other birds, helping to establish social familiarity, particularly in young chicks (Zajonc *et al.*, 1975; Riedstra and Groothuis, 2002).

If we take a broad view and consider the various types of non-aggressive inter-bird pecking that may take place – whether as gentle feather pecking or gentle pecking in the head region – it is still difficult to conclude that any of these behaviours is truly affiliative. The receiving birds seem highly tolerant of such pecks, but there is still no good evidence that they function to strengthen social relationships. The gentle head pecks noted by Dixon *et al.* (2006) and Dixon and Nicol (2008) tended to increase when the chickens experienced a change from a preferred diet to a less-preferred one, suggesting more of a foraging motivation, while other forms of gentle peck are directed towards unfamiliar birds as a form of exploration (Riedstra and Groothuis, 2002).

Evidence of affiliation could come from another source. If specific familiar individuals spent a disproportionate amount of time with each other, this could indicate a social preference. This would seem highly possible, as we have seen already that chickens prefer familiar companions and can also identify specific familiar individuals. Indeed, if two animals seek out each other's company and obtain specific benefits from that interaction, we could call them friends. In several social species, selected companions provide social support benefits, such that an individual will recover more quickly from stress if with a favoured companion than if alone or with a non-preferred companion (Kikusui *et al.*, 2006). Abeyesinghe *et al.* (2013) shed light on the question of whether hens have friends by housing birds in groups of 15 for 2 weeks before monitoring for dyadic relationships began. Using a social relationship analysis to investigate who associated with whom, no evidence was found for hens actively preferring others in their choice of resource area or in companion proximity. Most dyads were never observed roosting together, although some were shown to perch together frequently. The suggestion from this paper is that hens do *not* have specific friends. However hens clearly form tolerant relationships within small groups of familiar birds, and stable non-aggressive relationships in very large flocks.

Personality

One reason why hens do not appear to have friends might be because one chicken behaves very much like another, such that different companions are essentially interchangeable. However, this does not appear to be the case. In adult hens of the same age, strain and background, stable character differences or 'personalities' (henalities?) are apparent. Nicol *et al.* (2009) measured a wide range of characteristics in 60 hens over a period of many months. Statistical analysis revealed behavioural traits that were stable over time, and identified three types of hen. One group comprised individuals

that liked wood shavings and were highly food motivated, continuing to peck a dish that had contained mealworms even when all the mealworms were gone. A second group appeared more anxious, and these individuals preferred to be on a wire floor. A third group was less food motivated but still liked to forage in litter material, preferring environments with enriched litter material. In male birds, stable individual differences have also been detected. Whereas traits such as vigilance, activity and exploration were dependent on social status, cockerels varied in their boldness independently of their social position in the group (Favati *et al.*, 2014). These studies show that even closely related chickens are perfectly distinguishable in their behaviour if we play sufficiently close attention.

The fact that different stable behavioural traits influence the environmental preferences of chickens (Nicol *et al.*, 2009) shows that a bird's personality is an important factor in the decisions it makes, just as human personality affects the choices we make. It further suggests a role for housing systems that provide birds with a degree of choice over some aspects of their environment in securing the welfare of all birds.

The Foundations of Empathy

Empathy is a complex phenomenon of profound importance to human society. Given this, it is perhaps surprising that emerging evidence shows that chickens may possess some of the foundational mechanisms and capacities that support empathy. For any individual chicken, its experience of pain or distress is a private affair but one that may have wider consequences. An observer chicken, witnessing expressions of pain or distress in a companion, may experience a vicarious emotional reaction that is more appropriate to the companion than itself. Emotional transfer of this kind is thought to take place when human babies cry when they hear other infants crying, and emotional transfer is one of the foundations of more complex forms of empathic response (Edgar *et al.*, 2012a). Chickens clearly respond to others who may be in danger (e.g. by responding to alarm calls, as discussed above), but the emotional state of the birds in these studies has not been examined.

Evidence that there is a degree of emotional contagion between chickens comes from studies that show that when hens with a genetic tendency to fearfulness are housed with birds from a less-fearful line, the latter birds display increased signs of fear and stress (Cheng et al., 2002; Uitdehaag et al., 2008, 2009). Similarly, broiler chickens mounted a greater stress response to capture and caging if they had previously observed a conspecific being roughly handled than if they had observed gentler handling (Zulkifli and Azah, 2004). In both cases, increased fear or stress in subject birds may be a consequence of them simply using signs of fear or distress in others as a warning signal indicating their own potential danger. However, it is also possible that chickens could become frightened or distressed by the behaviour of their companions, even if they are clearly in no danger themselves. If observer birds find the distress of their companions aversive and upsetting, then this would suggest a degree of emotional empathy. Slightly different processes might underpin cognitive empathy, where evidence would be required that an observer understood something about the dangerous situation of its companion, even if this situation differed from that of its own (Edgar et al., 2012b).

As we have seen previously, female broody hens are committed parents, spending time with their newly hatched chicks, sheltering and keeping them warm, warning them of danger and closely guiding their feeding behaviour. The hens and chicks form social bonds and become distressed when separated, in ways not observed between related or unrelated adults. In searching for evidence of an empathic response in chickens, the maternal context seems the obvious place to start. In carefully controlled experimental work, we have shown that broody hens respond with a pronounced and consistent set of behavioural and physiological responses to the mild distress of their chicks. Mild distress was induced by directing a small puff of air to the chicks, while the hens were equipped with a small externally mounted heart-rate monitor situated within a small harness, and the surface temperature of their combs and eyes was measured using infrared thermography. When hens observed the mild distress of their chicks, there was a significant increase in their heart rate, a rapid reduction in core temperature detected at the eye (Fig. 6.2), increased standing alert, decreased preening and an increase in maternal vocalization relative to a

variety of control conditions such as the sound of the air puff on its own (Edgar et al., 2011).

Using the same procedure, Edgar et al. (2012b) found no evidence of any physiological or behavioural reaction when the same air puff was directed towards familiar but unrelated adult birds instead of chicks. A possible explanation for this apparent difference might be that chicks find the air puff more aversive than the adults and thus provide a clearer demonstration of distress to the observing hen. However, this possibility was rejected when a conditioned place preference test (see Chapter 4) showed that the chicks and the adults found the air puff equally (although mildly) aversive. Further evidence that broody hens are not responding solely to the reactions of their chicks comes from a subsequent experiment that independently manipulated both the hens' and the chicks' perception of the chicks' predicament (Edgar et al., 2013). On four consecutive days, chicks were placed alternately in boxes that were coloured red or yellow. One of the box colours (the two colours were systematically varied) signalled safety, while the other box colour signalled that an air puff would be delivered at regular intervals for 5 min. Hens were also placed in the coloured boxes but not at the same time as their chicks. Some hens received the same information about the link between box colour and safety or danger, while other hens received the opposite information; for this latter group, the box that the hens associated with safety was the box the chicks associated with danger and vice versa. Subsequently, some hens were exposed to the sight of their chicks in the coloured box that, to the hen, signalled danger. In this condition, the hens reacted with marked behavioural changes (increased vocalization, decreased preening), regardless of whether the chicks thought the box signalled safety or danger. However, physiological reactions by the hens occurred only when their information about the situation coincided with that of the chicks. In addition, the chicks appeared more agitated and gave more distress calls when their mother thought they were in danger, even if this conflicted with their own knowledge. Edgar et al. (2013) concluded that there was a 'combination of mediators of the maternal response in domestic chickens, with both learned associations and more simple distress cues influencing hen behaviour and physiology, and facilitating adaptive mirroring behaviour in the chicks'.

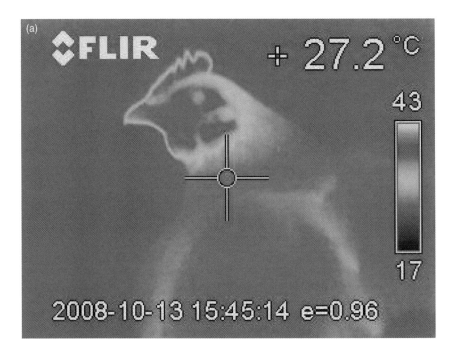

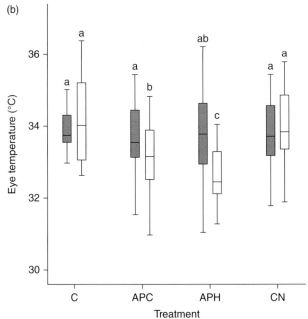

Fig. 6.2. Changes in eye temperature of broody hens observing mild distress of their chicks. (a) Thermal imaging was used to determine the core eye temperature. (b) The eye temperature of the hens was determined during a 10 min pre-treatment phase (grey bars) followed by a 10 min treatment phase (white bars). The hens' eye temperature dropped when they experienced an air puff themselves (APH) or when they witnessed their chicks being air puffed (APC) but not during a control treatment (C) or in response to the noise of the air puff only (CN). Different letters above bars indicate a statistically significant difference ($P<0.05$). (Image courtesy of Dr Joanne Edgar, University of Bristol, UK. Figure reproduced from Edgar *et al.* (2011).)

References

Abeyesinghe, S.M., McLeman, M.A., Owen, R.C., McMahon, C.E. and Wathes, C.M. (2009) Investigating social discrimination of group members by laying hens. *Behavioural Processes* 81, 1–13.

Abeyesinghe, S.M., Drewe, J.A., Asher, L., Wathes, C.M. and Collins, L.M. (2013) Do hens have friends? *Applied Animal Behaviour Science* 143, 61–66.

Albentosa M.J. and Cooper, J.J. (2005) Testing resource value in group-housed animals: an investigation of cage height preference in laying hens. *Behavioural Processes* 70, 113–121.

Alvino, G.M., Blatchford, R.A., Archer, G.S. and Mench, J.A. (2009) Light intensity during rearing affects the behavioural synchrony and resting patterns of broiler chickens. *British Poultry Science* 50, 275–283.

Appleby, M.C. (1983) The probability of linearity in hierarchies. *Animal Behaviour* 31, 600–608.

Appleby, M.C. (2004) What causes crowding? Effects of space, facilities and group size on behaviour, with particular reference to furnished cages for hens. *Animal Welfare* 13, 313–320.

Appleby, M.C., Mench, J.A. and Hughes, B.O. (2004) *Poultry Behaviour and Welfare*. CABI, Wallingford, UK.

Asher, L. and Collins, L.M. (2012) Assessing synchrony in groups: are you measuring what you think you are measuring? *Applied Animal Behaviour Science* 138, 162–169.

Asher, L., Collins, L.M., Pfeiffer, D.U. and Nicol, C.J. (2013) Flocking for food or flockmates? *Applied Animal Behaviour Science* 147, 94–103.

Bayly, K.L. and Evans, C.S. (2003) Dynamic changes in alarm call structure: a strategy for reducing conspicuousness to avian predators? *Behaviour* 140, 353–369.

Bayly, K.L., Evans, C.S. and Taylor, A. (2006) Measuring social structure: a comparison of eight dominance indices. *Behavioural Processes* 73, 1–12.

Bilcik, B. and Estevez, I. (2005) Impact of male–male competition and morphological traits on mating strategies and reproductive success in broiler breeders. *Applied Animal Behaviour Science* 92, 307–323.

Bilcik, B., Estevez, I. and Russek-Cohen, E. (2005) Reproductive success of broiler breeders in natural mating systems: the effect of male–male competition, sperm quality, and morphological characteristics. *Poultry Science* 84, 1453–1462.

Bokkers, E.A.M., de Boer, I.J.M. and Koene, P. (2011) Space needs of broilers. *Animal Welfare* 20, 623–632.

Bradshaw, R.H. (1992a) Conspecific discrimination and social preference in the laying hen. *Applied Animal Behaviour Science* 33, 69–75.

Bradshaw, R.H. (1992b) Individual attributes as predictors of social status in small groups of laying hens. *Applied Animal Behaviour Science* 34, 359–363.

Bradshaw, R.H. and Dawkins, M.S. (1993) Slides of conspecifics as representatives of real animals in laying hens (*Gallus domesticus*). *Behavioural Processes* 28, 165–172.

Buijs, S., Keeling, L.J., Vangestel, C., Baert, J. and Tuyttens, F.A.M. (2011) Neighbourhood analysis as an indicator of spatial requirements of broiler chickens. *Applied Animal Behaviour Science* 129, 111–120.

Campo, J.L., Gil, M.G. and Davila, S.G. (2005) Social aggressiveness, pecking at hands and its relationships with tonic immobility duration and heterophil to lymphocyte ratio in chickens of different breeds. *Archiv fur Geflugelkunde* 69, 11–15.

Carmichael, N.L., Walker, A.W. and Hughes, B.O. (1999) Laying hens in large flocks in a perchery system: influence of stocking density on location, use of resources and behaviour. *British Poultry Science* 40, 165–176.

Cheng, H.W., Singleton, P. and Muir, W.M. (2002) Social stress in laying hens: differential dopamine and corticosterone responses after intermingling different genetic strains of chickens. *Poultry Science* 81, 1265–1272.

Cloutier, S. and Newberry, R.C. (2000) Recent social experience, body weight and initial patterns of attack predict the social status attained by unfamiliar hens in a new group. *Behaviour* 137, 705–726.

Collias, N.E. (1952) The development of social behaviour in birds. *Auk* 69, 127–159.

Collias, N.E. and Collias, E. (1996) Social organization of a red junglefowl, *Gallus gallus*, population related to evolution theory. *Animal Behaviour* 51, 1337–1354.

Collias, N. and Joos, M. (1953) The spectrographic analysis of sound signals of the domestic fowl. *Behaviour* 5, 175–188.

Collins, L.M. and Sumpter, D.J.T. (2007) The feeding dynamics of broiler chickens. *Journal of the Royal Society Interface* 4, 65–72.

Collins, L.M., Asher, L., Pfeiffer, D.U., Browne, W.J. and Nicol, C.J. (2011) Clustering and synchrony in laying hens: the effect of environmental resources on social dynamics. *Applied Animal Behaviour Science* 129, 43–53.

Collins, S., Forkman, B., Kristensen, H.H., Sandoe, P. and Hocking P.M. (2011) Investigating the importance of vision in poultry: comparing the behaviour of blind and sighted chickens. *Applied Animal Behaviour Science* 133, 60–69.

Craig, J.V., Biswas, D.K. and Guhl, A.M. (1969) Agonistic behaviour influenced by strangeness, crowding and heredity in female domestic fowl (*Gallus gallus*). *Animal Behaviour* 17, 498–506.

Craig, J. V., Al-Rawi, B. and Kratzer, D.D. (1977) Social status and sex ratio effects on mating frequency of cockerels.
Poultry Science 56, 767–772.

Croney, C.C., Prince-Kelly, N. and Meller, C.L. (2007) A note on social dominance and learning ability in the domestic chicken (*Gallus gallus*). *Applied Animal Behaviour Science* 105, 254–258.

Dawkins, M.S. (1995) How do hens view other hens? The use of lateral and binocular visual fields in social recognition. *Behaviour* 132, 591–606.

Dawkins, M.S. (1996) Distance and social recognition in hens: implications for the use of photographs as social
stimuli. *Behaviour* 133, 663–680.

Dean, R., Nakagawa, S. and Pizzari, T. (2011) The risk and intensity of sperm ejection in female birds. *American
Naturalist* 178, 343–354.

D'Eath, R.B. and Keeling, L.J. (2003) Social discrimination and aggression by laying hens in large groups: from peck
orders to social tolerance. *Applied Animal Behaviour Science* 84, 197–212.

D'Eath, R.B. and Stone, R.J. (1999) Chickens use visual cues in social discrimination: an experiment with coloured
lighting. *Applied Animal Behaviour Science* 62, 233–242.

de Jong, I.C., Wolthuis-Fillerip, M. and van Emous, R.A. (2009) Development of sexual behaviour in
commercially-housed broiler breeders after mixing. *British Poultry Science* 50, 151–160.

Dixon, G. and Nicol, C.J. (2008) The effect of diet change on the behaviour of layer pullets. *Animal Welfare* 17,
101–109.

Dixon, G., Green, L.E. and Nicol, C.J. (2006) Effect of diet change on the behaviour of chicks of an egg-laying strain.
Journal of Applied Animal Welfare Science 9, 41–58.

Duncan, I.J.H., Widowski, T.M., Malleau, A.E., Lindberg, A.C. and Petherick, J.C. (1998) External factors and causation of dustbathing in domestic hens. *Behavioural Processes* 43, 219–228.

Edgar, J.L., Lowe, J.C., Paul, E.S. and Nicol, C.J. (2011) Avian maternal response to chick distress. *Proceedings of the
Royal Society B – Biological Sciences* 278, 3129–3134.

Edgar, J.L., Nicol, C.J., Clark, C.C.A. and Paul, E.S. (2012a) Measuring empathic responses in animals. *Applied Animal Behaviour Science* 138, 182–193.

Edgar, J.L., Paul, E.S., Harris, L., Penturn, S. and Nicol, C.J. (2012b) No evidence for emotional empathy in chickens
observing familiar adult conspecifics. *PLoS One* 7, e31542.

Edgar, J.L., Paul, E.S. and Nicol, C.J. (2013) Protective mother hens: cognitive influences on the maternal response.
Animal Behaviour 86, 223–229.

Emery, N.J., Seed, A.M., von Bayern, A.M.P. and Clayton, N.S. (2007) Cognitive adaptations of social bonding in
birds. *Philosophical Transactions of the Royal Society B – Biological Sciences* 362, 489–505.

Estevez, I., Newberry, R.C. and Arias de Reyna, L. (1997) Broiler chickens: a tolerant social system? *Etologia* 5,
19–29.

Estevez, I., Newberry, R.C. and Keeling, L.J. (2002) Dynamics of aggression in domestic fowl. *Applied Animal Behaviour Science* 76, 307–325.

Estevez, I., Keeling, L.J. and Newberry, R.C. (2003) Decreasing aggression with increasing group size in young domestic fowl. *Applied Animal Behaviour Science* 84, 213–218.

Evans, C.S. (1997) Referential signals. *Perspectives in Ethology* 12, 99–143.

Evans, C.S. and Evans, L. (1999) Chicken food calls are functionally referential. *Animal Behaviour* 58, 307–317.

Evans, C.S. and Evans, L. (2007) Representational signalling in birds. *Biology Letters* 3, 8–11.

Evans, C. S. and Marler, P. (1991) On the use of video images as social stimuli in birds: audience effects on alarm
calling. *Animal Behaviour* 41, 17–26.

Evans, C.S. and Marler, P. (1992) Female appearance as a factor in the responsiveness of male chickens during anti-
predator behaviour and courtship. *Animal Behaviour* 43, 137–145.

Evans, C.S. and Marler, P. (1994) Food calling and audience effects in male chickens, *Gallus gallus*: their relationships to food availability, courtship and social facilitation. *Animal Behaviour* 47, 1159–1170.

Evans, C.S., Evans, L. and Marler, P. (1993a) On the meaning of alarm calls: functional reference in an avian vocal
system. *Animal Behaviour* 46, 23–38.

Evans, C.S., Macedonia, J.M. and Marler, P. (1993b) Effects of apparent size and speed on the response of chickens,
Gallus gallus, to computer-generated simulations of aerial predators. *Animal Behaviour* 46, 1–11.

Favati, A., Leimar, O., Radesater, T. and Lovlie, H. (2014) Social status and personality: stability in social state can promote consistency of behavioural responses. *Proceedings of the Royal Society B – Biological Sciences* 281, 2013–2531.

Febrer, K., Jones, T.A., Donnelly, C.A. and Dawkins, M.S. (2006) Forced to crowd or choosing to cluster? Spatial
distribution indicates social attraction in broiler chickens. *Animal Behaviour* 72, 1291–1300.

Fisher, R.A. (1930) *The Genetical Theory of Natural Selection*. Clarendon Press, Oxford, UK.

Folstad, I. and Karter, A.J. (1992) Parasites, bright males and the immunocompetence handicap. *American Naturalist* 139, 603–622.

Forkman, B. (1996) The social facilitation of drinking: what is facilitated and who is affected? *Ethology* 102, 252–258.

Forkman, B. and Haskell, M.J. (2004) The maintenance of stable dominance hierarchies and the pattern of aggression: support for the suppression hypothesis. *Ethology* 110, 737–744.

Freire, R., Wilkins, L.J., Short, F. and Nicol, C.J. (2003) Behaviour and welfare of individual laying hens in a non-cage system. *British Poultry Science* 44, 22–29.

Ghareeb, K., Niebuhr, K., Awad, W.A., Waiblinger, S. and Troxler, J. (2008) Stability of fear and sociality in two strains of laying hens. *British Poultry Science* 49, 502–508.

Graves, H.B., Hable, C.P. and Jenkins, T.H. (1985) Sexual selection in *Gallus*: effects of morphology and dominance on female spatial behaviour. *Behavioural Processes* 11, 189–197.

Grigor, P.N., Hughes, B.O. and Appleby, M.C. (1995) Social inhibition of movement in domestic hens. *Animal Behaviour* 49, 1381–1388.

Guhl, A.M. (1958) The development of social organization in the domestic chick. *Animal Behaviour* 6, 92–111.

Guhl, A.M. (1964) Psychophysiological interrelations in the social behavior of chickens. *Psychology Bulletin* 61, 277–285.

Guhl, A.M., Collias, N.E. and Allee, W.C. (1945) Mating behaviour and the social hierarchy in small flocks of white leghorns. *Physiological Zoology* 18, 365–390.

Guzman, D.A. and Marin, R.H. (2008) Social reinstatement responses of meat-type chickens to familiar and unfamiliar conspecifics after exposure to an acute stressor. *Applied Animal Behaviour Science* 110, 282–293.

Gyger, M. and Marler, P. (1988) Food calling in the domestic fowl *Gallus gallus*: the role of external referents and deception. *Animal Behaviour* 36, 358–365.

Gyger, M., Marler, P. and Pickert, R. (1987) Semantics of an avian alarm call system: the male domestic fowl, *Gallus domesticus*. *Behaviour* 102, 15–40.

Gyger, M., Karakashian, S., Dufty, A.M. Jr and Marler, P. (1988) Alarm signals in birds: the role of testosterone. *Hormones and Behavior* 22, 305–314.

Hamilton, W.D. and Zuk, M. (1982) Heritable true fitness and bright birds: a role for parasites? *Science* 218, 384–387.

Hemsworth, P.H., Barnett, J.L. and Jones, R.B. (1993) Situational factors that influence the level of fear of humans by laying hens. *Applied Animal Behaviour Science* 36, 197–210.

Hogue, M. E., Beaugrand, J.P. and Lague, P.C. (1996) Coherent use of information by hens observing their former dominant defeating or being defeated by a stranger. *Behavioural Processes* 38, 241–252.

Hoppitt, W. and Laland, K.N. (2008) Social processes affecting feeding and drinking in the domestic fowl. *Animal Behaviour* 76, 1529–1543.

Hoppitt, W., Blackburn, L. and Laland, K.N. (2007) Response facilitation in the domestic fowl. *Animal Behaviour* 73, 229–238.

Hughes, B.O. (1971) Allelomimetic feeding in the domestic fowl. *British Poultry Science* 12, 359–366.

Hughes, B.O., Carmichael, N.L., Walker, A.W. and Grigor, P.N. (1997) Low incidence of aggression in large flocks of laying hens. *Applied Animal Behaviour Science* 54, 215–234.

Johnsen, T.S. and Zuk, M. (1998) Parasites, morphology and blood characters in male red junglefowl during development. *Condor* 100, 749–752.

Johnsen, T.S., Zuk, M. and Fessler, E.A. (2001) Social dominance, male behaviour and mating in mixed-sex flocks of red junglefowl. *Behaviour* 138, 1–18.

Johnson, K., Thornhill, R., Ligon, J.D. and Zuk, M. (1993) The direction of mothers and daughters preferences and the heritability of male ornaments in red junglefowl (*Gallus gallus*). *Behavioral Ecology* 4, 254–259.

Jones, M.E.J. and Mench, J. (1991) Behavioral correlates of male mating success in a multisire flock as determined by DNA fingerprinting. *Poultry Science* 70, 1493–1498.

Jones, R.B., Duncan, I.J.H. and Hughes, B.O. (1981) The assessment of fear in domestic hens exposed to a looming human stimulus. *Behavioural Processes* 6, 121–133.

Karakashian, S.J., Gyger, M. and Marler, P. (1988) Audience effects on alarm calling in chickens (*Gallus gallus*). *Journal of Comparative Psychology* 102, 129–135.

Keeling, L.J. and Hurnik, J.F. (1993) Chickens show socially facilitated feeding behaviour in response to a video image of a conspecific. *Applied Animal Behaviour Science* 36, 223–231.

Kikusui, T., Winslow, J.T. and Mori, Y. (2006) Social buffering: relief from stress and anxiety. *Philosophical Transactions of the Royal Society B – Biological Sciences* 361, 2215–2228.

Kim, T. and Zuk, M. (2000) The effects of age and previous experience on social rank in female red junglefowl, *Gallus gallus spadiceus*. *Animal Behaviour* 60, 239–244.

Knierim, U. (2000) [Degree of synchronous feeding behaviour of two types of laying hybrid hens in battery cages with a feeder space of 12 cm per hen]. *Deutsche Tierartzliche Wochenschrift* 107, 459–463 (in German).

Kokolakis, A., Smith, C.L. and Evans, C.S. (2010) Aerial alarm calling by male fowl (*Gallus gallus*) reveals subtle new mechanisms of risk managements. *Animal Behaviour* 79, 1373–1380.

Koshiba, M., Shirakawa, Y., Mimura, K., Senoo, A., Karino, G. and Nakamura, S. (2013) Familiarity perception call elicited under restricted sensory cues in peer-social interactions of the domestic chick. *PLoS One* 8, e58847.

Ligon, J.D. and Zwartjes, P.W. (1995) Female red junglefowl choose to mate with multiple males. *Animal Behaviour* 49, 127–135.

Ligon, J.D., Kimball, R. and Merola-Zwartjes, M. (1998) Mate choice by female red junglefowl: the issues of multiple ornaments and fluctuating asymmetry. *Animal Behaviour* 55, 41–50.

Leonard, M.L. and Horn, A.G. (1995) Crowing in relation to social status in roosters. *Animal Behaviour* 49, 1283–1290.

Leonard, M.L., Horn, A.G. and Fairfull, R.W. (1995) Correlates and consequences of allopecking in White Leghorn chickens. *Applied Animal Behaviour Science* 43, 17–26.

Lindberg A.C. and Nicol C.J. (1996) Effects of social and environmental familiarity on group preferences and spacing behaviour in laying hens. *Applied Animal Behaviour Science* 49, 109–123.

Lovlie, H. and Pizzari, T. (2007) Sex in the morning or in the evening? Females adjust daily mating patterns to the intensity of sexual harassment. *American Naturalist* 170, E1–E13.

Lovlie, H., Cornwallis, C.K. and Pizzari, T. (2005) Male mounting alone reduces female promiscuity in the fowl. *Current Biology* 15, 1222–1227.

Lundberg, A.S. and Keeling, L.J. (2003) Social effects on dustbathing behaviour in laying hens: using video images to investigate effect of rank. *Applied Animal Behaviour Science* 81, 43–57.

Marin, R.H., Jones, R.B., Garcia, D.A. and Arce, A. (1999) Early T-maze behaviour and subsequent growth in commercial broiler flocks. *British Poultry Science* 40, 434–438.

Marin, R.H., Freytes, P., Guzman, D. and Jones, R.B. (2001) Effects of an acute stressor on fear and on the social reinstatement responses of domestic chicks to cagemates and strangers. *Applied Animal Behaviour Science* 71, 57–66.

Marin, R.H., Satterlee, D.G., Castille, S.A. and Jones, R.B. (2003) Early T-maze behaviour and broiler growth. *Poultry Science* 82, 742–748.

Marler, P., Dufty, A. and Pickert, R. (1986a) Vocal communication in the domestic chicken: I. Does a sender communicate information about the quality of a food referent to a receiver? *Animal Behaviour* 34, 188–193.

Marler, P., Dufty, A. and Pickert, R. (1986b) Vocal communication in the domestic chicken: II. Is a sender sensitive to the presence and nature of a receiver. *Animal Behaviour* 34, 194–198.

McBride, G., Parer, I.P. and Foenander, F. (1969) The social organization and behaviour of the feral domestic fowl. *Animal Behaviour Monographs* 2, 126–181.

Millman, S.T. and Duncan, I.J.H. (2000a) Strain differences in aggressiveness of male domestic fowl in response to a male model. *Applied Animal Behaviour Science* 66, 217–233.

Millman, S.T. and Duncan, I.J.H. (2000b) Do female broiler breeder fowl display a preference for broiler breeder or laying strain males in a Y-maze test? *Applied Animal Behaviour Science* 69, 275–290.

Millman, S.T., Duncan, I.J.H. and Widowski, T.M. (2000) Male broiler breeder fowl display high levels of aggression toward females. *Poultry Science* 79, 1233–1241.

Moffatt, C.A. and Hogan, J.A. (1992) Ontogeny of chick responses to maternal food calls in Burmese Red Junglefowl (*Gallus gallus spadiceus*). *Journal of Comparative Psychology* 106, 92–96.

Nicol, C.J. (1989) Social influences on the comfort behaviour of laying hens. *Applied Animal Behaviour Science* 22, 75–81.

Nicol, C.J. (1995) The social transmission of information and behaviour. *Applied Animal Behaviour Science* 44, 79–98.

Nicol, C.J. and Pope, S.J. (1994) Social learning in small flocks of laying hens. *Animal Behaviour* 47, 1289–1296.

Nicol, C.J. and Pope, S.J. (1996) The maternal display of domestic hens is sensitive to perceived chick error. *Animal Behaviour* 52, 767–774.

Nicol, C.J. and Pope, S.J. (1999) The effects of demonstrator social status and prior foraging success on social learning in laying hens. *Animal Behaviour* 57, 163–171.

Nicol, C.J., Gregory, N.G., Knowles, T.G., Parkman, I.D. and Wilkins, L. (1999) Differential effects of increased stocking density, mediated by increased flock size, on feather pecking and aggression in laying hens. *Applied Animal Behaviour Science* 65, 137–152.

Nicol, C.J., Caplen, G., Edgar, J. and Browne, W.J. (2009) Associations between welfare indicators and environmental choice in laying hens. *Animal Behaviour* 78, 413–424.

O'Connor, E.A., Saunders, J.E., Grist, H., McLeman, M.A., Wathes, C.M. and Abeyesinghe, S.M. (2011) The relationship between the comb and social behaviour in laying hens. *Applied Animal Behaviour Science* 135, 293–299.

Oden, K., Vestergaard, K.S. and Algers, B. (2000) Space use and agonistic behaviour in relation to sex composition in large flocks of laying hen. *Applied Animal Behaviour Science* 67, 307–320.

Olsson, I.A.S., Duncan, I.J.J. and Keeling, L.J. (2002) How important is social facilitation for dustbathing in laying hens. *Applied Animal Behaviour Science* 79, 285–297.

Pagel, M. and Dawkins, M.S. (1997) Peck orders and group size in laying hens: 'futures contracts' for non-aggression. *Behavioural Processes* 40, 13–25.

Palleroni, A., Hauser, M. and Marler, P. (2005) Do responses of galliform birds vary adaptively with predator size? *Animal Cognition* 8, 200–210.

Parker, T.H. (2003) Genetic benefits of mate choice separated from differential maternal investment in red junglefowl (*Gallus gallus*). *Evolution* 57, 2157–2165.

Parker, T.H. and Ligon, J.D. (2003) Female mating preferences in red junglefowl: a meta-analysis. *Ethology, Ecology, and Evolution* 15, 63–72.

Pizzari, T. (2001) Indirect partner choice through manipulation of male behaviour by female fowl, *Gallus gallus domesticus*. *Proceedings of the Royal Society B – Biological Sciences* 268, 181–186.

Pizzari, T. (2003) Food, vigilance and sperm: the role of male direct benefits in the evolution of female preference in a polygamous bird. *Behavioural Ecology* 14, 593–601.

Pizzari, T. and Birkhead, T.R. (2000) Female feral fowl eject sperm of subdominant males. *Nature* 405, 787–789.

Pizzari, T. and Birkhead, T.R. (2001) For whom does the hen cackle? The function of postoviposition cackling. *Animal Behaviour* 61, 601–607.

Pizzari, T., Cornwallis, C.K., Lovlie, H., Jakobsson, S. and Birkhead, T.R. (2003) Sophisticated sperm allocation in male fowl. *Nature* 426, 70–74.

Porter, R.H., Roelofsen, R., Picard, M. and Arnould, C. (2005) The temporal development and sensory mediation of social discrimination in domestic chicks. *Animal Behaviour* 70, 359–364.

Porter, R.H., Arnould, C., Simac, L. and Hild, S. (2006) Retention of individual recognition in chicks and the effects of social experience. *Animal Behaviour* 72, 707–712.

Prokop, Z.M., Michalczyk, L., Drobniak, S.M., Herdegen, M. and Radwan, J. (2012) Meta-analysis suggests choosy females get sexy sons more than "good genes". *Evolution* 66, 2665–2673.

Riedstra, B. and Groothuis, T.G.G. (2002) Early feather pecking as a form of social exploration: the effect of group stability on feather pecking and tonic immobility in domestic chicks. *Applied Animal Behaviour Science* 77, 127–138.

Rodenburg, T.B. and Koene, P. (2007) The impact of group size on damaging behaviours, aggression, fear and stress in farm animals. *Applied Animal Behaviour Science* 103, 205–214.

Rodriguez-Aurrekoetxea, A. and Estevez, I. (2014) Aggressiveness in the domestic fowl: distance versus 'attitude'. *Applied Animal Behaviour Science* 153, 68–74.

Rushen, J. (1982a) The peck orders of chickens: how do they develop and why are they linear? *Animal Behaviour* 30, 1129–1137.

Rushen, J. (1982b) Development of social behaviour in chickens: a factor analysis. *Behavioural Processes* 7, 319–333.

Rushen, J. (1984) Frequencies of agonistic behaviours as measures of aggression in chickens: a factor analysis. *Applied Animal Behaviour Science* 12, 167–176.

Ryan, C.M.E. (1982) Concept formation and individual recognition in the domestic chicken (*Gallus gallus*). *Behaviour Analysis Letters* 2, 213–220.

Ryan, C.M.E. and Lea, S.E.G. (1994) Images of conspecifics as categories to be discriminated by pigeons and chickens: slides, video tapes, stuffed birds and live birds. *Behavioural Processes* 33, 155–175.

Savory, C.J. and Mann, J.S. (1997) Behavioural development in groups of pen-housed pullets in relation to genetic strain, age and food form. *British Poultry Science* 38, 38–47.

Savory, C.J., Wood-Gush, D.G.M. and Duncan, I.J.H. (1978) Feeding behaviour in a population of domestic fowls in the wild. *Applied Animal Ethology* 4, 13–27.

Schjelderup-Ebbe, T. (1922) Beitrage zur Sozialpsycholgie des Haushuhns. *Zeitschrift Psychologie* 88, 225–252.

Schwean-Lardner, K., Fancher, B.I. and Classen, H.L. (2014) Effect of day length on flock behavioural patterns and melatonin rhythms in broilers. *British Poultry Science* 55, 21–30.

Shimmura, T., Eguchi, Y., Uetake, K. and Tanaka, T. (2007) Differences of behaviour, use of resources and physical conditions between dominant and subordinate hens in furnished cages. *Animal Science Journal* 78, 307–313.

Shimmura, T., Azuma, T., Hirahara, S., Eguchi, Y., Uetake, K. and Tanaka, T. (2008) Relation between social order and use of resources in small and large furnished cages for laying hens. *British Poultry Science* 49, 516–524.

Shimmura, T., Nakamura, T., Azuma, T., Eguchi, Y., Uetake, K. and Tanaka, T. (2010) Effects of social rank and familiarity on dustbathing in domestic fowl. *Animal Welfare* 19, 67–73.

Smith, C.L. and Evans, C.S. (2008) Multimodal signalling in fowl, *Gallus gallus*. *Journal of Experimental Biology* 211, 2052–2057.

Smith, C.L. and Evans, C.S. (2009) Silent tidbitting in male fowl, *Gallus gallus*: a referential visual signal with multiple functions. *Journal of Experimental Biology* 212, 835–842.

Smith, C.L. and Evans, C.S. (2011) Exaggeration of display characteristics enhances detection of visual signals. *Behaviour* 148, 287–305.

Smith, C.L., van Dyk, D.A., Taylor, P.W. and Evans, C.S. (2009) On the function of an enigmatic ornament: wattles increase the conspicuousness of visual displays in male fowl. *Animal Behaviour* 78, 1433–1440.

Smith, C.L., Taylor, A. and Evans, C.S. (2011) Tactical multimodal signalling in birds: facultative variation in signal modality reveals sensitivity to social costs. *Animal Behaviour* 82, 521–527.

Tolman, C.W. (1967a) The feeding behaviour of domestic chicks as a function of rate of pecking by a surrogate companion. *Behaviour* 29, 57–62.

Tolman, C.W. (1967b) The effects of tapping sounds on feeding behaviour of domestic chicks. *Animal Behaviour* 15, 134–142.

Tolman, C.W. (1968) The varieties of social stimulation in the feeding behaviour of domestic chicks. *Behaviour* 30, 275–286.

Tolman, C.W. and Wilson, G.F. (1965) Social feeding in domestic chicks. *Animal Behaviour* 13, 134–142.

Uitdehaag, K.A., Rodenburg, T.B., van Hierden, Y.M., Bolhuis, J.E., Toscano, M.J., Nicol, C.J. and Komen, J. (2008) Effects of mixing housing of birds from two genetic lines of laying hens on open field and manual restraint responses. *Behavioural Processes* 79, 13–18.

Uitdehaag, K.A., Rodenburg, T.B., Bolhuis, J.E., Decuypere, E. and Komen, J.(2009) Mixed housing of different genetic lines of laying hens negatively affects feather pecking and fear related behaviour. *Applied Animal Behaviour Science* 116, 58–66.

van Kampen, H.S. (1994) Courtship food-calling in Burmese red junglefowl: I. The causation of female approach. *Behaviour* 131, 261–275.

van Kampen, H.S. (1999) Courtship food-calling in Burmese red junglefowl: II. Sexual conditioning and the role of the female. *Behaviour* 134, 775–787.

van Kampen, H.S. and Hogan, J.A. (2000) Courtship food-calling in Burmese red junglefowl: III. Factors influencing the male's behaviour. *Behaviour* 137, 1191–1209.

Vonschantz, T., Tufvesson, M., Goransson, G., Grahn, M., Wilhelmson, M. and Wittzell, H. (1995) Artificial selection for increased comb size and its effects on other sexual characters and viability in *Gallus domesticus* (the domestic chicken). *Heredity* 75, 518–529.

Wauters, A.M. and Richard-Yris, M.A. (2001) Experience modulates emission of food calls in broody hens. *Comptes Rendue de l'Academie des Sciences Serie III* 324, 1021–1027.

Wauters, A.M. and Richard-Yris, M.A. (2002) Mutual influence of the maternal hen's food calling and feeding behaviour on the behaviour of her chicks. *Developmental Psychobiology* 41, 25–36.

Wauters, A.M. and Richard-Yris, M.A. (2003) Maternal food calling in domestic hens: influence of feeding context. *Comptes Rendus Biologies* 326, 677–686.

Wauters, A.M., Richard-Yris, M.A., Pierre, J.S., Lunel, C. and Richard, J.P. (1999a) Influence of chicks and food quality on food calling in broody domestic hens. *Behaviour* 136, 919–933.

Wauters, A.M., Richard-Yris, M.A., Richard, J.P. and Forasté, M. (1999b) Internal and external factors modulate food-calling in domestic hens. *Animal Cognition* 2, 1–10.

Webster, A.B. and Hurnik, J.F. (1994) Synchronization of behaviour among laying hens in battery cages. *Applied Animal Behaviour Science* 40, 153–165.

Wennrich, G. (1981) Zum lautinventar bei haushühnern (*Gallus f. domesticus*). *Berliner Münchener Tierärztlich Wochenschrift* 94, 90–95.

Wilson, D.R. and Evans, C.S. (2008) Mating success increases alarm-calling effort in male fowl, *Gallus gallus*. *Animal Behaviour* 76, 2029–2035.

Wilson, D.R. and Evans, C.S. (2010) Female fowl (*Gallus gallus*) do not prefer alarm-calling males. *Behaviour* 147, 525–552.

Wilson, D.R., Bayly, K.L., Nelson, X.J., Gillings, M. and Evans, C.S. (2008) Alarm calling best predicts mating and reproductive success in ornamented male fowl, *Gallus gallus*. *Animal Behaviour* 76, 543–554.

Wilson, D.R., Nelson, X.J. and Evans, C.S. (2009) Seizing the opportunity: subordinate male fowl respond rapidly to variation in social context. *Ethology* 115, 995–1004.

Wood-Gush, D.G.M. (1971) *The Behaviour of the Domestic Fowl*. Heinemann Studies in Biology 7. Heineman Educational Books, London.

Zajonc, R.B., Wilson, W.R. and Rajecki, D.W. (1975) Affiliation and social discrimination produced by brief exposure in day-old domestic chicks. *Animal Behaviour* 23, 131–138.

Zimmerman, P.H., Lindberg, A.C., Pope, S.J., Glen, E., Bolhuis, J.E. and Nicol, C.J. (2006) The effect of stocking density, flock size and modified management on laying hen behaviour and welfare in a non-cage system. *Applied Animal Behaviour Science* 101, 111–124.

Zuk, M. and Johnsen, T.S. (2000) Social environment and immunity in male red junglefowl. *Behavioral Ecology* 11, 146–153.

Zuk, M., Johnson, J., Thornhill, R. and Ligon, J.D. (1990) Mechanisms of female choice in red junglefowl. *Evolution* 44, 477–485.

Zuk, M., Ligon, J.D. and Thornhill, R. (1992) Effects of experimental manipulation of male secondary sex characters on female mate preference in red junglefowl. *Animal Behaviour* 44, 999–1006.

Zuk, M., Johnsen, T.S. and Maclarty, T. (1995a) Endocrine-immune interactions, ornaments and mate choice in red junglefowl. *Proceedings of the Royal Society B – Biological Sciences* 260, 205–210.

Zuk, M., Popma, S.L. and Johnsen, T.S. (1995b) Male courtship displays, ornaments and female mate choice in captive red junglefowl. *Behaviour* 132, 821–836.

Zulkifli, I. and Azah, A.S.N. (2004) Fear and stress reactions, and the performance of commercial broiler chickens subjected to regular pleasant and unpleasant contacts with human beings. *Applied Animal Behaviour Science* 88, 77–87.

7

Learning, Intelligence and Cognition

We have already seen that chickens perform complex sequences of behaviour and possess finely honed senses and sophisticated social structures, but this does not necessarily mean that they are acting in an intelligent way. Some very complex behaviours can be thought of as evolved 'rules of thumb' such that a trigger leads to a particular response; for example, if you see A, do behaviour X. Such rules can lead to complex sequences of behaviour (e.g. in situation A do X; if this leads to situation B, then do Y), but these apparently complex sequences may not require a lot of mental processing. In other cases, significant brain power may be required to perform behaviour, but most of the processing may occur in 'lower' brain regions, which, in humans at least, are not associated with thought. For example, significant brain processing is required for a chicken to fly down from a perch, but this is not generally regarded as a sign of intelligence.

In this chapter, we will explore the remarkable learning and cognitive abilities of chickens. Many people are interested in whether chickens are intelligent birds (in comparison with other species), but making such direct comparisons is fraught with difficulty (Nicol, 1996). Without doubt, chickens are successful and speedy learners, and are problem solvers in their natural environment and in the experimental psychology laboratory. Their associative and discrimination learning abilities have been put to good use in tests of their sensory perception and welfare, and the emergent properties of these apparently simple processes can result in reasoning. In contrast, it is

noteworthy that young chicks hatch with highly developed brains capable of feats of spontaneous cognition that require little or no training or experience. Their earliest experiences of learning (imprinting, food recognition) produce stable memories that persist throughout life, and yet they remain capable of continuing and reversible learning in other contexts. Cognitive abilities of older birds depend on combinations of innate aptitude and learning or training. This chapter will show that both unlearned and learned behaviour in chickens can be complex and flexible.

To distinguish intelligent behaviour from evolved responses, comparative psychologists suggest using a general problem-solving task in an artificial scenario, where simple evolved rules of thumb cannot be used. Discrimination or maze-running tasks (to reach a food reward) are often used, and during early trials each separate task can be solved by simple trial-and-error learning over a reasonable period of time. To show intelligence, the individual would be expected to learn new exemplars of each task at a faster rate – in other words, to show evidence of 'learning to learn', something shown in young chickens over 50 years ago (Alpert *et al.*, 1962). However, there is a potential difficulty in testing animals in artificial settings, with a risk of posing problems in an unfair manner (Shettleworth, 2003). For example, a child who does not want to work or who cannot see the blackboard properly may appear stupid; equally, if a chicken is scared or distracted, it will be less likely to perform a task than one that is not.

We have to be careful not to confuse motivation with intelligence and to ensure that all tests take full account of the differing sensory worlds of humans and chickens.

To take an overview on the question of chicken intelligence, we will also consider the flexibility of their behaviour and the ways in which their learning and cognitive abilities support their complex social, communicative, spatial and navigational skills. Finally, some of the implications of these findings for chicken welfare will be considered.

Associative Learning

The first and critical learning event for most chickens occurs when they discover, encode and remember the characteristics of their mother (or, in experiments, an artificial object) during the imprinting process reviewed in Chapter 3. One reason why chicks are used so widely in cognition research is because their desire to regain visual contact with the imprinted goal object requires no further training and remains strong for many days. Their spontaneous tendency to prefer a larger group of objects they have been reared with has allowed an assessment of their mathematical abilities, reviewed later.

Thereafter and throughout life, chickens continually form associations between events, and use this learning to guide their behaviour. Chickens detect contingencies or predictive relationships between events via classical or Pavlovian processes. For example, in commercial systems, chickens become expert at detecting the subtlest signs that predict the arrival of food. The detection of a conditioned stimulus such as the feed chain motor switching on can set off a chain of anticipatory responses and vocalizations. Classical conditioning allows chickens to predict events (the motor has started, so food will arrive shortly) but gives them little control or influence over the timing of food delivery. A second form of associative learning is called instrumental or operant conditioning, where the first event is a response made by the chicken and the second event is the associated reinforcing consequence. In this case, the chicken gains not only the ability to predict what might happen next but also the ability to control the timing of the reward. An example would be the ease with which chickens learn to use operant feeders to obtain food rewards in experimental settings. Many of their pheasant cousins, reared for shooting purposes, obtain all of their food by learning (via instrumental conditioning) to peck at a spiral or a nozzle situated below the main body of the outdoor feeder. However, despite its widespread use in the production of game birds, operant feeding is not used in commercial chicken systems.

As we saw in Chapter 4, using instrumentally acquired responses to control the environment may be an important component of good chicken welfare. Certainly, chickens prefer to obtain some food by performing a simple operant response, even in the presence of free food, provided the work demands are not too high (Duncan and Hughes, 1972; Taylor et al., 2001). In theory, there could be many benefits to allowing chickens first to learn, and subsequently to use, an operant response pattern to obtain their feed. Unlike wild jungle fowl, commercial chickens do not have the problem of too much to do with too little time but rather the problem of filling too much time from a relatively limited number of behavioural options available (Hughes and Duncan, 1988). The use of operant feeders could therefore in theory fill this 'spare' time productively and harmlessly, possibly even preventing the development of unwanted behaviours such as feather pecking. When operant feeders have been tested with small groups of laying hens, the birds rapidly acquire the necessary learned responses (Lindberg and Nicol, 1994; Taylor et al., 2001). Continuing to make operant pecks also leads to an increase in the overall time spent on feeding-related behaviour (Lindberg and Nicol, 1994). However, the exclusive use of operant feeders was not a success when this was the only way in which food could be acquired (Lindberg and Nicol, 1994). Under these conditions, the hens appeared to become frustrated, either because they had difficulty accessing the operant feeders, which were defended by conspecifics, or because they were overexcited by the very visible release of food falling to the ground. For these reasons, Lindberg and Nicol (1994) did not advocate the exclusive use of such feeders. However, when used in conjunction with the *ad libitum* provision of a mash diet, supplementary operant feeding could still have a role in keeping flocks productively occupied (Taylor et al., 2001).

The ability of chickens to associate their behavioural responses with specific outcomes means that operant tests have been used to measure the extent to which chickens will work to obtain or avoid specific resources and stimuli, an approach reviewed in Chapter 4. Chickens will work to obtain positive reinforcers such as food, perches, nests and foraging substrates. They will also work to achieve the removal of negative reinforcers such as noise or vibration. These situations are reinforcing because the chicken obtains a positive outcome or avoids a negative outcome by performing an active response. In contrast, punishing outcomes (e.g. pecking at an unpalatable or sickness-inducing food item) lead to a decrease in active behaviour. If chickens eat a novel coloured or textured food that results in sickness many hours later, they will subsequently avoid that food in the future (Martin and Bellingham, 1979), showing that just one encounter is sufficient to establish learning in this biologically important context. Similarly, young chicks take just one trial to learn *not* to peck at a bead coated with an aversive substance such as quinine or methyl anthranilate (MeA). This passive avoidance procedure forms the basis for much work examining the neural changes that accompany learning.

The brains of chicks can be examined before and after imprinting (see Chapter 3) or passive avoidance learning events have taken place. The clearly identifiable learning experience is associated with a chain of events in regions such as the lobus parolfactorius and the intermediate and medial mesopallium (previously termed the intermediate and medial hyperstriatum ventral) (Stewart *et al.*, 1987; Bourne *et al.*, 1991; Rose, 2000; McCabe, 2013). Increased glutamate release is followed by upregulation of dopamine-1 and N-methyl-D-aspartate (NMDA) glutamate receptors, the synthesis of synaptic membrane proteins, cell proliferation and increases in synaptic density in these brain regions (Steele *et al.*, 1995; Stewart *et al.*, 1996; Rose, 2000). Correlations between neural events, such as neuronal firing in the intermediate and medial mesopallium, and the strength of behavioural imprinting have been detected in individual chicks (Bradford and McCabe, 1994). More recently, upregulation of glutamate receptor subunits has been demonstrated in association with taste avoidance learning in older chickens (Atkinson *et al.*, 2008).

Generalization, Discrimination and Categorization

The ability of animals to categorize stimuli is an important question for animal intelligence as remarkable abilities, including concept formation, can emerge from these processes (Nicol, 1996). The basis of categorization is generalization and discrimination between stimuli. Generalized stimuli will all produce the same response, as occurs when chickens that have been habituated to one person by regular handling show reduced fear of other people (Jones, 1994).

In nature, chickens discriminate all the time. Some groupings are based on unlearned categories (e.g. moving versus static), but many are learned from the first few days of life onwards. Young chicks learn to discriminate between the calls of their mother and those of other hens because their mother's calls are associated with the reward of her visual appearance (Cowan, 1974). Evans (1972) showed that chicks would discriminate between two similar biologically relevant sounds (the author repeating the word 'cluck' or the word 'brupp') to regain access to a white disc with red markings on which they had been imprinted. Discrimination learning is also employed when learning what to eat and what to avoid. Learning to avoid a bitter or harmful substance is easier when the visual features of that substance are novel (Shettleworth, 1972). When familiar coloured water is contaminated with a sickness-inducing substance, chicks ingest lethal doses (Hayne *et al.*, 1996) and do not show food aversion learning of the type noted by Martin and Bellingham who used novel food types (1979). Avoidance learning is also easier when object features stand out from the background (Roper and Wistow, 1986). Chicks initially peck more at conspicuous than at cryptic distasteful food items, but subsequently they learn to avoid the contrasting items more quickly. Roper and Wistow (1986) proposed that this could explain why many natural but distasteful prey items have bright coloration. Some stimulus features seem to be naturally easier to learn than others due to long periods of co-evolution between predators and prey. Striped (wasp-like) patterns (Haugland *et al.*, 2006), bright colours (Aronsson and Gamberale-Stille, 2008) and the odour of pyrazine (Siddall and Marples, 2008) have all been shown to enhance learning and memory.

The ability of chickens to discriminate among stimuli has been used to investigate their sensory and perceptual abilities. We saw in Chapter 2 that discrimination tasks could be used to reveal visual thresholds of detection of light wavelength, flicker and acuity (Demello *et al.*, 1993; Prescott and Wathes, 1999; Jarvis *et al.*, 2002, 2009; Lisney *et al.*, 2011), auditory thresholds of detection (Temple *et al.*, 1984) and their perception of video images (Patterson-Kane *et al.*, 1997; Railton *et al.*, 2010). This research is not designed to explore the chickens' capacity to learn; rather, it uses learning as a way of uncovering the perceptual abilities of chickens as the sample and comparison stimuli become ever harder to detect or distinguish.

Successful discrimination of objects, sounds or individuals does, of course, depend on the relevant differences falling within the chickens' perceptual range. However, even within this range, there may be innate or learned preferences to pay more attention to some features than to others, and this can greatly influence success in discrimination tasks. In goal-detection tasks, individuals of many species predominantly and preferentially use spatial rather than object cues in discrimination. However, because chickens are highly visual foragers, they appear to form a combined representation of spatial and object features within a test arena. Even when trained to learn a spatial task, chicks remember and encode much information about the visual details of the objects in the environment (Vallortigara and Zanforlin, 1989), and the stability of their memory for feature cues is greater than for spatial cues (van Kampen and Devos, 1992). Chicks, especially females, preferentially use information about object features when information from spatial and object cues conflicts (Vallortigara *et al.*, 1990; Vallortigara, 1996). If landmarks are present within an enclosure, then chicks will preferentially use these local features to guide their food-searching behaviour (Della Chiesa *et al.*, 2006), making use of information from separate landmarks rather than extracting information from the overall shape of the landmark array (Pecchia and Vallortigara, 2010). Chicks are also better at encoding absolute information about stimulus colour than about stimulus size (Hauf *et al.*, 2008). However, in the absence of information from local features or objects, chicks can use 'pure' spatial information to find food. Once trained to find food in the centre of either a small (0.7 × 0.7 m) or large (1.4 × 1.4 m)

test arena, they can extrapolate the centre rule and use it to search the centre of the large or smaller arena not previously encountered (Tommasi and Vallortigara, 2000) and they can also encode information about the relative lengths and angles of walls (Tommasi and Polli, 2004). The coding of overall geometric information that allows a chick to detect the centre of an arena takes place in the right hemisphere. It is complemented by a degree of learning about where to search based on coding absolute distances from the wall in the training arena, and encoded in the left hemisphere.

The learning involved in discrimination tasks can be subtly different. In 'go/no-go' tests, both stimuli are presented together (for example, a lit and an unlit panel), with one stimulus designated as positive so that a peck to that panel results in a food reward (Prescott and Wathes, 1999). In 'two alternative choices' tests, a sample stimulus is presented and the chicken must make a choice between two responses based on the characteristics of this sample (for example, peck left if the sample is plain, peck right if the sample is grated; Demello *et al.*, 1993). In two alternative choices tests, attention to the original sample is important and can be improved by increasing the length of time the sample is present (Foster *et al.*, 1995) or by requiring chickens to make a number of pecks to the sample before proceeding to choose their response (Demello *et al.*, 1993). Discrimination performance in chickens can also be improved if the alternative responses are rewarded differently (Poling *et al.*, 1996). The learning required to solve these types of go/no-go and two alternative choices discrimination tasks is the relatively straightforward acquisition of an 'if... then' rule (Wright, 2001). Reward centres in the lobus parolfactorius brain region (homologous to the nucleus accumbens) are implicated in this type of rewarded discrimination learning (Matsushima *et al.*, 2003).

A further type of discrimination task, matching to sample (MTS), has the potential to tell us much more about chicken learning and cognitive abilities. In an MTS test, a sample stimulus is presented on a middle key or panel for a certain amount of time, or until it is pecked. In either case, two comparison stimuli are then revealed and the chicken must peck the comparator that matches the sample to obtain reinforcement. In the simplest form of this test, the criterion for matching is simple physical resemblance (for example, if the sample is green, peck the green comparator;

if sample is red, peck the red comparator), and this is well within the learning capability of laying hens (Foster *et al.*, 1995; Poling *et al.*, 1996; Nakagawa *et al.*, 2004). It was immediately apparent that, after a period of training, chickens could perform well on an MTS task using 'if ... then' learning of a limited number of rules, and this is probably the case in the studies mentioned. However, MTS tests can also be solved using more complex cognitive processing such that birds come to appreciate that there is an abstract distinction between stimuli that are the 'same as' or 'different from' the sample. The best way of deciding whether a bird is using high-level categorization is to present it with novel stimulus pairs. Unfortunately, this has not yet been investigated in chickens. Pigeons can use this degree of abstract cognition in MTS tests but only if they are given sufficient exposure to many initial training stimuli. If trained on only two stimulus pairs, then pigeons do not form a general 'same/different' category and do not perform above chance levels when presented with novel stimuli. However, when trained with 152 initial stimulus pairs, these birds are able to form a same/different category and apply it immediately to solving novel stimulus problems (Wright, 2001).

Working Memory and Object Permanence

One way of examining the working memory abilities of chickens is to develop the MTS test so that the initial sample stimulus disappears before the comparators appear. Laying hens with excellent discriminative performance at a delay of 0.25 s showed a slight drop in performance when the delay was increased to 2 or 8 s, and a further drop (to marginally above-chance levels) with a delay of 16 s (Nakagawa *et al.*, 2004). The delay between sample removal and comparator appearance is not the only influence – a delay in reward presentation after a comparator has been selected also reduces discrimination performance (Nakagawa *et al.*, 2004). As expected, retaining a memory for the sample stimulus becomes more difficult as a function of time since it was last seen. A more surprising finding is that training experience with particular delay intervals primes birds to pay more attention at these precise delay intervals in subsequent tests.

This type of working memory helps chickens in their appreciation that when a predator, prey or conspecific moves out of sight it is likely still to exist. Chicks of just a few days of age move readily towards hidden prey items or social partners. When imprinted on an artificial object such as a ball, chicks will successfully make detours around obstacles to rejoin the object (Regolin *et al.*, 1994, 1995). Having accomplished a detour once, they remember the route taken 24 h later (Regolin and Rose, 1999). Success in a simple detour task may be accomplished by using an evolutionary rule of thumb, such as going to the position where the companion was last seen (Freire and Nicol, 1999). However, it appears that a more detailed mental representation of the vanished object is formed. Young chicks will extrapolate the complete shape of an object that is partially hidden (Regolin and Vallortigara, 1995). Other evidence that chickens possess an object permanence capacity comes from studies showing that if an imprinting object disappears behind one of two screens of different sizes and shapes, chicks consistently choose the screen compatible with the presence of the object behind it. The chicks did not need to interact with the screens beforehand, suggesting that they could make intuitive inferences about possible and impossible hiding places (Chiandetti and Vallortigara, 2011).

When assessing memory using searching tasks that are perhaps more natural than an MTS procedure, it has been shown that chicks encode the location of a social stimulus more readily than that of a food reward (Regolin *et al.*, 1995). When an imprinted goal object is hidden behind one of two different opaque screens, chicks will search for and correctly locate the object after delays of 30 s or more (Vallortigara *et al.*, 1998; Regolin *et al.*, 2005a). As the delay increases from 10 to 180 s, correct responses remain higher for an imprinted goal object than for a food reward (Regolin *et al.*, 2005b). Even when chicks were unable to view the correct location during the delay period, their performance remained at above-chance levels over delays of 3 min when searching for the social goal (Regolin *et al.*, 2005b). As the authors state, this is an impressive demonstration of working memory for a young bird.

Chicks do not appear to be able to use their representation to predict where a moving object might reappear after vanishing behind a screen (Freire and Nicol, 1999). Perhaps their representation does not encode features such as movement, speed or direction, or perhaps the chicks in this study were reared under artificial conditions that did not permit full cognitive development. During development, chicks show a sudden peak in movement out of sight of their mother at about day 11 of life (Vallortigara et al., 1997), and experience of the disappearance and reappearance of objects during this sensitive phase for the development of spatial memory is crucial. Freire et al. (2004) imprinted chicks on a ball and then housed the chicks in pairs between 8 and 12 days of age in pens that provided either transparent or opaque barriers. Chicks reared with opaque barriers that enabled experience of object occlusion showed better retrieval of the hidden imprinting ball and made fewer errors in a spatial detour test.

Behavioural Flexibility

There is now good agreement that behavioural flexibility is one measure of intelligence (Roth and Dicke, 2005) as it has the potential to disprove the idea that chickens are working to simple rules of thumb. An apparently simple rule of thumb, for example 'when a predator is seen, give an alarm call', is revealed as an inadequate explanation for the sophisticated alarm-calling strategies we saw in Chapter 6. In that chapter, the extent to which chickens can adjust their referential communication to a range of specific and unique circumstances was reviewed and is highly relevant to questions of chicken intelligence. Consider the fact that non-broody hens do not give alarm calls to small hawks, whereas broody hens do. In addition, the broody hens reduce their alarm calling to small hawks as their chicks get older and bigger, but they do not reduce their alarm calling to medium-sized hawks that remain a threat (Palleroni et al., 2005). Such flexible adjustment of behaviour requires mental processing of environmental cues, the individual's needs and the task posed, and is one example of what we might call 'thought' in humans. Another example of the broody hen's ability to use the information around her to produce a context-specific behaviour was provided by Nicol and Pope (1996). In their study, 24 hens were housed together with their chicks, except at meal times when they were fed separately. The mother hens were taught that food within a feeder of a certain colour was palatable, while the food in a feeder of a different colour was unpalatable. The chicks were offered the same

options, but for some chicks the colours were the opposite to those of their mothers. This meant that they were trained to eat from the feeder that the mother hen was trained to avoid. Once this had been learned, the mother observed her chicks through wire mesh while they were eating. The mother went to one colour while in a separate pen the chicks went to the other. When the mother hen observed her chicks eating unpalatable food, her maternal display increased in response to her assessment of their mistake (Nicol and Pope, 1996). To assess the chicks' choice of food, relate it to what she knows and then encourage the chicks to change their feeding preference requires great cognitive ability as the hen must synthesize all of this information and respond in a scenario that she has not encountered previously. It is even possible that the hens attempt to 'teach' their chicks what is correct based on some understanding of the perspective of her offspring, although further work would be required to establish this.

Using a related experimental design (Edgar et al., 2011), broody hens and their chicks were trained independently and separately that a compartment of one colour was associated with the occurrence of an air puff every 30 s for 5 min, while an identical compartment of a different colour was safe. Broody hens were then given the opportunity to observe their chicks in compartments of each colour. Hens showed increased agitation (indicated by increased walking and reduced preening) and also performed vocalizations for a greater proportion of time when they perceived their chicks to be threatened compared with when they thought the chicks were safe. The fascinating aspect of this result, and one that mirrors the results of Nicol and Pope (1996), is that the hens responded to chicks that they perceived to be threatened even when those same chicks had been trained that they were safe (Edgar et al., 2013). This shows that the hen's response to chick distress is not primarily mediated as a direct response to distress cues from her chicks but is instead controlled by her own knowledge, which is then applied to a novel situation.

Representation, Expectation and Abstract Thought

The object permanence abilities of chickens hint at their capacity to form mental representations.

One corollary of this is that chickens may be able to develop expectations about their environment in a way that would allow a rapid response to change. For example, if chickens expect to find high-quality food at the end of a runway but then unexpectedly encounter poor-quality food, two outcomes are possible. If the chickens rely on associative learning processes, they will gradually traverse the runway more slowly as the association between a rapid approach and obtaining a high reward weakens. However, a mental representation of the reward could be updated after the first mismatch is detected, and the chickens could slow or halt their runway performance on the very next trial. Although Haskell *et al.* (2001) found no evidence of an immediate response to downgraded food reward mediated by expectations in broiler chickens, more recent work on cognitive bias suggests that chickens can form expectancies under some conditions. The fact that ambiguous stimuli are interpreted more negatively after a short period of isolation suggests an emotional component to the expectations formed by young chicks (Salmeto *et al.*, 2011).

Mathematics and Physics

Competence in symbolic mathematics is the preserve of (most) adult humans, but it will surprise many to learn that chickens possess some kinds of numerical abilities. Not only that, but these numerical abilities appear almost immediately after hatching, without any training or formal education. Perhaps the simplest ability is being able to discriminate small sets of identical stimuli where sets differ only in number. One of the first studies to investigate this in chickens found that chicks could distinguish stimulus sets of one versus two, and two versus three on the basis of number, i.e. the experimenters were careful to control for the overall size and area of the stimuli available so that chicks could not simply choose the set with the biggest surface area (Rugani *et al.*, 2008).

In this first study, chicks appeared to have difficulty in distinguishing larger sets, such as four versus five or four versus six, on the basis of number, although they could do so by using overall size and area cues. However, a second study showed that chicks can keep track of numbers up to five. In this study, chicks were imprinted on a set of five

identical objects (plastic Kinder Surprise capsules), which they were subsequently very motivated to follow. The experimenters then asked what the chicks would do if the set of five capsules was split into a subset of three and a subset of two. Almost without exception, the chicks would approach the larger subset, even when both were hidden behind screens. This allowed the experimenters to probe the chicks' numerical abilities in more depth. Initially, the chick observed its set of five capsules split into a subset of four, which moved behind one screen, and a subset of one, which moved behind a second screen. Following this, two capsules were visibly moved from behind screen A to behind screen B, so that just two remained behind screen A, with three now behind screen B. Amazingly, chicks were able to compute the end result of these various manoeuvres, and when allowed to choose which screen to approach, they headed for screen B with the larger number of capsules. This shows not simply number discrimination ability but a capacity for basic arithmetic (Rugani *et al.*, 2009). Unfortunately, their arithmetic abilities seem limited to these small numbers as chicks fail in similar tasks with larger numbers, such as six versus nine (Rugani *et al.*, 2011), unless quantitative cues (cumulative area of volume of the stimuli) are also available. However, when numbers differ by a larger ratio, such as five versus ten or ten versus 20, then the problems can be solved with or without the presence of quantitative cues (Rugani *et al.*, 2013).

A slightly different numerical ability is the ability to order numbers, something else that very young chicks can master (Rugani *et al.*, 2007). Using apparatus that offered the chicks ten holes in a line with only one containing food, the ability of the chicks to learn the correct hole in the sequence was studied. On day 4 of their lives, they were trained to peck a hole that offered food to them, and this was either the third, fourth or sixth hole in the apparatus (depending on the chick's reinforcement position). All other holes offered no food. On the test day, on day 5 of their lives, the chicks were placed into the apparatus and underwent a test to discover whether they could immediately choose the correct hole in the line. After the first peck, the test was considered to be over. The chicks rapidly approached the third, fourth or sixth position. Further experiments ruled out the use of distance cues, showing that chicks possess an ability to sequence numbers. More recently, it has been shown that, as in humans, chicks seem to represent their number line as proceeding from left to right. In a similar test to the above where chicks had to find food at the fourth or sixth position in a line of 16, they could do so if they started from the left but not if they started from the right (Rugani *et al.*, 2010). Kundey *et al.* (2010) examined whether adult chickens could use sequencing to predict when a particularly large or small food reward might be available on repeated trials. They found that when food amounts varied monotonically (e.g. in the order four, three, two, one) chickens significantly slowed their running speed on the fifth trial, correctly predicting the zero reward, compared with situations where the food amount varied (four, one, three, two), or where a larger amount of food was available (eliminating satiety as an explanation for slower running) but where the sequence was weaker (four, three, three, one).

It also seems that chicks are born not only with a rudimentary understanding of numbers but also an innate ability in physics, particularly structural engineering. They show a significant preference to approach a two-dimensional diagram of a possible object (a cube with shadows indicating realistic structural elements) rather than a diagram of an impossible object (a cube with inconsistent structural information) (Regolin *et al.*, 2011) (Fig. 7.1).

Fig. 7.1. Newly hatched, naïve chicks are presented with two-dimensional images that could potentially be perceived as three-dimensional objects. The chicks show a spontaneous preference for the image of the structurally 'possible' cube over the image of the structurally 'impossible' cube. (Photo courtesy of Professor Lucia Regolin, University of Padova, Italy.)

Spatial Cognition, Orientation and Navigation

The spatial abilities of chicks in object permanence, arena goal-finding and detour tests have already been described. However, Lipp *et al.* (2001) found that chickens were not as proficient as rats or crows at locating food in the arms of a large octagonal maze when consistent external maze cues were provided, although they did learn the task and performed better than guinea fowl or geese (Pleskacheva *et al.*, 2001). Comparisons of species performance in tests of this kind are also fraught with confounding factors to do with perceptual, locomotor and motivational differences. In many birds, resource location over small distances is guided by a hierarchy of preference for global, landmark and local feature cues, not all of them visual (Chappell and Guilford, 1995). Thus, chickens may not have performed to their full capability in maze tests because their preferred cues were not available.

In many bird species, the sun provides the preferred global cue used for resource location. To examine whether chickens could use information from the sun's position to locate food, Zimmerman *et al.* (2003) designed an experiment where the sun was the only consistent cue available. An eight-armed maze was constructed and placed in an outdoor location away from external landmark cues. Chickens had to find food placed in just one arm of the maze in a compass direction that was consistent within individuals but that varied among subjects. Every day the maze was placed in a different part of the field and was rotated. Food residue was placed in pots in all arms so that direct visual or olfactory cues associated with food could not be used to solve the task. When tested on sunny days, seven out of eight birds consistently found food with a high success rate. As the tests were conducted at different times of day, this could mean that the chickens were using the sun as a time-compensated compass (accounting for its change in azimuth over the day). Alternatively, they may have used the magnetic sense that was described in Chapter 2. To distinguish which cues were being used, Zimmerman *et al.* (2009) performed two further manipulations. First, they subjected chickens to a clock-shift procedure by housing them indoors on a light schedule that resulted in them being 6 h ahead of normal time. If, for example, one of these chickens was trained to find food due south at 6 a.m., after clock shifting it would consider the time of day to be 12 noon, and would expect the sun to be due south and should head towards the sun. In reality, at this time of day the sun will still be in the east and so the chicken will be wrong in its orientation by around 90°. The second manipulation was to fit the chickens with small (2 × 3 mm) powerful magnets glued to their head feathers. The clock-shifting procedure disrupted the birds' goal orientation (although not quite as neatly as predicted by the theory), but the magnetic disruption had no effect on goal orientation. This suggests that chickens are able to account for the changing position of the sun throughout the day and use this as a cue in local navigation tasks.

Social Learning

Social learning is an important and widely used cognitive ability in group-living chickens and is a way in which they can avoid the mistakes and dangers of individual associative learning when in a novel environment or situation. Chickens often use their observations of conspecifics to guide their behaviour and avoid the cost of the trial and error of individual learning (Nicol, 2004).

In Chapters 3 and 6, we saw how chicks are sensitive to social guidance about which foods they should and should not ingest, and we described the role of the mother hen in attracting their young to eat the correct food. As they get older, chicks rely more on their flockmates for social learning. Day-old chicks will avoid pecking at an aversive stimulus after observing the disgust responses of another chick (Johnston et al., 1998). Young chicks may be particularly attentive to the behaviours of conspecifics and less attentive to their own experiences of ingestion than older birds. Sherwin et al. (2002) found that pullets obtained at 9 weeks of age did not learn to avoid an unpalatable food by watching a demonstrator showing a disgust reaction.

As chicks become more independent, feeding influences shift towards a dual role for both social transmission and individual associative learning. Gajdon et al. (2001) seeded a test arena with hidden food, which could be located by noting the location of coloured markers. They found that groups of chicks containing a knowledgeable

demonstrator developed more successful foraging behaviour than groups of chicks with a naïve demonstrator with no prior knowledge of the significance of the coloured markers. Junglefowl chicks of a few weeks of age will approach specific marked food bowls 2 days after having observed conspecifics (or videos of conspecifics) feeding from the same type of dish or in the same location (McQuoid and Galef, 1992, 1993). Sherwin et al. (2002) used a circular apparatus that allowed eight observers at one time to view the feeding behaviour of a demonstrator placed at the centre, and exposed observers to a demonstrator not feeding, feeding on normal food or feeding on 'highly palatable' feed. There was no treatment effect on the total amount of food consumed by observers, but there was a positive correlation between the pecking rate of the demonstrators and the proportion of red-coloured feed consumed. As the observers otherwise tended to avoid red food, this suggested that social learning might have a role in overcoming unlearned aversions to particular colours (Nicol, 2004).

So far, we have reviewed how social factors influence the development of natural feeding behaviour in young birds. The possibility that chickens might learn new or novel patterns of behaviour via social learning was examined in an operant experiment by Johnson et al. (1986). They found that chickens that had observed a trained demonstrator responded earlier and more frequently on a subsequent auto-shaping operant test than chickens that had observed only correlations between key-light and food-hopper operation. However, this experiment did not include a control for the mere presence of another chicken during observation sessions, which may have reduced fear levels.

Nicol and Pope (1992) showed that, having watched a trained demonstrator, naïve hens were better able to learn the behaviour they had just seen than naïve hens that had watched untrained demonstrators (controlling for mere presence effects) or no demonstrator. This was done using a duplicate cage set-up where both hens could view each other through a Perspex partition (Fig. 7.2). The experienced demonstrator performed a key-pecking discrimination task to obtain food, and the observer was able to translate what she had seen to produce the same behaviour herself.

In a second experiment, the effects of observing a live, trained demonstrator were compared with those of observing an artificial rod 'pecking'

Fig. 7.2. Duplicate cage box used in social learning experiments. The demonstrator hen on the left is trained to peck one of the coloured keys to obtain food. The observer watches. When the observer is moved to the left-hand chamber, she is much faster at doing the task than hens that have only watched an empty box or an untrained demonstrator. (Photo courtesy of Stuart Pope, University of Bristol, UK.)

the key and opening the door to the food hopper. This showed that birds that had observed the rod appeared to learn nothing about the task at hand, and that a social model is critical (Nicol and Pope, 1992). It might be thought that removal of the artificial barriers separating demonstrators and observers could facilitate social learning, with observers obtaining a better view of demonstrator behaviour. However, in hens, partition removal simply meant that the response rates of the demonstrators declined as they spent time defending the area around the pecking key (Nicol and Pope, 1999).

The identity of the social demonstrator is highly relevant in the context of social learning (Nicol and Pope, 1994). Using dominant, subordinate or unfamiliar demonstrators, it was found that observers learned the most when viewing the dominant bird complete a task. This was initially thought to be because dominance was correlated with success in tasks such as foraging. However, it was later shown that dominant hens foraged no better than any other type of hen (Nicol and Pope, 1999), and no difference has been detected in the ability of dominant and subordinate chickens to learn a simple visual discrimination task (Croney *et al.*, 2007). Furthermore, prior foraging

success had no significant effect on the influence of birds as demonstrators (Nicol and Pope, 1999). An alternative reason for the greater effectiveness of dominant hens might be that they simply provide a more striking or noticeable presence during the performance of specific behaviours. For example, dominant birds might be bigger, adopt a taller body posture or peck with greater force. However, this line of reasoning was not supported by the inclusion, in a second experiment (Nicol and Pope, 1999), of trained cockerels as demonstrators. The cockerels were larger and appeared to peck the key more forcefully, but this resulted in little social learning by the hens. The influence of the dominant hens is therefore more likely to be due to the greater attention paid to them by subordinates seeking to avoid confrontation.

Planning Ahead

The question of whether an animal is able to think about the past or the future is of great importance when considering their intelligence. Without such abilities, the individual lives only in the present and will maximize its immediate benefit at the cost of a possibly greater future benefit. If the animal has the ability to consider its actions and their consequences, it has the opportunity to make larger gains in the long run. This requires an individual to perceive and process time and apply this to its current situation; furthermore, it must impose a certain amount of self-control on itself. In children, it has been shown that a high degree of self-control is correlated with cognitive competence later in life (Metcalfe and Mischel, 1999). After listening to an adult explain that waiting will result in more sweets or better crayons, some children will dutifully sit out the waiting period while others will choose instead to take a smaller immediately available reward. Children are better at waiting as they age, and their ability also increases with a greater reward.

Experiments testing animals define self-control as choosing a large reward available only after a 'long' delay over a small reward available straight away or after a 'short' delay (Logue, 1988). The first relevant work in this area was by Taylor et al. (2002), who showed that hens were able to predict a time interval when given a reliable signal. The hens would view a computer screen, which would indicate the start of the fixed interval of 6 min. The first hen peck after this time had elapsed resulted in a light coming on, the screen going off and the food being delivered. When testing the hens to see if they had learned the time at which they would receive a reward, it was found that pecking increased gradually, reaching a maximum at 6min. This revealed that domestic hens could estimate the occurrence of an event several minutes in advance of the first signal. Building on this, the next step was to show that hens could not only estimate the passing of time but also show self-control to receive a greater reward. Using a set-up that gave hens the option of two keys – one that had a 2 s delay followed by 3 s of access to a feeder and another that had a 6 s delay followed by 7 s of access to a feeder – Abeyesinghe et al. (2005) found that only 22% of hens showed the self-control needed to gain access to the better resource. This suggested that the hens were not showing restraint and were choosing the key that gave them the immediate reward. However, when the experimenters upped the reward for a 'jackpot' group that had a 6 s delay followed by 22 s of access to a feeder, 93% of hens chose to exhibit self-control and receive the larger reward. This provided clear evidence

that self-control was possible in hens while also demonstrating the sophisticated trade-off calculations made by the hens in deciding what level of reward was worthwhile.

Transitive Inference: The Logical Chicken

In Chapter 6, we reviewed the ways in which chickens assess social status and develop dominance hierarchies when housed in small groups where individual recognition is a possibility. In a revealing study, it was shown that when a hen sees the result of a conspecific's fight with a stranger, it evaluates its own potential to defeat the stranger from what it has seen (Hogue et al., 1996). With this ability, an individual has the huge benefit of avoiding the risks of fights that it is likely to lose. Such reasoning may occur through transitive inference, where if you know A>B and B>C, you know that A>C. The study set up different conditions for assessment of a bystander's social learning ability. The first showed individuals their prior dominant being defeated by a stranger and then introduced them to that stranger. The second showed them their prior dominant defeating a stranger. In the first condition, the hen should infer that, because the dominant hen was defeated, they would also be defeated. This was exactly the case, with none of the bystanders initiating an attack on this 'powerful' stranger. If the stranger initiated an attack, then these hens would immediately show signs of submission. In the second condition, the bystander cannot be sure whether it is able to defeat the stranger or not. It could be said that it has a 50/50 chance of winning or losing. The hens were found to attack the stranger 50% of the time and win 50% of the time. These conditions both showed clearly that the bystanders were able to learn from what they had witnessed and act accordingly. Keeping track of the social position of others within the group requires significant cognitive ability, and it seems that this ability may be underpinned by a degree of logical reasoning.

Daisley et al. (2010) recently investigated the logical abilities of chickens using a learning task where chicks were trained to peck at a small visual stimulus to receive a food reward. The chicks were then presented with paired presentations of stimuli that would permit them to learn that the stimuli could be ranked in a hierarchy. For example, a dot could be rewarded over a square, a square over a diamond, a diamond over a cross and a cross over a star, thus A>B>C>D>E.

However, the chicks were only ever trained with adjacent pairs e.g. A>B or C>D. A test of the chicks' reasoning ability came when untrained pairs of stimuli were presented for the first time. When A and E were presented together, the chicks chose A, but this could be because A had always been associated with a food reward and E had never been associated with food in previous training. A tougher test of reasoning came from presenting the chicks with the stimuli B and D, both of which had been equally rewarded and unrewarded in the past. To solve this problem, the chicks needed to appreciate and remember the original stimulus hierarchy, something they did at significantly better than chance levels.

It is also worth noting that chickens show transitivity in their long-term preferences. If they prefer environment A over B, and B over C, they also tend to prefer A over C (Browne et al., 2010). In this case, formal logic does not have to be involved, as the environments differ in many respects and the chickens simply need to be consistent in their preferences. It does, however, show that chickens know what they want and then stick to it, sometimes over many months of testing.

Environmental Influences on Chicken Cognition

Considerable work in mammals has shown that development within a more complex environment results in a more complex, flexible brain, even one that is more resistant to developing stereotypic behaviour (Tanimura et al., 2008). The same depth of work does not exist for chickens, but the results that are available suggest that a complex environment also enhances brain development and cognition in chickens. Pre-hatch auditory stimulation with chicken vocalizations or with music (specifically, in one study, the sitar) led to increased hippocampal development and synaptic density, as well as improved spatial learning performance in a T-maze (Chaudhury et al., 2010; Kauser et al., 2011). When chicks were reared in cages either with or without many wooden

egg-shaped and cube-shaped objects and were compared with chicks reared in semi-natural pens, no difference in discrimination ability was detected (although all chicks detected size differences in cubes more easily than size differences in eggs) (Zoeke et al., 1980). Krause et al. (2006) found that just 1 week's access to a simple grass outdoor area increased the speed of learning by young layer-strain pullets in a Y-maze test, although it had no effect on their subsequent memory for the task. The effects of environmental complexity on neural development are, surprisingly perhaps, not limited to young birds. In one study, hens housed in free-range systems from the point of lay developed larger cells in some hippocampal brain regions and a greater fibre density in their serotonin pathways than caged hens. These changes were described as relatively mild, but it is significant that they occurred at all (Patzke et al., 2009).

Performance in learning or cognition tasks can be greatly affected by the internal motivational state or fear level of the chickens. In many learning experiments, chickens are mildly food deprived because motivational state partially determines the stimuli to which animals will attend (Dorrance and Zentall, 2001). However, in a social learning context, food deprivation seems to result in chickens paying more attention to the food than to the method being used to obtain that food and can therefore be counterproductive (Nicol and Pope, 1993). Chronic and stressful levels of hunger may impair learning more seriously. Feed-restricted broiler breeders were unable to associate simple black and white cues with differential food rewards (Buckley et al., 2011). Severely restricted birds performed worse, but all subjects had low success rates (in marked contrast to the success of non-food-deprived chicks and laying-strain birds on a variety of visual discrimination tasks outlined above).

A study by Nordquist et al. (2011) studied two lines of hens that differed in their basic fear and anxiety levels (assessed in an open-field test and in a test of readiness to approach a human) and observed how this affected T-maze performance. The less fearful hens approached a human more readily, were far more likely to leave the T-maze start box before the 10 min testing period had elapsed, and were significantly more likely to enter the rewarded compartment. Thus, the less fearful birds had a greater exploratory tendency and showed a greater ability to navigate the maze. Similarly, stress induced by housing under an unpredictable light/dark schedule impaired spatial learning in both white leghorn chickens and red junglefowl (Lindqvist and Jensen, 2009).

Welfare Implications

It may seem that studying chicken learning and cognition has little relevance to practical questions about their welfare, but nothing could be further from the truth. This knowledge can help us identify the situations in which chickens may suffer and can help us to improve our techniques of assessing welfare, design better housing systems and control unwanted behaviour. For example, if a hen has no appreciation of numbers above four or five, she is unlikely to suffer if one of her ten chicks is removed, provided of course that she does not react to a reduction in the overall mass, visual area occupied, heat or sound generated by her brood. A chicken that can remember events over a period of many days or weeks has at least the potential to suffer from memories of past ill treatment. The demonstrated ability of chickens to form mental representations and expectations only increases the possibility that they may suffer from the absence of resources such as nests or dustbathing substrates as was proposed in Chapter 4. Knowledge of the way in which chickens integrate different cues to orient and navigate could be used to facilitate a more uniform distribution of birds in large houses or free-range systems, and much greater use could be made of predictive cues

that would make chicken husbandry procedures more predictable and hence less stressful.

The capacity of chickens to both learn by observation and, to an extent that we are only beginning to discern, influence each other's emotional state has important practical implications. Behaviours such as smothering, feather pecking and cannibalism often spread rapidly. In the case of feather pecking, social learning may play a part in its transmission within a flock (Zeltner et al., 2000), although this study did not exclude the possibility that clusters of birds were exposed to the same environmental causes, or direct effects (e.g. pecked feathers are more attractive for further pecking). However, in a controlled experiment, Cloutier et al. (2002) found that the tendency to peck through a membrane to access blood was facilitated by observational learning in hens. Consideration should be given to the placement of visual barriers that could narrow this route of transmission.

Finally, studies of chicken learning, cognition and intelligence can also help us rethink our ethical stance and relationship with these birds. Once we appreciate that these birds do not simply respond to their environments, to each other or to us with a set of simple, fixed or 'unthinking' responses, we may decide that they merit a different position within an ethical framework. We may admire and appreciate their complexity just as we might admire great paintings or diverse and complex landscapes, and perhaps accord them greater respect or protection on these grounds alone. Additionally, it may be that some aspects of learning and cognition turn out to be linked to the capacity to suffer.

References

Abeyesinghe, S.M., Nicol, C.J., Hartnell, S.J. and Wathes, C.M. (2005) Can domestic fowl, *Gallus gallus domesticus*, show self-control? *Animal Behaviour* 70, 1–11.

Alpert, M., Schein, M.W., Beck, C.H. and Warren, M. (1962) Learning set formation in young chickens. *American Zoologist* 2, 502.

Aronsson, M. and Gamberale-Stille, G. (2008) Domestic chicks primarily attend to colour not pattern when learning an aposematic coloration. *Animal Behaviour* 75, 417–423.

Atkinson, R., Migues, P.V., Hunter, M. and Rostas, J.A.P. (2008) Molecular changes in the intermediate medial mesopallium after a one trial avoidance learning in immature and mature chickens. *Journal of Neurochemistry* 104, 891–902.

Bourne, R.C., Davies, D.C., Stewart, M.G., Csillag, A. and Cooper, M. (1991) Cerebral glycoprotein synthesis and long-term memory formation in the chick (*Gallus domesticus*) following passive-avoidance training depends on the nature of the aversive stimulus. *European Journal of Neuroscience* 3, 243–248.

Bradford, C.M. and McCabe, B.J. (1994) Neuronal activity related to memory in the intermediate and medial part of the hyperstriatum ventrale of the chick brain. *Brain Research* 640, 11–16.

Browne, W.J., Caplen, G., Edgar, J., Wilson, L.R. and Nicol, C.J. (2010). Consistency, transitivity and inter-relationships between measures of choice in environmental preference tests with chickens. *Behavioural Processes* 83, 72–78.

Buckley, L.A., McMillan, L.M., Sandilands, V., Tolkamp, B.J., Hocking, P.M. and D'Eath, R.B. (2011) Too hungry to learn? Hungry broiler breeders fail to learn a Y-maze food quantity discrimination task. *Animal Welfare* 20, 469–481.

Chappell, J. and Guilford, T. (1995) Homing pigeons primarily use the sun compass rather than fixed directional visual cues in an open-field arena food-searching task. *Proceedings of the Royal Society B – Biological Sciences* 260, 59–63.

Chaudhury, S., Jain, S. and Wadhwa, S. (2010). Expression of synaptic proteins in the hippocampus and spatial learning in chicks following prenatal auditory stimulation. *Developmental Neuroscience* 32, 114–124.

Chiandetti, C. and Vallortigara, G. (2011) Intuitive physical reasoning about occluded objects by inexperienced chicks. *Proceedings of the Royal Society B – Biological Sciences* 278, 2621–2627.

Cloutier, S., Newberry, R.C., Honda, J. and Alldredge, J.R. (2002) Cannibalistic behaviour spread by social learning. *Animal Behaviour* 63, 1153–1162.

Cowan, P.J. (1974) Selective responses to parental calls of different individual hens by young *Gallus gallus* – auditory discrimination learning vs auditory imprinting. *Behavioural Biology* 10, 541–545.

Croney, C.C., Prince-Kelly, N. and Meller, C.L. (2007) A note on social dominance and learning ability in the domestic chicken (*Gallus gallus*). *Applied Animal Behaviour Science* 105, 254–258.

Daisley, J.N., Vallortigara, G. and Regolin, L. (2010). Logic in an asymmetrical (social) brain: transitive inference in the young domestic chick. *Social Neuroscience* 5, 309–319.

Della Chiesa, A., Pecchia, T., Tommasi, L. and Vallortigara, G. (2006) Multiple landmarks, the encoding of environmental geometry and the spatial logics of a dual brain. *Animal Cognition* 9, 281–293.

Demello, L.R., Foster, T.M. and Temple, W. (1993) The effect of increased response requirements on discriminative performance of the domestic hen in a visual-acuity task. *Journal of the Experimental Analysis of Behavior* 60, 595–609.

Dorrance, B.R. and Zentall, T.R. (2001) Imitative learning in Japanese quail (*Coturnix japonica*) depends on the motivational state of the observer quail at the time of observation. *Journal of Comparative Psychology* 115, 62–67.

Duncan, I.J.H. and Hughes, B.O. (1972) Free and operant feeding in domestic fowls *Animal Behaviour* 20, 775–777.

Edgar, J.L., Lowe, J.S., Paul, E.S. and Nicol, C.J. (2011) Avian maternal response to chick distress. *Proceedings of the Royal Society B – Biological Sciences* 278, 3129–3134.

Edgar, J.L., Paul, E. and Nicol, C.J. (2013) Protective mother hens: cognitive influences on the maternal response. *Animal Behaviour* 86, 223–229.

Evans, R.M. (1972) Development of an auditory discrimination in domestic chicks (*Gallus gallus*). *Animal Behaviour* 20, 77–82.

Foster, T.M., Temple, W., MacKenzie, C., Demello, L.R. and Poling, A. (1995) Delayed matching-to-sample performance of hens – effects of sample duration and response requirements during the sample. *Journal of the Experimental Analysis of Behavior* 64, 19–31.

Freire, R. and Nicol, C.J. (1999) Effect of experience of occlusion events on the domestic chick's strategy for locating a concealed imprinting object. *Animal Behaviour* 58, 593–599.

Freire, R., Cheng, H.W. and Nicol, C.J. (2004) Development of spatial memory in occlusion-experienced domestic chicks. *Animal Behaviour* 67, 141–150.

Gajdon, G.K., Hungerbuhler, N. and Stuffacher, M. (2001). Social influence on early foraging of domestic chicks (*Gallus gallus*) in a near-to-nature procedure. *Ethology* 107, 913–937.

Haskell, M.J., Vilarion, M., Savina, M., Atamna, J. and Picard, M. (2001) Do broiler chicks have a cognitive representation of food quality? Appetitive, behavioural and ingestive responses to a change in diet quality. *Applied Animal Behaviour Science* 72, 63–77.

Hauf, P., Prior, H. and Sarris, V. (2008) Generalization gradients and representational modes after absolute and relative discrimination learning in young chickens. *Behavioural Processes* 93, 93–99.

Haugland, K., Hagen, S.B. and Lampe, H.M. (2006) Responses of domestic chicks (*Gallus gallus domesticus*) to multimodal aposematic signals. *Behavioural Ecology* 17, 392–398.

Hayne, H., Rovee-Collier, C., Collier, G., Tudor, L. and Morgan, C.A. (1996) Learning and retention of conditioned aversions by freely feeding chicks. *Developmental Psychobiology* 29, 417–431.

Hogue, M.E., Beaugrand, J.P. and Laguë, P.C. (1996) Coherent use of information by hens observing their former dominant defeating or being defeated by a stranger. *Behavioural Processes* 38, 241–252.

Hughes, B.O. and Duncan, I.J.H. (1988) Behavioral needs – can they be explained in terms of motivational models. *Applied Animal Behaviour Science* 19, 352–355.

Jarvis, J.R., Taylor, N.R., Prescott, N.B., Meeks, I. and Wathes, C.M. (2002) Measuring and modelling the photopic flicker sensitivity of the chicken (*Gallus g. domesticus*). *Vision Research* 42, 99–106.

Jarvis, J.R., Abeyesinghe, S.M., McMahon, C.E. and Wathes, C.M. (2009) Measuring and modelling the spatial contrast sensitivity of the chicken (*Gallus g.domesticus*). *Vision Research* 49, 1448–1454.

Johnson, S.B., Hamm, R.J. and Leahey, T.H. (1986) Observational learning in *Gallus gallus domesticus* with and without a conspecfic model. *Bulletin of the Psychonomic Society* 24, 237–239.

Johnston, A.N.B., Burne, T.H.J. and Rose, S.P.R. (1998) Observational learning in day-old chicks using a one-trial passive avoidance learning paradigm. *Animal Behaviour* 56, 1347–1353.

Jones, R.B. (1994) Regular handling and the domestic chick's fear of human beings: generalisation of response. *Applied Animal Behaviour Science* 42, 129–144.

Kauser, H., Roy, S., Pal, A., Sreenivas, V., Mathur, R., Wadhwa, S. and Jain, S. (2011) Prenatal complex rhythmic music sound stimulation facilitates postnatal spatial learning but transiently impairs memory in the domestic chick. *Developmental Neuroscience* 33, 48–56.

Krause, E.T., Naguib, M., Trillmich, F. and Schrader, L. (2006) The effects of short term enrichment on learning in chickens from a laying strain (*Gallus gallus domesticus*). *Applied Animal Behaviour Science* 101, 318–327.

Kundey, S.M.A., Strandell, B., Mathis, H. and Rowan, J.D. (2010) Learning of monotonic and nonmonotonic sequences in domesticated horses (*Equus caballus*) and chickens (*Gallus domesticus*). *Learning and Motivation* 41, 213–223.

Lindberg, A.C. and Nicol, C.J. (1994) An evaluation of the effect of operant feeders on welfare of hens maintained on litter. *Applied Animal Behaviour Science* 41, 211–227.

Lindqvist, C. and Jensen, P. (2009) Domestication and stress effects on contrafreeloading and spatial learning performance in red jungle fowl (*Gallus gallus*) and White Leghorn layers. *Behavioural Processes* 81, 80–84.

Lipp, H.P., Pleskacheva, M.G., Gossweiler, H., Ricceri, L., Smirnova, A.A., Garin, N.N., Perepiolkina, O.P., Voronkov, D.N., Kuptsov, P.A. and Dell'Omo, G. (2001) A large outdoor radial maze for comparative studies in birds and mammals. *Neuroscience Review* 25, 83–99.

Lisney, T.J., Rubene, D., Rozsa, J., Lovlie, H., Hastad, O. and Odeen, A. (2011). Behavioural assessment of flicker fusion frequency in chicken *Gallus gallus domesticus*. *Vision Research* 51, 1324–1332.

Logue, A.W. (1988) Research on self-control: an integrating framework. *Behavioral and Brain Sciences* 11, 665–679.

Martin, G.M. and Bellingham, W.P. (1979) Learning of food aversions by chickens (*Gallus gallus*) over long delays. *Behavioural and Neural Biology* 25, 58–68.

Matsushima, T., Izawa, E.I., Aoki, N. and Yanagihara, S. (2003) The mind through chick eyes: memory, cognition and anticipation. *Zoological Science* 20, 395–408.

McCabe, B.J. (2013) Imprinting. *Wiley Interdisciplinary Reviews – Cognitive Science* 4, 375–390.

McQuoid, L.M. and Galef, B.G. (1992) Social influences on feeding site selection by Burmese fowl (*Gallus gallus*). *Journal of Comparative Psychology* 106, 137–141.

McQuoid, L.M. and Galef, B.G. (1993) Social stimuli influencing feeding behaviour of Burmese fowl: a video analysis. *Animal Behaviour* 46, 13–22.

Metcalfe, J. and Mischel, W. (1999) A hot/cool-system analysis of delay of gratification: dynamics of willpower. *Psychological Review* 106, 3–19.

Nakagawa, S., Etheredge, R.J.M., Foster, T.M., Sumpter, C.E. and Temple, W. (2004) The effects of changes in consequences on hens' performance in delayed-matching-to-sample tasks. *Behavioural Processes* 67, 441–451.

Nicol, C.J. (1996) Farm animal cognition. *Animal Science* 62, 375–391.

Nicol, C.J. (2004) Development, direction and damage limitation: social learning in domestic fowl. *Learning and Behavior* 32, 72–81.

Nicol, C.J. and Pope, S.J. (1992) Effects of social learning on the acquisition of discriminatory keypecking in hens. *Bulletin of the Psychonomic Society* 30, 293–296.

Nicol, C.J. and Pope, S.J. (1993) Food deprivation during observation reduces social learning in hens. *Animal Behaviour* 45, 193–196.

Nicol, C.J. and Pope, S.J. (1994) Social learning in small flocks of laying hens. *Animal Behaviour* 47, 1289–1296.

Nicol, C.J. and Pope, S.J. (1996) The maternal feeding display of domestic hens is sensitive to perceived chick error. *Animal Behaviour* 52, 767–774.

Nicol, C.J. and Pope, S.J. (1999) The effects of demonstrator social status and prior foraging success on social learning in laying hens. *Animal Behaviour* 57, 163–171.

Nordquist, R., Heerkens, J.L.T., Rodenburg, T.B., Boks, S., Ellen, E.D. and van der Staay, F.J. (2011) Laying hens selected for low mortality: behaviour in tests of fearfulness, anxiety and cognition. *Applied Animal Behaviour Science* 131, 110–122.

Palleroni, A., Hauser, M. and Marler, P. (2005) Do responses of galliform birds vary adaptively with predator size? *Animal Cognition* 8, 200–210.

Patterson-Kane, E., Nicol, C.J., Foster, T.M. and Temple, W. (1997) Limited perception of video images by domestic hens. *Animal Behaviour* 53, 951–963.

Patzke, N., Ocklenburg, S., van der Staay, F.J., Gunturkun, O. and Manns, M. (2009) Consequences of different housing conditions on brain morphology in laying hens. *Journal of Chemical Neuroanatomy* 37, 141–148.

Pecchia, T. and Vallortigara, G. (2010) Reorienting strategies in a rectangular array of landmarks by domestic chicks (*Gallus gallus*). *Journal of Comparative Psychology* 124, 147–158.

Pleskacheva, M.G., Dell'Omo, G., Kuptsov, P.A., Voronkov, D.N., Garin, N.N. and Lipp, H.P. (2001) Domestic birds in a giant radial maze: spatial learning in chickens, guinea fowls and geese. *Society for Neuroscience Abstracts* 27, 1417.

Poling, A., Temple, W. and Foster, T.M. (1996) The differential outcomes effect: a demonstration in domestic chickens responding under a titrating-delayed-matching-to-sample procedure. *Behavioural Processes* 36, 109–115.

Prescott, N.B. and Wathes, C.M. (1999) Spectral sensitivity of the domestic fowl (*Gallus g. domesticus*). *British Poultry Science* 40, 332–339.

Railton, R.C.R., Foster, T.M. and Temple, W. (2010) Transfer of stimulus control from a TFT to CRT screen. *Behavioural Processes* 85, 111–115.

Regolin, L. and Rose, S.P.R. (1999) Long-term memory for a spatial task in young chicks. *Animal Behaviour* 57, 1185–1191.

Regolin, L. and Vallortigara, G. (1995) Perception of partially occluded objects by young chicks. *Perception and Psychophysics* 57, 971–976.

Regolin, L., Vallortigara, G. and Zanforlin, M. (1994) Perceptual and motivational aspects of detour behaviour in young chicks. *Animal Behaviour* 47, 123–131.

Regolin, L., Vallortigara, G. and Zanforlin, M. (1995) Detour behaviour in the domestic chick: searching for a disappearing prey or a disappearing social partner. *Animal Behaviour* 50, 203–211.

Regolin, L., Garzatto, B., Rugani, R., Pagni, P. and Vallortigara, G. (2005a) Working memory in the chick: parallel and lateralized mechanisms for encoding of object- and position-specific information. *Behavioural Brain Research* 157, 1–9.

Regolin, L., Rugani, R., Pagni, P. and Vallortigara, G. (2005b) Delayed search for social and non-social goals by young domestic chicks, *Gallus gallus domesticus*. *Animal Behaviour* 70, 885–864.

Regolin, L., Rugani, R., Stancher, G. and Vallortigara, G. (2011) Spontaneous discrimination of possible and impossible objects by newly hatched chicks. *Biology Letters* 7, 654–657.

Roper, T.J. and Wistow, R. (1986) Aposematic coloration and avoidance-learning in chicks. *Quarterly Journal of Experimental Psychology Section B – Comparative and Physiological Psychology* 38, 141–149.

Rose, S.P.R. (2000) God's organism? The chick as a model system for memory studies. *Learning and Memory* 7, 1–17.

Roth, G. and Dicke, U. (2005) Evolution of the brain and intelligence. *Trends in Cognitive Sciences* 9, 250–257.

Rugani, R., Regolin, L. and Vallortigara, G. (2007) Rudimental numerical competence in 5-day-old domestic chicks (*Gallus gallus*): identification of ordinal position. *Journal of Experimental Psychology: Animal Behavior Processes* 33, 21–31.

Rugani, R., Regolin, L. and Vallortigara, G. (2008) Discrimination of small numerosities in young chicks. *Journal of Experimental Psychology – Animal Behavior Processes* 34, 388–399.

Rugani, R., Fontanari, L., Simoni, E., Regolin, L. and Vallortigara, G. (2009) Arithmetic in newborn chicks. *Proceedings of the Royal Society B: Biological Sciences* 276, 2451–2460.

Rugani, R., Kelly, D.M., Szelest, I. and Vallortigara, G. (2010) Is it only humans that count from left to right? *Biology Letters* 6, 290–292.

Rugani, R., Regolin, L. and Vallortigara, G. (2011) Summation of large numerousness by newborn chicks. *Frontiers in Psychology* 2, 179.

Rugani, R., Vallortigara, G. and Regolin, L. (2013) Numerical abstraction in young domestic chicks (*Gallus gallus*). *PLoS One* 8, e65262.

Salmeto, A.L., Hymel, K.A., Carpenter, E.C., Brilot, B.O., Bateson, M. and Sufka, K.J. (2011) Cognitive bias in the chick anxiety-depression model. *Brain Research* 1373, 124–130.

Sherwin, C.M., Heyes, C.M. and Nicol, C.J. (2002) Social learning influences the preferences of domestic hens for novel food. *Animal Behaviour* 63, 933–942.

Shettleworth, S.J. (1972) Role of novelty in learned avoidance of unpalatable prey by domestic chicks (*Gallus gallus*). *Animal Behaviour* 20, 29–34.

Shettleworth, S.J. (2003) Memory and hippocampal specialization in food-storing birds: challenges for research on comparative cognition. *Brain, Behavior and Evolution* 62, 108–116.

Siddall, E.C. and Marples, N.M. (2008) Better to be bimodal: the interaction of color and odor on learning and memory. *Behavioural Ecology* 19, 425–432.

Steele, R.J., Stewart, M.G. and Rose, S.P.R. (1995) Increases in NMDA receptor binding are specifically related to memory formation for a passive avoidance task in the chick – a quantitative autoradiographic study. *Brain Research* 674, 352–356.

Stewart, M.G., Csillag, A. and Rose, S.P.R. (1987) Alterations in synaptic structure in the paleostriatal complex of the domestic chick, *Gallus domesticus*, following passive avoidance training. *Brain Research* 426, 69–81.

Stewart, M.G., Kabai, P., Harrison, E., Steele, R.J., Kossut, M., Gierdalski, M. and Csillag, A. (1996) The involvement of dopamine in the striatum in passive avoidance training in the chick. *Neuroscience* 70, 7–14.

Tanimura, Y., Yang, M.C. and Lewis, M.H. (2008) Procedural learning and cognitive flexibility in a mouse model of restricted, repetitive behaviour. *Behavioural Brain Research* 189, 250–256.

Taylor, P.E., Coerse, N.C.A. and Haskell, M. (2001) The effects of operant control over food and light on the behaviour of domestic hens. *Applied Animal Behaviour Science* 71, 319–333.

Taylor, P.E., Haskell, M., Appleby, M.C. and Waran, N.K. (2002) Perception of time duration by domestic hens. *Applied Animal Behaviour Science* 76, 41–51.

Temple, W., Foster, T. and O'Donnell, C.S. (1984) Behavioural estimates of auditory thresholds in hens. *British Poultry Science* 25, 487–493.

Tommasi, L. and Polli, C. (2004) Representation of two geometric features of the environment in the domestic chick (*Gallus gallus*). *Animal Cognition* 7, 53–59.

Tommasi, L. and Vallortigara, G. (2000) Searching for the centre: spatial cognition in the domestic chick (*Gallus gallus*). *Journal of Experimental Psychology – Animal Behavior Processes* 26, 477–486.

Vallortigara, G. (1996) Learning of colour and position cues in domestic chicks: males are better at position, females at colour. *Behavioural Processes* 36, 289–296.

Vallortigara, G. and Zanforlin, M. (1989) Place and object learning in chicks (*Gallus gallus domesticus*). *Journal of Comparative Psychology* 103, 201–209.

Vallortigara, G., Zanforlin, M. and Pasti, G. (1990) Geometric modules in animals spatial representations – a test with chicks (*Gallus gallus domesticus*). *Journal of Comparative Psychology* 104, 248–254.

Vallortigara, G., Andrew, R.J., Serori, L. and Regolin, L. (1997) Sharply timed behavioural changes during the first 5 weeks of life in the domestic chick (*Gallus gallus*). *Bird Behaviour* 12, 29–40.

Vallortigara, G., Regolin, L., Rigoni, M. and Zanforlin, M. (1998) Delayed search for a concealed imprinted object in the domestic chick. *Animal Cognition* 1998, 17–24.

van Kampen, H.S. and Devos, G.J. (1992) Memory for the spatial position of an imprinting object in jungle fowl chicks. *Behaviour* 122, 26–40.

Wright, A.A. (2001) Learning strategies in matching-to-sample. In: Cook, R.G. (ed.) *Avian Visual Cognition.* http://www.pigeon.psy.tufts.edu/avc/wright/.

Zeltner, E., Klein, T. and Huber-Eicher, B. (2000) Is there social transmission of feather pecking in groups of laying hen chicks? *Animal Behaviour* 60, 211–216.

Zimmerman, P.H., Pope, S.J., Guilford, T. and Nicol CJ. (2003) Navigational ability in the domestic fowl (*Gallus gallus domesticus*). *Applied Animal Behaviour Science* 80, 327–336.

Zimmerman, P.H., Pope, S.J., Guilford, T. and Nicol, C.J. (2009) Involvement of the sun and the magnetic compass of domestic fowl in its spatial orientation. *Applied Animal Behaviour Science* 116, 204–210.

Zoeke, B., Althoff, K.P., Judt, P. and Krebs, U. (1980) The influence of object type and early experience on the discrimination ability of domestic chickens. *Zeitschrift fur Tierpsycholgie* 52, 149–170.

8

Applied Ethology of Broilers and Broiler Breeders

Broilers

In many European countries, broiler production is increasingly the preserve of specialist integrated companies with a controlling hand in all aspects of the production chain. In the UK, for example, broiler production is dominated by just four such companies, which produce over 80% of birds. Average flock sizes have increased markedly in the past decade, and in the UK the majority of birds are now reared on premises holding more than 100,000 birds. Most broilers are reared in indoor floor systems, with just 5–6% of birds kept in free-range systems and 1–2% in organic.

The broiler production chain begins when chicks are transported, in controlled-environment lorries, from the hatchery to the rearing house. On arrival, small pellets or crumbs are spread on paper so that the chicks learn how to feed. For the first week, the house lights are on for 23 h out of every 24 h, to encourage the new arrivals to feed and drink. By 3 or 4 days of age, chicks learn to feed from automated feeders and the paper is removed. Chicks are kept warm either by heating the whole house to an initial temperature of approximately 30°C or by the provision of spot brooders. After 7 days, the EU Broiler Directive (Council Directive 2007/43/CE) requires that growing birds are allowed a total of 6 h darkness in every 24 h to enable rest. Careful control of house temperature, ventilation and humidity is critical and has major implications for bird health, as will be discussed later. The birds grow rapidly on a sequence of starter, grower and finisher diets, and are usually slaughtered between 5 and 10 weeks of age. Many flocks (79% in a recent Belgian survey; Tuyttens et al., 2014) are partially depopulated or 'thinned' during the rearing period, with approximately 20% of birds removed for early slaughter. This allows higher numbers of chicks to be placed initially without subsequently exceeding legal stocking density limits as the birds near slaughter weight.

The increasing consumption of meat from broiler chickens in the EU and worldwide shows no sign of slowing, but it is accompanied by a sense of unease in many consumers. In a comprehensive survey of EU citizens (Eurobarometer, 2007), 42% of respondents chose broiler chickens as one of the farm animals 'whose current level of welfare should be improved the most'. Whereas animal welfare experts rate welfare problems such as leg health and hock burn ahead of stocking density (Haslam and Kestin, 2003), many members of the public are concerned that broiler chickens are unable to express their normal behaviour and that they are kept in overcrowded conditions (de Jonge and van Trijp, 2013). The basis for these concerns and the relationships between them are considered next.

Space use and stocking density

It is only since 2007 that legislation in Europe has been implemented to address the question of stocking density for broiler chickens. The EU Broiler

---

Directive referred to above was designed to encourage better animal welfare; it does so in the conventional way – by specifying regulations about the physical environment – but also by permitting consideration of the condition of the chickens themselves. In this sense, the nature of the legislation is unusual as it addresses both husbandry inputs and bird welfare outcomes (animal-based measures). Essentially the Directive states that chickens should not be stocked at more than 33 kg m^{-2} unless producers can show that they have complied with guidelines on the house temperature, humidity and ammonia level, and reported rates of on-farm mortality. If these conditions are satisfied, and if the official veterinarian at the slaughterhouse does not observe a high prevalence of conditions such as contact dermatitis (see below), then birds can be stocked at up to 42 kg m^{-2}. In some countries, levels of dermatitis are monitored formally at the slaughterhouse to inform these decisions. In England, 39 kg m^{-2} is the maximum stocking density currently allowed, although in most other EU member states, birds can be stocked at up to 42 kg m^{-2} if producers can show that cumulative mortality rates in preceding flocks have been sufficiently low. If one takes adult broiler bodyweight to be approximately 2.2 kg, then stocking densities of 33, 39 and 42 kg m^{-2} equate approximately to 15, 17.7 and 19.1 birds m^{-2}, respectively. Notwithstanding the legal position, some reports indicate that stocking densities of 47–48 kg m^{-2} remain in use in member states such as Belgium (Verspecht et al., 2011) and Poland (Utnik-Banas et al., 2014).

The practice of thinning to maintain legal stocking densities must also be taken into consideration as it can have substantial effects on flock (and human) health and welfare. Biosecurity is, for example, often compromised during thinning, and the practice is therefore linked with a higher risk of Campylobacter contamination, which can cause serious food poisoning in people (Hue et al., 2010). It has also been suggested that the stress of thinning (catching teams, machinery, noise and disruption) may increase the susceptibility to infection of the birds that remain (Bull et al., 2008). In contrast, the benefit of a lower stocking density at key points in the growing period is that it can reduce the risk of skin dermatitis (de Jong et al., 2012). Overall, however, there is very little information and research on the implications of partial depopulation for bird behaviour and welfare.

The emphasis of the EU Directive on measuring the health outcomes for broiler chickens, and allowing some discretion over stocking density, fits with the view of scientists who consider that if the environment is well controlled, stocking density per se is not a major issue for the health of broiler chickens. Estevez (2007) wrote 'the welfare of broilers can be ensured at a range of (reasonable) densities, as long as the requirements for environmental quality are fulfilled', while Dawkins et al. (2004) concluded that 'differences among producers in the environment that they provide for chickens have more impact on welfare than has stocking density itself.' Dawkins et al. (2004) based this conclusion on a study in which commercial producers agreed to stock 114 broiler houses at densities ranging between 30 and 46 kg. It was found that much of the variation in broiler health and welfare at high stocking densities depended on the extent to which producers were able to control temperature and relative humidity (Jones et al., 2005). Their ability to achieve tight control depended in turn on factors such as house age, the number and type of drinkers, the provision of fan ventilation, and the number of stockpersons and daily inspection visits.

Even though environmental control and management may be of primary importance for achieving good flock health, a sharp focus on stocking density looks set to continue. Reasons for this include the fact that good environmental management is harder to achieve when stocking densities are high (in that sense, a high stocking density is a proxy indicator that a problem might arise), that stocking density is relatively easy to manipulate within existing houses by altering the number of birds housed, and that new work is emerging on the behavioural effects of different stocking densities and the spatial preferences of broiler chickens. In addition, sophisticated statistical analyses that control for differences and confounding variables among producers continue to identify high stocking density as a risk factor for leg problems. Knowles et al. (2008) found that for every 1 kg m^{-2} increase in stocking density measured at the time of a leg health assessment, across a range from 16 to 46 kg m^{-2}, there was a 0.013 deterioration in flock gait score, which was assessed for 51,000 individual birds on a scale from 0 (smooth, fluid locomotion) to 5 (incapable of sustained walking). For all of these reasons, current and future EU legislation, as well as voluntary assurance standards, seems likely to be framed around stocking density for some time to come.

In considering the effects of stocking density on bird behaviour, overhead video analysis has been a useful technique. This has revealed that male broiler chickens at 6 weeks of age, occupy an average of 687 cm^2 when standing idle, and 704 cm^2 when preening, when housed at low stocking densities. If housed at higher densities of 16 birds m^{-2} (approximately 35–39 kg m^{-2}), then the space occupied while performing these activities is less, with birds occupying just 628 and 613 cm^2, respectively. Similar results were found for female birds. The authors considered that the reduced space occupied at higher densities could be explained by closely packed birds squashing each other's soft tissues and plumage (Bokkers et al., 2011).

An analysis of space occupied is useful, but it does not tell us how broilers perceive spatial restriction or high stocking densities. Some information on this aspect can be gleaned from studies of spatial distribution where birds have an element of choice and can be observed in situ. Patterns of random distribution can be compared with situations where birds are closer together than expected by chance – clumped (with a high variation in inter-individual distance) or more evenly dispersed than expected by chance (with a low variation in inter-individual distance). Factors encouraging birds to stay in close proximity include shared preferences for resources such as food or heat that are located in certain regions of the house, and a desire to reduce predation risk. Competition for resources provides a counter-motivation encouraging dispersion, while thermoregulatory needs can encourage either clumping to keep warm or dispersion to cool down. Broiler distribution has been studied in small groups at relatively low stocking densities. Under these conditions (e.g. at a stocking density of 6.7 birds m^{-2}, approximately 15 kg m^{-2}), broilers tend to be more dispersed than would be expected if they were adopting random, or indifferent, spatial positions (Leone et al., 2010). As pen size increases, the broilers move yet farther apart (Leone et al., 2010), although it is worth reporting that at no point in this study were the birds found at the maximum possible inter-individual distances. Similar results were obtained in another study using small experimental groups housed at stocking densities that varied from 6 to 56 kg m^{-2} (Buijs et al., 2011b).

These authors found evidence, using a variety of different measures of bird association and distribution, of a tendency to disperse once stocking density reached 15 kg m^{-2}. In commercial broiler houses, however, slightly different results have been obtained. Studying commercial flocks at stocking densities ranging between 30 and 46 kg m^{-2}, Febrer et al. (2006) found a tendency for birds to cluster, even at the highest stocking densities. This non-random distribution occurred despite greatly increased incidences of jostling and disturbance at the higher stocking densities. In addition, when walking en route to feeders, broilers did not avoid or move around clumps of other birds and the distance travelled was not affected by stocking density (Collins, 2008). Thus, broilers with a clear goal were not averse to the close proximity of other birds and would sometimes step or walk over resting birds in their path. Unlike laying hens, there is little evidence that broiler chicks engage in overt social competition (Chapter 6), or that threats or aggression prevent birds from moving freely. One reason for the differing results from small-scale and large-scale studies may be that birds in commercial systems are clustering not because of a social attraction but because they share preferences for certain resources (heaters, vertical walls, areas of dry litter) that are in more limited supply in commercial houses than in experimental pens. For example, broilers tend to congregate against the house walls or in corners (Newberry and Hall, 1990) as they are less disturbed by the movement of other birds when they have their backs to a wall (Buijs et al., 2010).

Examining bird distribution is one way to infer whether there is a latent demand for additional space, but there is a paucity of more formal preference or demand experiments addressing this question. Recently, however, Buijs et al. (2011a) tested the demand of broilers to access pen areas of lower stocking density. Broilers had to cross a barrier of variable height (7–28 cm) to move from one side of a pen to the other. An equal number of birds was placed on each side of the barrier at the start of the test, but the stocking density on one side was adjusted by increasing or decreasing the pen area. Broilers were highly motivated to access areas of stocking density of less than 14.7 birds m^{-2} (equivalent to 40 kg m^{-2}) and were willing to cross the high barrier to do so. When food deprivation was instead used as a yardstick, many birds (almost a quarter of the study population) did not cross the high barriers, even after 6 h of food deprivation. Given that broilers have a very high energy demand, this comparison shows that the motivation to reach a relatively low stocking density is strong.

A series of studies of small groups of broiler chickens (up to 30 birds) has been conducted to partition the effects of enclosure size, stocking density and group size. Studies examined the effects of altering available area in both rectangular- and square-shaped pens as these have different implications for the amount of available perimeter wall per unit area. Generally, positive effects of housing broilers in larger pens were detected (Leone and Estevez, 2008a) including greater overall movement in square pens (Mallapur et al., 2009) and increased distances between birds at equivalent stocking densities in larger pens (Leone and Estevez, 2008a; Leone et al., 2010). Increased stocking density negated some of the benefits of larger pens, while increased group size had few adverse effects, with the possible exception of some increased disturbance and jostling between birds. These small-scale studies demonstrate important principles, but the extent to which the results can be generalized to commercial situations is not known. Small-scale studies have also demonstrated how inexpensive alterations to the pen environment could be used to encourage a more uniform distribution of birds. For example, Cornetto and Estevez (2001a) placed vertical panels in the centre of pens, which attracted broiler chicks away from the pen walls and corners and resulted in more even spacing.

The legal framework governing the production of broiler chickens provides a bottom line below which chicken welfare is likely to be compromised. Organic standards for indoor space allowance are much higher, with birds to be kept at no more than 21 kg m^{-2} or 10 birds m^{-2}. At the time of writing, the cost of an organic whole chicken in the UK is nearly twice that of a bird reared under standard production and so, despite a loyal consumer base for organic produce, sales of organic chicken account for less than 1% of all UK chicken production. Given the limited market for fully organic chicken, many producers choose to subscribe to other consumer-assurance and labelling initiatives that set standards intermediate between conventional and organic. In the UK, such schemes include Red Tractor, which covers 90% of UK broiler production and requires birds to be kept at no more than 38 kg m^{-2}. More demanding standards are set by Freedom Food (the RSPCA's farm-assurance and food-labelling scheme). This scheme suggests that the flock size of indoor-reared broiler chickens should not exceed 30,000, and it requires birds to be kept at no more than 30 kg m^{-2} or approximately 13.6 birds m^{-2}. However, only 2% of UK chicken production is currently raised to the Freedom Food criteria.

Importance of activity for broiler welfare

Another issue that concerns consumers is the lack of activity shown by broiler chickens as they approach slaughter weight. As we have seen, the behaviour of chickens changes rapidly during early development (Chapter 3), but it is greatly limited in broiler chickens from around 3 or 4 weeks of age through rapidly increasing bodyweight and, for many birds, difficulties in walking and locomotion. Body weight becomes the major factor governing walking behaviour in heavier birds, even those without obvious leg problems (Bokkers and Koene, 2004). In contrast to young pullets, broilers show steady but substantial reductions in perching and locomotion (Chapter 5) and foraging (Cornetto and Estevez, 2001b) as they age. The distance covered during the day declines (Newberry and Hall, 1990) as walking becomes tiring (Corr et al., 2003). Compared with the efficient gait of the junglefowl, the leg and back movements of non-lame broiler chickens are exaggerated in compensation for the large breast mass (Caplen et al., 2012). The gait of lame chickens shows further deviation from the junglefowl baseline, and walking is likely to be a painful experience as it can be improved temporarily by the use of analgesic drugs (Caplen et al., 2013). Drugs such as carprofen and meloxicam, which are both non-steroidal anti-inflammatory agents, increased the birds' walking speed, although (at higher doses and greater speeds) at the expense of good balance. No one is suggesting that birds should be given these drugs to assist their locomotion. Rather, these experimental studies present evidence that the declining activity of broiler birds is not entirely voluntary and is, to some extent, a symptom of pain.

If moving is painful and difficult for the birds, it might be thought, short of a better long-term solution, that the kindest thing to do would be to leave them in peace and allow them to be inactive for the last few weeks of their life. However, inactivity can result in a worsening of the situation. As exercise provides the essential forces that stimulate proper bone growth and muscular development (Sherlock et al., 2010), it is probably worth trying to keep birds moving for as long as possible to maintain the locomotor abilities they still possess. This idea gains support when some of the other consequences of inactivity are considered.

Ulcerative skin conditions such as footpad dermatitis, hock burn and breast burn are commonly found within commercial broiler flocks. Moderate to severe footpad dermatitis was recorded at an average level of 11%, and moderate to severe hockburn at an average level of 1.3%, on 149 UK broiler farms that volunteered to take part (Haslam et al., 2007). Levels of these conditions on other farms and in other countries can be much higher (e.g. moderate to severe footpad lesions were detected in more than 90% of birds from farms in France and in 38.4% of birds from flocks in The Netherlands; Allain et al., 2009; de Jong et al., 2012). Breast burns occur far more rarely. Ulcerative dermatitis results primarily from the burning effects of ammonia, which arise when high concentrations of urea are present in the litter. Maintaining litter quality by proper drinker design and good control of house ventilation, temperature and humidity are therefore of the utmost importance to prevent these conditions. However, the risks of developing these different, although correlated (Haslam et al., 2006), forms of contact dermatitis are also influenced by bird activity. This is shown first by the fact that birds with walking difficulties and high gait scores are more likely to suffer from contact dermatitis (Sorensen and Kestin, 1999; Su et al., 1999), specifically hock burn (Haslam et al., 2006), and secondly that smaller, lighter birds are at lower risk of hock burn than birds with higher bodyweight at slaughter age (Haslam et al., 2006; Hepworth et al., 2010) or than birds with other serious conditions such as ascites (an accumulation of fluid in the abdominal cavity) and septicaemia (Haslam et al., 2008; Hepworth et al., 2011). Presumably, the more active, healthy birds spend less time with their hocks in contact with the litter. Increasing bird activity

has a lesser effect on footpad dermatitis as the birds' feet are in constant contact with the litter whether they are moving or sitting. For this reason, at a farm level, correlations between litter quality and footpad dermatitis are stronger than for hock burn or breast burn (Haslam et al., 2006, 2007). Inactivity can also result in reduced feathering on the breast due to constant friction from sitting on the litter, and in breast blisters (a condition distinct from breast burns where fluid accumulates in sacs along the breast bone) (Allain et al., 2009).

These additional findings add support to the idea that it could be beneficial to encourage greater activity in broiler chickens. Indeed, chickens forced to be more active with treadmill exercise have been reported to have better leg health (Reiter, 2006). However, in a commercial environment, increased activity cannot be forced and carrots rather than sticks are required. Possible incentives to increase activity are adjustments to lighting (reviewed in Chapter 2) or environmental enrichment programmes, which are discussed next.

Environmental enrichment to increase broiler activity

Environments may be altered by the provision of resources, materials or toys, but they are enriching to chickens only if they have a demonstrable effect on behaviour and only if that effect is one that improves welfare (Newberry, 1995). It is not sufficient that an environment containing toys (for example) appears to be improved from a human perspective. Objective evidence that changes to the environment have increased bird activity, and thereby lowered the prevalence of walking problems and inflammatory skin conditions, is required.

In practice, the success of environmental enrichment strategies in increasing bird activity has been mixed. More complex environments do not always encourage more locomotion or increase the overall distance that broilers (aged 3–5 weeks) walk (Estevez et al., 2010). Broilers make more use of low perches (8.5 cm) than higher perches, even when lower perches are arranged in steps to make access as easy as possible (LeVan et al., 2000), but the number of birds culled for leg problems in this study did not differ between groups with or without perches. As we saw in Chapter 5, low-level barriers placed between food and water have had a degree of success in increasing broiler activity levels but not to the extent that they reliably improve walking ability (Bizeray et al., 2002a,b; Ventura et al., 2010), although perches in combination with sand have been reported to reduce levels of hockburn (Simsek et al., 2009). Other manipulations that one might hope would encourage movement, such as providing moving spot lights (Bizeray et al., 2002a) or string for pecking (Arnould et al., 2004), have had little or no observable effect on leg health. Even attempts to promote an increased intensity or duration of normal behaviours such as dustbathing (which involves a certain amount of energetic leg movement) have met with limited success. As broilers prefer sand for dustbathing over other substrates such as wood shavings, rice hulls or paper bedding (Shields et al., 2004), it was logical to determine whether the overall amount of dustbathing could be increased by the provision of sand. Unfortunately, although the preference for sand was confirmed (Shields et al., 2005), access to this preferred substrate was insufficient to provoke an overall increase in active behaviours such as dustbathing or locomotion. There is therefore no hint that this method of stimulating activity could lead to improved leg health in a commercial environment.

One attempt to increase locomotion associated with foraging behaviour has been to scatter grain (additional to the normal food ration) in the litter, but this practice has not generally had the desired effect (Bizeray et al., 2002a; Jordan et al., 2011). In contrast, if feed troughs are removed and the normal daily ration of food pellets is scattered in the litter, then broiler walking, scratching and pecking activity is increased significantly (Jordan et al., 2011). However, in this laboratory study, this treatment also resulted in some food loss and a lower body weight by 6 weeks of age, factors unlikely to be accepted in commercial practice.

One promising idea, especially applicable in warmer climates, is to use cooled perches. As we saw in Chapter 5, broilers continue to use these perches with associated improvements in leg health (Zhao et al., 2012). Perhaps the most encouraging results have come from studies of enrichment on commercial farms. In one study, two commercial farms in the UK each contributed two (essentially identical) houses, one of which acted as a control,

while the other was provisioned with straw bales (81 on one farm, 118 on the other). The provision of straw bales on both farms resulted in a significant increase in locomotion (from 4.95 to 8.63% across all ages), and reductions in resting and sitting (Kells et al., 2001). A major advantage of the straw bales was that they were used for many different purposes – perching, pecking and foraging and support for resting. A further study examined combinations of straw bales (30 per house of 23,000 birds) and natural light provision in a 2 × 2 experiment and also directly measured gait score and other indicators of leg health (Bailie et al., 2013). Flocks housed without natural light or straw bales had higher (worse) gait scores than all other treatments. Other interactions were more difficult to interpret, but the greatest beneficial effects (e.g. increased latency to lie down) were associated primarily with the provision of natural light.

The importance of light in increasing broiler activity was reviewed in Chapter 2. In practice, the Freedom Food scheme requires natural light to be provided once the birds are 7 days old. In addition, from 2014, the Red Tractor scheme also requires newly built broiler houses to admit natural light.

Strategies to reduce fearfulness

The appearance and behaviour of human stockpersons can trigger fear responses in commercial broiler flocks during the growing period and especially during procedures such as thinning and catching for transportation or slaughter. Rough handling induces fear (Jones 1992). In contrast, regular gentle handling can have beneficial effects in reducing fear-related behaviour (Jones and Waddington, 1992), while also improving disease resistance (Gross and Siegel, 1982) and, in some studies, production (Hemsworth et al., 1994). Commercially, the production of flocks of broilers with a high tendency to avoid humans is reduced in comparison with broilers that are more tolerant of human proximity. In one study, the withdrawal tendency of broiler chickens explained no less than 28% of variation in food conversion across 22 commercial farms (Hemsworth et al., 1994). This suggests that finding ways to mitigate the birds' fear of humans could have welfare and production advantages.

Regular, gentle handling of birds has a significant effect in reducing their fearfulness and tendency to withdraw from humans (Hemsworth *et al.*, 1994). However, birds can form expectancies about their environment and can react badly if these are violated. For example, in one experiment, birds that experienced gentle handling were more fearful after 60 min of transportation than birds that had received no human handling at all (Nicol, 1992). Similarly, Kannan and Mench (1996, 1997) found that gentle handling reduced stress in comparison with rough handling when birds were kept in their home pens, but if placed in transport crates or handled more roughly at slaughter age, the nature of the prior handling did not affect their stress levels. In addition, it is clearly not feasible to handle every bird in a flock of many thousands. For this reason, some studies have examined whether increased visual contact with humans is sufficient to attenuate fear. Zulkifli *et al.* (2002) found that 10 min of visual contact with a human from 0 to 3 weeks, or from 0 to 6 weeks of age, significantly reduced the fear and stress responses of broiler chickens when they were caught and placed in transport crates at the age of 42 days. Zulkifli and Azah (2004) also found that observation of one conspecific (from a group of 30) being gently handled reduced underlying fearfulness of the whole group in comparison with unhandled controls.

More positive attitudes expressed by stockpeople towards chickens in a questionnaire correlated with reduced bird withdrawal (Cransberg *et al.*, 2000). However, in contrast with studies on pig and dairy farms, it has proved difficult to identify correlations between the expressed attitude of a stockperson and his or her actual behaviour when in the broiler house, or to identify any features of stockperson behaviour that are strongly associated with fearfulness in commercial broiler flocks (Cransberg *et al.*, 2000).

Environmental enrichment (Altan *et al.*, 2013), or even imprinting on a complex stimulus (Gvaryhu *et al.*, 1989) can also reduce underlying fearfulness in broilers. Perhaps of most direct commercial relevance, simply allowing broiler chickens to get more sleep can reduce their fear levels. Birds housed with no dark period show greater fear responses than birds housed under dark periods of 2–12 h (Zulkifli *et al.*, 1998; Sanotra *et al.*, 2002). This was confirmed in a survey by Bassler *et al.* (2013), who found enormous variation in the length of the dark period provided for intensive broiler flocks in four EU countries (with a range of 0–12 h). Flocks where birds were able to get more rest and sleep were also the ones where individual chickens showed less avoidance of human contact. Birds reared to the RSPCA Freedom Food standard must have a minimum of 6 h and a maximum of 12 h of continuous dark to enable proper resting.

Broilers and range use

According to the British Poultry register, fewer than 8% of broiler chickens have access to an outdoor area as part of free-range or organic production. The corresponding numbers in many other countries are likely to be much lower. Unlike consumers, farmers generally perceive outdoor access as something that could be detrimental to broiler welfare as well as being more costly and time-consuming to manage (Tuyttens *et al.*, 2014). It is undoubtedly the case that some of the potential benefits of access to an outdoor range for broiler chickens will depend on the birds' leg health. Fast-growing, commercial strains of broiler chicken may make little use of outdoor facilities (Weeks *et al.*, 1994; Nielsen *et al.*, 2003; Jones *et al.*, 2007), and their activity levels may be no higher than for birds reared indoors (Weeks *et al.*, 1994). Many of these birds experience difficulties in moving, and this is likely to be one reason why outdoor areas are not greatly used by the birds. In a study of 40 flocks of 20,000 commercial free-range broilers, very few if any birds went outside in winter (Dawkins *et al.*, 2003). Even in summer, it was rare for more than 5% of the birds to be seen outside during daylight hours, although Dawkins *et al.* (2003) reported that many more birds went out at dusk. In contrast, slow- and medium-growing hybrid broilers kept in small organic flocks can make good use of an outdoor range and show a broader behavioural repertoire and more activity than indoor birds (Zupan *et al.*, 2005; Sosnowka-Czajka *et al.*, 2007). While this increased activity may reduce walking difficulties, a potential disadvantage is that levels of the skin condition footpad dermatitis have been reported to be higher in free-range and organic broilers than in birds kept indoors (Pagazaurtundua and Warriss, 2006).

In smaller flocks, variable numbers of birds are seen outside; factors that tend to increase ranging are increasing bird age (Mirabito and Lubac, 2001), warmer weather (Jones *et al.*, 2007), smaller flocks (Sosnowka-Czajka *et al.*, 2007) and the presence of trees or shrubs on the range (Sosnowka-Czajka *et al.*, 2007; Dal Bosco *et al.*, 2014). As with the large commercial flocks, ranging behaviour is more apparent at dawn and dusk than during the middle of the day (Nielsen *et al.*, 2003).

If slow-growing breeds are retained for a longer growing period, then ranging behaviour appears to increase further, with 68.6% of slow-growth broilers and 39.9% of medium-growth broilers observed outdoors between 11 and 16 weeks of age (Almeida *et al.*, 2012). Broilers in organic production that are fed a protein-restricted diet during the latter part of rear may be especially motivated to access the range to forage for insects or other high-protein snacks (Almeida *et al.*, 2012).

Hunger in broiler chickens

Broiler chickens have a high genetic growth potential and when fed *ad libitum* they will reach slaughter weight in just 35–42 days, often with the associated leg and health problems discussed at length in preceding sections. In an attempt to counter the commercial pressure for such rapid weight gain and young age at slaughter, and to ensure that consumers are not misled about the animal welfare, legal standards require free-range chickens to reach an age of 56 days and standard breeds of organic chickens an age of 81 days before slaughter. The minimum age does not apply if slow-growing strains of broiler are used in organic production (interpreted as a weight gain of less than 45 g day^{-1} (Defra) or less than 35 g day^{-1} (Soil Association)), but such birds are likely to take around 70 days to reach slaughter weight. If these longer lifespans are achieved by using strains of bird with a lower genetic growth potential, then the intended welfare benefits are likely to be realized. However, it is possible to restrict the growth of standard breeds by reducing the amount of feed provided at key periods during rearing. If growth rate is restricted in this way, the birds are likely to experience chronic hunger, as they have been bred with a large appetite that then remains unsatisfied. In an attempt to avoid

this scenario, the RSPCA Freedom Food scheme has developed a protocol to assess the welfare of different breeds under specified conditions on approved establishments (Cooper, 2013). Only breeds that maintain high welfare under these conditions will be permitted to be used for Freedom Food production. By including an assessment of the time taken for a breed to reach a target weight of 2.2 kg when food is provided *ad libitum*, it is hoped this new scheme will increase the pressure on breeders to produce slower-growing birds. This revolutionary proposal could provide a counter-force that opposes the direction of genetic travel that has taken place over the past few decades, but it will require more farms to join the scheme, and we will have to await the outcome of trials due to start in 2015 before assessing its impact.

This same issue – a mismatch between appetite and food provision – also affects the parent birds that produce the fertile eggs for broiler production. These broiler breeder chickens will be considered next.

Broiler Breeders

There are approximately 7 million broiler breeder birds in the UK. These birds are reared in floor pens containing around 1000 birds, usually with male and female birds kept separately. The birds are weighed and graded at key points in development so that feeding can be adjusted and growth controlled to strict targets. Feed is often scattered by hand, or broadcast from above using automated spin feeders. Perches can be introduced during rear to encourage use of raised structures, which aids the subsequent use of nest boxes by female birds. Light is increased to bring the birds into reproductive condition, and the males and females are then transferred to the laying unit and housed together at a ratio of approximately eight males to every 100 females. During the laying period, the males and females continue to be fed separately via the use of female feeders (with narrow grids or roll bars that exclude males with wider heads) and male feeders that are too high for the females to reach. Male birds may be progressively removed from the flock as their condition declines and, after peak lay, a small proportion of younger males may be added to restimulate fertility and mating activity in the older males.

The fertile eggs are collected and sent to hatcheries. A female bird will produce approximately 150 fertile eggs before being sent for slaughter at around 60 weeks of age.

Hunger in broiler breeders

Broiler breeders possess the large appetite and potential for rapid growth that are such notable characteristics of all broilers. However, if fed *ad libitum* before they reach sexual maturity, broiler breeder birds become obese resulting in health problems (thermal discomfort and panting, skeletal and metabolic disorders), reduced fertility due to multiple ovulation and increased mortality (Mench, 2002; van Krimpen and de Jong, 2014). For these reasons, young broiler breeder birds are fed only 30–40% of the food they would eat under unrestricted conditions (Hocking *et al.*, 1993). Generally, feed is scattered on the ground in one daily meal, which is consumed very quickly (Savory *et al.*, 1996), often within minutes. In some countries, birds may be fed only on alternate days, or on 4 or 5 days out of 7, although this is not permitted in the UK, Sweden, Norway or Denmark. Adult females may also be restricted, albeit to a lesser extent, during the laying period and particularly after peak lay (Hocking *et al.*, 2002). Although they appear healthy, these birds are chronically hungry, and this can lead to frustration, problems of aggression and feather pecking, fearfulness and raised stress levels (Mench, 1991; Savory *et al.*, 1993; Hocking *et al.*, 1997). Hocking *et al.* (1996) examined the effects of restricting birds' food during the rearing period so that they gained between 25 and 85% of the body weight of birds fed *ad libitum*. Restriction to 40% or less was associated with the greatest signs of stress assessed by heterophil:lymphocyte ratio and plasma corticosterone. The problem has become even more severe due to further increases in growth potential in recent years (van Krimpen and de Jong, 2014). For the breeding birds, this means yet greater food restriction, to as little as 25% of free-feeding intake, and rising stress levels (van Emous *et al.*, 2014).

The question of just how hungry broiler breeders are when kept under different levels of food restriction is therefore highly relevant (D'Eath *et al.*, 2009) and can be addressed by quantifying the birds' motivation to obtain food or to avoid places associated with food restriction, using the methods described in Chapter 4. For example, a very high motivation to feed is detected when restricted birds are allowed subsequent access to *ad libitum* food, with rates of intake three times that of non-restricted birds (de Jong *et al.*, 2003). However, somewhat unexpectedly, conditioned place aversion experiments (Chapter 4) have not shown that broiler breeders avoid places (colour-cued pens) associated with severe levels of food restriction (Dixon *et al.*, 2013). This may be because hunger triggers a strong motivation to search for food, and the primary behaviour shown by these birds was a tendency to visit the most novel pen and forage. In this context, a conditioned place aversion experiment measures exactly the wrong thing. When a different test of motivation was devised, such that birds had to cross a water bath to gain access to a raised foraging arena, the restricted-fed birds were twice as likely to cross the water as birds given two or three times more food for their daily ration (Dixon *et al.*, 2014). As no food was actually forthcoming during the experiment, the birds were not just responding to food cues. Rather, this study demonstrates that broiler breeders cannot simply switch off and think about something else – they must deal continuously with feelings of hunger.

Moderate or substantial food restriction also greatly increases the performance of oral behaviours such as excessive drinking, and pecking at drinkers, empty feeders, floor litter or spots on the wall (Savory *et al.*, 1992; Savory and Maros, 1993). These active behaviours occur at higher rates post-feeding than in the intervals between or preceding meals. For young broiler birds on a daily ration of just 25% of *ad libitum* intake, meal size has a paradoxical effect. A relatively large meal provokes a decline in drinking but an increase in post-feeding pecking (Savory and Mann, 1999). These authors suggested that the larger meal provided more positive feedback than the smaller meal, and this feedback (possibly mediated via dopamingeric pathways) led to a sustained high level of feeding motivation after the meal was finished. Arousal associated with high feeding motivation may also stimulate post-feeding pecking (Savory and Kostal, 1996). The performance of post-feeding pecking is linked (although in a complex manner) with subsequent reductions in arousal, as measured by heart rate and slow-wave EEG (Savory and Kostal, 2006).

Repetitive pecking at feeders and drinkers occurred more often during cold weather and at higher stocking densities in open-sided houses in Israel (Spinu *et al.*, 2003), so is affected by factors other than unfulfilled feeding motivation. However, frustrated motivation to feed is clearly the underlying causal factor, and the oral behaviour of broiler breeders can reasonably be described as stereotypic. Stereotypic behaviour has traditionally been defined as repetitive, invariant and apparently functionless (Mason, 1991) or, more recently, as behaviour induced by frustration, repeated attempts to cope and/or central nervous system dysfunction (Mason, 2006). The repetitive pecking of broiler breeders fits both definitions.

The management of broiler breeder birds presents a welfare dilemma. Ideally, a way forward should be found so that the birds' health could be protected without causing the birds to suffer continuous hunger (Decuypere *et al.*, 2006). It has been suggested that the degree of food restriction should be limited to perhaps 50–80% of free-feeding weight gain (Hocking *et al.*, 1996), but this has not been adopted by the industry. Experimental dwarf genotypes that require a lesser degree of feed restriction have also been trialled (Jones *et al.*, 2004; Bruggeman *et al.*, 2005), but unfortunately their rate of lay is lower than that of a fully restricted commercial genotype. Scattering food in the litter, or dividing food into two meals per day instead of one, are also practices that have little effect on most indices of hunger (de Jong *et al.*, 2005a). A different approach, and one receiving increasing attention, is to consider qualitative changes to the birds' diet such as the addition of bulking or diluting agents, the use of appetite suppressants or feeding low-protein diets. Some of the first attempts to restrict birds' growth in these ways resulted in uneven growth rates (Savory *et al.*, 1996), a particular problem when the modified diet is offered *ad libitum*. However, diets and feeding schedules that achieve uniform and target weight gain have now been formulated (Tolkamp *et al.*, 2005; Sandilands *et al.*, 2006; Nielsen *et al.*, 2011). These diets are fed in restricted amounts, although at a lesser degree of restriction than is necessary with a commercial diet.

Studies examining the effects of dietary dilution have usually found that it prolongs feeding time (Zuidhof *et al.*, 1995; Savory *et al.*, 1996; Hocking *et al.*, 2004; de Jong *et al.*, 2005b;

Sandilands *et al.*, 2005). A diet with lower levels of protein provided throughout rear similarly increased feeding time, although only at the ages of 12 and 17 weeks (van Emous *et al.*, 2014). However other effects on behaviour have been mixed and sometimes contradictory. Zuidhof *et al.* (1995) found repetitive object pecking behaviour was unaffected when ground oat hulls were added to the diet, but de Jong *et al.* (2005b) found diets diluted with palm kernel meal, wheat bran, lucerne, wheat gluten and sunflower seed meal reduced object pecking during the rearing period. Similar results have been obtained by adding ground oat hulls with an appetite suppressant (Sandilands *et al.*, 2005; Morrissey *et al.*, 2014a) and by lowering protein levels (van Emous *et al.*, 2014). Careful interpretation of these results is required. A redirection of pecking activity from an empty feeder or drinker towards other substrates such as litter or the (possibly unpalatable) new diet itself does not necessarily indicate an improvement in welfare. Often the overall total amount of pecking is very similar between control groups (on commercial quantitative restriction) and treatment groups (given modified diets), suggesting that, in growth-restricted birds, pecks are simply directed to the substrates in the environment that most resemble food particles. Indeed, a similar redirection of pecking behaviour can be achieved by housing restricted-fed broiler breeders on litter rather than plastic floors (Hocking *et al.*, 2005). It is also possible that appetite suppressants could cause a general sense of malaise or nausea in birds, rather than simply reducing hunger. Further evidence of beneficial effects is required before we can be convinced that modified diets have improved broiler breeder welfare.

Diet dilution can reduce feather pecking and cannibalism (Hocking *et al.*, 2004) and (in combination with the inclusion of an appetite suppressant) can improve plumage cover (Morrissey *et al.*, 2014b), which is clearly an indirect welfare benefit to members of the flock. However, physiological indicators of stress and immune function are not necessarily improved by a modified diet (Savory *et al.*, 1996; Sandilands *et al.*, 2006; Hocking, 2006, reviewed by D'Eath *et al.*, 2009; van Emous *et al.*, 2014). The real test that dietary dilution has improved welfare would be a demonstration of reduced feeding motivation. This can be assessed by making food available after a period of restriction and assessing compensatory feeding, or by assessing how hard birds are prepared to work for food using operant tests. Neither method is without difficulties (e.g. animals on an *ad libitum* but poor-quality diet will be unable to show any compensatory feeding; high-quality feeds may have to be offered in smaller amounts in operant tests to provide similar reward per unit of effort and this may itself affect bird motivation) and, so far, evidence in this area is contradictory. Some authors report no reduction in food motivation between birds whose growth had been restricted using qualitative versus quantitative methods (Savory and Lariviere, 2000), but others have found reductions in feeding motivation for birds on high-fibre diets (de Jong *et al.*, 2005b; Nielsen *et al.*, 2011) or diets that contain both additional fibre and an appetite suppressant (Sandilands *et al.*, 2005) compared with quantitatively restricted birds. In this latter experiment, the pre-test fasting period of treatment and control birds varied, and Sandilands *et al.* (2005) were therefore cautious in their interpretations regarding feeding motivation. Nielsen *et al.* (2011) controlled for pre-test fasting and found that birds fed a high-fibre diet with a relatively high proportion of insoluble fibre were less food motivated than birds fed a commercial diet. However, in this as in other experiments, while the feeding motivation of birds on modified diets is somewhat reduced in comparison with commercial practice, it is not reduced to anything like the level seen in birds fed *ad libitum*. These various findings suggest that dietary dilution may have some positive effects in increasing time spent feeding, promoting short-term satiety through increased distension of the gastrointestinal tract and partially reducing residual feeding motivation. While accepting that, we still do not know enough about how food intake goals are set in broiler breeder chickens, and whether these goals are fixed or dependent on the available food quality (D'Eath *et al.*, 2009). Overall, however, it seems clear that the positive effects of modified diets remain insufficient to fully address the hunger problem.

As we saw in Chapter 5, chickens in semi-natural conditions spend more than 50% of their daylight hours eating and foraging – approximately 42 h a week. In contrast, the commercially restricted broiler breeder spends less than 2 h a week actually feeding. Commercial practice is beginning to shift slightly in light of the scientific findings presented above, but with limited goals (e.g. of permitting birds to feed for 40–60 min on 4 or 5 days out of 7; Hubbard, 2011). As selection for broiler growth rate intensifies further, perhaps the only sustainable solution will be for concerned consumers to press for assurance schemes to demand the adoption of approved (less-productive) breeds of bird.

Aggression and fatigue in broiler breeders

Broiler strain males have not traditionally been known for their aggressiveness (Mench, 1988), so it was a surprise when outbreaks of extreme aggression towards females were reported in North American flocks in the 1990s (Duncan, 2009). The rate of aggressive acts directed towards females was five to ten times higher for two strains of broiler breeder than seen in layer strain males (Millman *et al.*, 2000). In commercial flocks, this aggression resulted in females being restricted to certain areas of the house, injured, attacked or even killed.

The hyperaggressive broiler breeders appeared to have specific deficits in their courtship behaviour that led them to attack females. This is a strange pattern of behaviour as even game birds, deliberately selected for their male–male fighting ability, rarely attack females (Millman and Duncan, 2000a). While Mench (1988) had detected slight increases in male aggression associated with food restriction, this could not explain the hyperaggressive behaviour in this case. The level of feed restriction during lay (Millman *et al.*, 2000) or during rear (Millman and Duncan, 2000b) was not a significant influence. If anything, males given more generous rations became even more aggressive (Millman *et al.*, 2000). It is possible that selection for a trait designed to improve male libido, such as rapid approach towards a female, has inadvertently altered normal male courtship behaviour (Duncan, 2009), but management factors are also important. de Jong *et al.* (2009) more recently reported that over 80% of matings in Dutch breeder flocks were forced, with females often struggling or injured, as shown in Fig. 8.1. They suggested that management factors such as separate rearing of males and females might interfere with the development of proper courtship

Fig. 8.1. Broiler breeder female showing feather loss and scratch marks due to male aggression. (Photo courtesy of Dr Ingrid de Jong, Wageningen UR, The Netherlands.)

behaviour (see also de Jong and Guemene, 2011). It is also possible that too much is expected of the females. In Chapter 6, we found that chickens evolved to be receptive to occasional matings from just one or two cockerels during a relatively short breeding season. In contrast, broiler breeder females are subject to frequent copulation attempts during the late afternoon and evening periods (de Jong et al., 2009). Add to this the energetic demands of high rates of lay (around five eggs per week) and it is reasonable to suppose that fatigue and exhaustion may play a part in reduced female receptivity to the point where mating can be achieved only through aggression.

Reducing stress levels in broiler breeders

Reducing stress levels in parent flocks is a topic of increasing interest, not only because it is ethical to consider the welfare of the parent birds but also because of an increased understanding that maternal stress levels can influence offspring development (see Chapters 1 and 3). We have seen above that severe feed restriction and the demands of aggressive males are prime causes of high stress levels in broiler breeder flocks (de Jong and Guemene, 2011; van Emous et al., 2014). However, even against this difficult background, it is possible that environmental enrichment could be used to reduce stress and fearfulness and improve bird welfare.

Hocking and Jones (2006) added string (for pecking) or bales of wood shavings during the rearing period for commercial broiler breeders, either from day 1 or from when the birds were 8 weeks of age. The bales were used extensively (the string less so), but in neither case was there any effect on the aggression levels of the young breeder birds when they were assessed at 16 weeks of age. However, broiler breeders reared with perches have been reported to be less fearful as adults (Brake et al., 1994), while birds reared in an environment where bales of wood shavings were added subsequently maintained better eggshell quality during their laying period (Edmond et al., 2005). Another manipulation trialled with significant success on commercial farms housing 7000 female and 800 male birds was the provision of regions of cover (Leone and Estevez, 2008b). Mesh-covered semi-opaque panels (70 × 70 cm) were placed vertically approximately 4.5 m apart on the litter area of each shed. The reproductive performance of each flock was improved significantly, for reasons that were thought to include reduced levels of female stress.

Another beneficial effect of the provision of panels was a significant reduction in floor eggs, particularly on the one commercial farm that had the biggest problem (Leone and Estevez, 2008b). Other possible influences on floor eggs have been explored in broiler breeder flocks. For example, it had been proposed that feeding these hungry birds during the morning nesting period might cause females to leave their nests and not return again to lay. However, Sheppard and Duncan (2011) found that although females would leave their nests to eat, the vast majority returned again to lay, and feeding time did not have a significant influence on the incidence of floor eggs (Sheppard and Duncan, 2011).

Conclusions

The behaviour of broiler chickens is relatively unconstrained for the first few weeks of life but, from around 4 weeks of age, increasing weight and bulk make normal activities more difficult and tiring to perform. In addition, significant proportions of broiler chickens are affected by painful conditions that cause lameness, despite more attention now being paid to this issue by breeding companies. Attempts to stimulate activity by providing environmental enrichment and natural lighting are worthwhile but do not provide a full solution. The welfare of broiler chickens remains a focus of attention for the public and animal welfare non-governmental organizations, and initiatives to approve only certain breeds for high welfare assurance schemes may have growing influence in the future. There is also a need to address the mismatch between consumer concerns about the importance of natural behaviour and the views of farmers who see welfare more in terms of physical health (Tuyttens et al., 2014). Both are crucial for good welfare.

In comparison with broiler chickens, the behaviour and welfare of their parent flocks has received far less attention. Consumers cannot directly influence the management of these birds by their buying decisions, and there is little public awareness of the degree of food restriction and aggression experienced by these birds. A full picture of the welfare of these birds is yet to emerge.

References

Allain, V., Mirabito, L., Arnould, C., Colas, M., Le Bouquin, S., Lupo, C. and Michel, V. (2009) Skin lesions in broiler chickens measured at the slaughterhouse: relationships between lesions and between their prevalence and rearing factors. *British Poultry Science* 50, 407–417.

Almeida, G.F.D., Hinrichsen, L.K., Horsted, K., Thamsborg, S.M. and Hermansen, J.E. (2012) Feed intake and activity level of two broiler genotypes foraging different types of vegetation in the finishing period. *Poultry Science* 91, 2105–2113.

Altan, O., Seremet, C. and Bayraktar, H. (2013) The effects of early environmental enrichment on performance, fear and physiological responses to acute stress of broilers. *Archiv für Geflügelkunde* 77, 23–28.

Arnould, C., Bizeray, D., Faure, J.M. and Leterrier, C. (2004) Effects of the addition of sand and string to pens on use of space, activity, tarsal angulations and bone composition in broiler chickens. *Animal Welfare* 13, 87–94.

Bailie, C.L., Ball, M.E.E. and O'Connell, N.E. (2013) Influence of the provision of natural light and straw bales on activity levels and leg health in commercial broiler chickens. *Animal* 7, 618–626.

Bassler, A.W., Arnould, C., Butterworth, A., Colin, L., de Jong, I.C., Ferrante, V., Ferrari, P., Haslam, S., Wemelsfelder, F. and Blokhuis, H.J. (2013) Potential risk factors associated with contact dermatitis, negative emotional state, and fear of humans in broiler chicken flocks. *Poultry Science* 92, 2811–2826.

Bizeray, D., Estevez, I., Leterrier, C. and Faure, J.M. (2002a) Effects of increasing environmental complexity on the physical activity of broiler chickens. *Applied Animal Behaviour Science* 79, 27–41.

Bizeray, D., Estevez, I., Letterier, C. and Faure, J.M. (2002b) Influence of increased environmental complexity on leg condition, performance and level of fearfulness in broilers. *Poultry Science* 81, 767–773.

Bokkers, E.A.M. and Koene, P. (2004) Motivation and ability to walk for a food reward in fast- and slow-growing broilers to 12 weeks of age. *Behavioural Processes* 67, 121–130.

Bokkers, E.A.M., de Boer, I.J.M. and Koene, P. (2011) Space needs of broilers. *Animal Welfare* 20, 623–632.

Brake, J., Keeley, T.P. and Jones, R.B. (1994) Effect of age and presence of perches during rearing on tonic immobility fear reactions of broiler breeder pullets. *Poultry Science* 73, 1470–1474.

Bruggeman, V., Onagbesan, O., Ragot, O., Metayer, S., Cassy, S., Favreau, F., Jego, Y., Trevidy, J.J., Tona, K., Williams, J., Decuypere, E. and Picard, M. (2005) Feed allowance – genotype interactions in broiler breeder hens. *Poultry Science* 84, 298–306.

Buijs, S., Keeling, L.J., Vangestel, J., Baert, J., Vangeyte, J. and Tuyttens, F.A.M. (2010) Resting or hiding? Why broiler chickens stay near walls and how density affects this. *Applied Animal Behaviour Science* 124, 97–103.

Buijs, S., Keeling, L.J. and Tuyttens, F.A.M. (2011a) Using motivation to feed as a way to assess the importance of space for broiler chickens. *Animal Behaviour* 81, 145–151.

Buijs, S., Keeling, L.J., Vangestel, C., Baert, J. and Tuyttens, F.A.M. (2011b) Neighbourhood analysis as an indicator of spatial requirements of broiler chickens. *Applied Animal Behaviour Science* 129, 111–120.

Bull, S.A., Thomas, A., Humphrey, T., Ellis-Iversen, J., Cook, A.J., Lovell, R. and Jorgensen, F. (2008) Flock health indicators and *Campylobacter* spp. in commercial housed broilers reared in Great Britain. *Applied and Environmental Microbiology* 74, 5408–5413.

Caplen, G., Hothersall, B., Murrell, J.C., Nicol, C.J., Waterman-Pearson, A.E., Weeks, C.A. and Colborne, G.R. (2012) Kinematica analysis quantifies gait abnormalities associated with lameness in broiler chickens and identifies evolutionary gait differences. *PLoS One* 7, e40800.

Caplen, G., Colborne, G.R., Hothersall, B., Nicol, C.J., Waterman-Pearson, A.E., Weeks, C.A. and Murrell, J.C. (2013) Lame broiler chickens respond to non-steroidal anti-inflammatory drugs with objective changes in gait function: a controlled clinical trial. *Veterinary Journal* 196, 477–482.

Collins, L.M. (2008) Non-intrusive tracking of commercial broiler chickens in situ at different stocking densities. *Applied Animal Behaviour Science* 112, 94–105.

Cooper, M.D. (2013) The development of a protocol to assess the welfare of meat chicken breeds. In: *Proceedings of the 11th World Conference on Animal Production*, Beijing, 15–20 October 2013.

Cornetto, T. and Estevez, I. (2001a) Influence of vertical panels on use of space by domestic fowl. *Applied Animal Behaviour Science* 71, 141–153.

Cornetto, T. and Estevez, I. (2001b) Behavior of the domestic fowl in the presence of vertical panels. *Poultry Science* 80, 1455–1462.

Corr, S.A., Gentle, M.J., McCorquodale, C.C. and Bennett, D. (2003). The effect of morphology on walking ability in the modern broiler. A gait analysis study. *Animal Welfare* 12, 159–171.

Cransberg, P.H., Hemsworth, P.H. and Coleman, G.J. (2000) Human factors affecting the behaviour and productivity of commercial broiler chickens. *British Poultry Science* 41, 272–279.

Dal Bosco, A., Mugnai, C., Rosati, A., Paoletti, A., Caporali, S. and Castellini, C. (2014) Effect of range enrichment on performance, behaviour and forage intake of free-range chickens. *Journal of Applied Poultry Research* 23, 137–145.

Dawkins, M.S., Cook, P.A., Whittingham, M.J., Mansell, K.A. and Harper, A.E. (2003) What makes free-range broiler chickens range? In situ measurement of habitat preference. *Animal Behaviour* 66, 151–160.

Dawkins, M.S., Donnelly, C.A. and Jones, T.A. (2004) Chicken welfare is influenced more by housing conditions than by stocking density. *Nature* 427, 342–344.

D'Eath, R.B., Tolkamp, B.J., Kyriazakis, I. and Lawrence, A.B. (2009) Freedom from hunger and preventing obesity: the animal welfare implications of reducing food quantity or quality. *Animal Behaviour* 77, 275–288.

Decuypere, E., Hocking, P.M., Tona, K., Onagbesan, O., Bruggeman, V., Jones, E.K.M., Cassy, S., Rideau, N., Metayer, S., Jego, Y., Putterflam, J., Tessseraud, S., Colin, A., Duclos, M., Trevidy, J.J. and Williams, J. (2006) Broiler breeder paradox: a project report. *World's Poultry Science Journal* 62, 443–453.

de Jong, I.C. and Guemene, D. (2011) Major welfare issues in broiler breeders. *World's Poultry Science Journal* 67, 73–81.

de Jonge, J. and van Trijp, H.C.M. (2013) The impact of broiler production system practices on consumer perceptions of animal welfare. *Poultry Science* 92, 3080–3095.

de Jong, I.C., van Voorst, S. and Blokhuis, H.J. (2003) Parameters for quantification of hunger in broiler breeders. *Physiology and Behaviour* 78, 773–783.

de Jong, I.C., Fillerup, M. and Blokhuis, H.J. (2005a) Effect of scattered feeding and feeding twice a day during rearing on indicators of hunger and frustration in broiler breeders. *Applied Animal Behaviour Science* 92, 61–76.

de Jong, I.C., Enting, H., van Voorst, A. and Blokhuis, H.J. (2005b) Do low-density diets improve broiler breeder welfare during rearing and laying? *Poultry Science* 84, 194–203.

de Jong, I.C., Wolthuis-Fillerup, M. and van Emous, R.A. (2009) Development of sexual behaviour in commercially-housed broiler breeders after mixing. *British Poultry Science* 50, 151–160.

de Jong, I.C., van Harn, J., Gunnink, H., Hindle, V.A. and Lourens, A. (2012) Footpad dermatitis in Dutch broiler flocks: prevalence and factors of influence. *Poultry Science* 91, 1569–1574.

Dixon, L.M., Sandilands, V., Bateson, M., Brocklehurst, S., Tolkamp, B.J. and D'Eath, R.B. (2013) Conditioned place preference or aversion as animal welfare assessment tools: limitations in their application. *Applied Animal Behaviour Science* 148, 164–176.

Dixon, L.M., Brocklehurst, S., Sandilands, V., Bateson, M., Tolkamp, B.J. and D'Eath, R.B. (2014) Measuring motivation for appetitive behaviour: food-restricted broiler breeder chickens cross a water barrier to forage in an area of wood shavings without food. *PLoS One* 9, e102322.

Duncan, I.J.H. (2009) Mating behaviour and fertility. In: Hocking, P. (ed.) *Biology of Breeding Poultry*. CABI, Wallingford, UK, pp. 111–131.

Edmond, A., King, L.A., Solomen, S.E. and Bain, M.M. (2005) Effect of environmental enrichment during the rearing phase on subsequent eggshell quality in broiler breeders. *British Poultry Science* 46, 182–189.

Estevez, I. (2007) Density allowances for broilers: where to set the limits. *Poultry Science* 86, 1265–1272.

Estevez, I., Mallapur, A., Miller, C. and Christman, M.C. (2010) Short- and long-term movement patterns in complex confined environments in broiler chickens: the effects of distribution of cover panels and food resources. *Poultry Science* 89, 643–650.

Eurobarometer (2007) http://ec.europa.eu/food/animal/welfare/survey/sp_barometer_fa_en.pdf.

Febrer, K., Jones, T.A., Donnelly, C.A. and Dawkins, M.S. (2006) Forced to crowd or choosing to cluster? Spatial distribution indicates social attraction in broiler chickens. *Animal Behaviour* 72, 1291–1300.

Gross, W.B. and Siegel, E.B. (1982) Socialization as a factor in resistance to infection, feed efficiency and response to antigen in chickens. *American Journal of Veterinary Research* 43, 2010–2012.

Gvaryhu, G., Cunningham, D.L. and van Tienhoven, A. (1989) Filial imprinting, environmental enrichment and music application effects on behaviour and performance of meat strain chicks. *Poultry Science* 68, 211–217.

Haslam, S.M. and Kestin, S.C. (2003) Use of conjoint analysis to weight welfare assessment measures for broiler chickens in UK husbandry systems. *Animal Welfare* 12, 669–675.

Haslam, S.M., Brown, S.N., Wilkins, L.J., Kestin, S.C., Warriss, P.D. and Nicol, C.J. (2006) Preliminary study to examine the utility of using foot burn or hock burn to assess aspects of housing conditions for broiler chicken. *British Poultry Science* 47, 13–18.

Haslam, S.M., Knowles, T.G., Brown, S.N., Wilkins, L.J., Kestin, S.C., Warriss, P.D. and Nicol, C.J. (2007) Factors affecting the prevalence of foot pad dermatitis, hock burn and breast burn in broiler chicken. *British Poultry Science* 48, 264–275.

Haslam, S.M., Knowles, T.G., Brown, S.N., Wilkins, L.J., Kestin, S.C., Warriss, P.D. and Nicol, C.J. (2008) Prevalence and factors associated with it, of birds dead on arrival at the slaughterhouse and other rejection conditions in broiler chickens. *British Poultry Science* 49, 685–696.

Hemsworth, P.H., Coleman, G.J., Barnett, J.L. and Jones, R.B. (1994) Behavioural responses to humans and the productivity of commercial broiler chickens. *Applied Animal Behaviour Science* 41, 101–114.

Hepworth, P.J., Nefedov, A., Muchnik, I.B. and Morgan, K.L. (2010) Early warning indicators for hock burn in broiler flocks. *Avian Pathology* 39, 405–409.

Hepworth, P.J., Nefedov, A.V., Muchnik, I.B. and Morgan, K.L. (2011) Hock burn: an indicator of broiler flock health. *Veterinary Record* 168, 303.

Hocking, P.M. (2006) High-fibre pelleted rations decrease water intake but do not improve physiological indexes of welfare in food-restricted female broiler breeders. *British Poultry Science* 47, 19–23

Hocking, P.M. and Jones, E.K.M. (2006) On-farm assessment of environmental enrichment for broiler breeders. *British Poultry Science* 47, 418–425.

Hocking, P.M., Maxwell, M.H. and Mitchell, M.A. (1993) Welfare assessment of broiler breeder and layer females subjected to food restriction and limited access to water during rearing. *British Poultry Science* 34, 443–458.

Hocking, P.M., Maxwell, M.H. and Mitchell, M.A. (1996) Relationships between the degree of food restriction and welfare indices in broiler breeder females. *British Poultry Science* 37, 263–278.

Hocking, P.M., Hughes, B.O. and Keerkeer, S. (1997) Comparison of food intake, rate of consumption, pecking activity and behaviour in layer and broiler breeder males. *British Poultry Science* 38, 237–240.

Hocking, P.M., Maxwell, M.H., Robertson, G.W. and Mitchell, M.A. (2002) Welfare assessment of broiler breeders that are food restricted after peak rate of lay. *British Poultry Science* 43, 5–15.

Hocking, P.M., Zaczek, V., Jones, E.K.M. and MacLeod, M.G. (2004) Different concentrations and sources of dietary fibre may improve the welfare of female broiler breeders. *British Poultry Science* 45, 9–19.

Hocking, P.M., Jones, E.K.M. and Picard, M. (2005) Assessing the welfare consequences of providing litter for feed-restricted broiler breeders. *British Poultry Science* 46, 545–552.

Hubbard (2011) Hubbard Management Guide. http://www.hubbardbreeders.com/media/breeder_nutrition_guide_final2011__016567500_0945_07012015.pdf.

Hue, O., Le Bouquin, S., Laisney, M.J., Allain, V., Lalande, F., Petetin, I., Rouxel, S., Quesne, S., Gloaguen, P.Y., Picherot, M., Santolini, J., Salvat, G., Bougeard, S. and Chemaly, M. (2010) Prevalence of and risk factors for *Campylobacter* spp. contamination of broiler chicken carcasses at the slaughterhouse. *Food Microbiology* 27, 992–999.

Jones, E.K.M., Zaczek, V., MacLeod, M. and Hocking, P.M. (2004) Genotype, dietary manipulation and food allocation affect indices of welfare in broiler breeders. *British Poultry Science* 45, 725–737.

Jones, R.B. (1992) The nature of handling immediately prior to test affects tonic immobility fear reactions in laying hens and broilers. *Applied Animal Behaviour Science* 34, 247–254.

Jones, R.B. and Waddington, D.G. (1992) Modification of fear in domestic chicks, *Gallus gallus domesticus*, via regular handling and early environmental enrichment. *Animal Behaviour* 43, 1021–1033.

Jones, T., Feber, R., Hemery, G., Cook, P., James, K., Lamberth, C. and Dawkins, M. (2007) Welfare and environmental benefits of integrating commercially viable free-range broiler chickens into newly planted woodland: a UK case study. *Agricultural Systems* 94, 177–188.

Jones, T.A., Donnelly, C.A. and Dawkins, M.S. (2005) Environmental and management factors affecting the welfare of chickens on commercial farms in the United Kingdom and Denmark stocked at five densities. *Poultry Science* 84, 1155–1165.

Jordan, D., Stuhec, I. and Bessei, W. (2011) Effect of whole wheat and feed pellets distribution in the litter on broilers' activity and performance. *Archiv fur Geflugelkunde* 75, 98–103.

Kannan, G. and Mench, J.A. (1996) Influence of different handling methods and crating periods on plasma cortico-sterone concentrations in broilers. *British Poultry Science* 37, 21–31.

Kannan, G. and Mench, J.A. (1997) Prior handling does not significantly reduce the stress response to pre-slaughter handling in broiler chickens. *Applied Animal Behaviour Science* 51, 87–99.

Kells, A., Dawkins, M.S. and Borja, M.C. (2001) The effect of a 'freedom food' enrichment on the behaviour of broilers on commercial farms. *Animal Welfare* 10, 347–356.

Knowles, T.G., Kestin, S.C., Haslam, S.M., Brown, S.N., Green, L.E., Butterworth, A., Pope, S.J., Pfeiffer, D. and Nicol, C.J. (2008) Leg disorders in broiler chickens: prevalence, risk factors and prevention. *PLoS One* 3, e1545.

Leone, E.H. and Estevez, I. (2008a) Use of space in the domestic fowl: separating the effects of enclosure size, group size and density. *Animal Behaviour* 76, 1673–1682.

Leone, E.H. and Estevez, I. (2008b) Economic and welfare benefits of environmental enrichment for broiler breeders. *Poultry Science* 87, 14–21.

Leone, E.H., Christman, M.C., Douglass, L. and Estevez, I. (2010) Separating the impact of group size, density, and enclosure size on broiler movement and space use at a decreasing perimeter to area ratio. *Behavioural Processes* 83, 16–22.

LeVan, N.F., Estevez, I. and Stricklin, W.R. (2000) Use of horizontal and angled perches by broiler chickens. *Applied Animal Behaviour Science* 65, 349–365.

Mallapur, A., Miller, C., Christman, M.C. and Estevez, I. (2009) Short-term and long-term movement patterns in confined environments by domestic fowl: Influence of group size and enclosure size. *Applied Animal Behaviour Science* 117, 28–34.

Mason, G.J. (1991) Stereotypies: a critical review. *Animal Behaviour* 41, 1015–1037.

Mason, G.J. (2006) Stereotypic behaviour in captive animals: fundamentals, and implications for welfare and beyond. In: Mason, G.J. (ed) *Stereotypic Animal Behaviour: Fundamentals and Applications to Welfare.* CABI, Wallingford, pp 325–256.

Mench, J.A. (1988) The development of aggressive behaviour in male broiler chicks – a comparison with laying-type males and the effects of feed restriction. *Applied Animal Behaviour Science* 21, 233–242.

Mench, J.A. (1991) Research note: feed restriction in broiler breeders causes a persistent elevation in cortico-sterone secretion that is modulated by dietary tryptophan. *Poultry Science* 70, 2547–2550.

Mench, J.A. (2002) Broiler breeders: feed restriction and welfare. *World's Poultry Science Journal* 58, 23–29.

Millman, S.T. and Duncan, I.J.H. (2000a) Strain differences in aggressiveness of male domestic fowl in response to a male model. *Applied Animal Behaviour Science* 66, 217–233.

Millman, S.T. and Duncan, I.J.H. (2000b) Effect of male-to-male aggressiveness and feed-restriction during rearing on sexual behaviour and aggressiveness towards females by male domestic fowl. *Applied Animal Behaviour Science* 70, 63–82.

Millman, S.T., Duncan, I.J.H. and Widowski, T.M. (2000) Male broiler breeder fowl display high levels of aggression towards females. *Poultry Science* 79, 1233–1241.

Mirabito, L., and Lubac, L. (2001) Descriptive study of outdoor run occupation by 'Red Label' type chickens. *British Poultry Science* 42, S16–S17.

Morrissey, K.L.H., Widowski, T., Leeson, S., Sandilands, V., Arnone, A. and Torrey, S. (2014a) The effect of dietary alter-ations during rearing on growth, productivity and behaviour in broiler breeder females. *Poultry Science* 93, 285–295.

Morrissey, K.L.H., Widowski, T., Leeson, S., Sandilands, V., Arnone, A. and Torrey, S. (2014b) The effect of dietary alterations during rearing on feather condition in broiler breeder females. *Poultry Science* 93, 1636–1643.

Newberry, R.C. (1995) Environmental enrichment – increasing the biological relevance of captive environments. *Applied Animal Behaviour Science* 44, 229–243.

Newberry, R.C. and Hall, J.W. (1990) Use of pen space by broiler chickens: effects of age and pen size. *Applied Animal Behaviour Science* 25, 125–136.

Nicol, C.J. (1992) Effects of environmental enrichment and gentle handling on behaviour and fear responses of transported broilers. *Applied Animal Behaviour Science* 33, 367–380.

Nielsen, B.L., Thomsen, M.G., Sorensen, P. and Young, J.F. (2003) Feed and strain effects on the use of outdoor areas by broilers. *British Poultry Science* 44, 161–169.

Nielsen, B.L. Thodberg, K., Malmkvist, J. and Steenfeldt, S. (2011) Proportion of insoluble fibre in the diet affects behaviour and hunger in broiler breeder growing at similar rates. *Animal* 5, 1247–1258.

Pagazaurtundua, A. and Warriss, P.D. (2006) Levels of foot pad dermatitis in broiler chickens reared in 5 different systems. *British Poultry Science* 47, 529–532.

Reiter, K. (2006) Behaviour and welfare of broiler chicken. *Archiv fur Geflugelkunde* 70, 208–215.

Sandilands, V., Tolkamp, B.J. and Kyriazakis, I. (2005) Behaviour of foot restricted broilers during rearing and lay – effects of an alternative feeding method. *Physiology and Behaviour* 85, 115–123.

Sandilands, V., Tolkamp, B.J., Savory, C.J. and Kyriazakis, I. (2006) Behaviour and welfare of broiler breeders fed qualitatively restricted diets during rearing: are there viable alternatives to quantitative restriction? *Applied Animal Behaviour Science* 96, 53–67.

Sanotra, G.S., Lund, J.D. and Vestergaard, K.S. (2002) Influence of light–dark schedules and stocking density on behaviour, risk of leg problems and occurrence of chronic fear in broilers. *British Poultry Science* 43, 344–354.

Savory, C.J. and Kostal, L. (1996) Temporal patterning of oral stereotypies in restricted-fed fowls: investigations with a single daily meal. *International Journal of Comparative Psychology* 9, 117–139.

Savory, C.J. and Kostal, L. (2006) Is expression of some behaviours associated with de-arousal in restricted-fed chickens? *Physiology and Behaviour* 88, 473–478.

Savory, C.J. and Lariviere, J.M. (2000) Effects of qualitative and quantitative food restriction treatments on feeding motivational state and general activity level of growing broiler breeders. *Applied Animal Behaviour Science* 69, 135–147.

Savory, C.J. and Mann, J.S. (1999) Stereotyped pecking after feeding by restricted-fed fowls is influenced by meal size. *Applied Animal Behaviour Science* 62, 209–217.

Savory, C.J. and Maros, K. (1993) Influence of degree of food restriction, age and time of day on behaviour of broiler breeder chickens. *Behavioural Processes* 29, 179–190.

Savory, C.J., Seawright, E. and Watson, A. (1992) Stereotyped behaviour in broiler breeders in relation to husbandry and opioid receptor blockade. *Applied Animal Behaviour Science* 32, 349–360.

Savory, C.J., Carlisle, A., Maxwell, M.H., Mitchell, M.A. and Robertson, G.W. (1993) Stress, arousal and opioid peptide-like immunoreactivity in restricted-fed and ad-lib-fed broiler breeder fowls. *Comparative Biochemistry and Physiology A – Physiology* 106, 587–594.

Savory, C.J., Hocking, P.M., Mann, J.S. and Maxwell, M.H. (1996) Is broiler breeder welfare improved by using qualitative rather than quantitative food restriction to limit growth rate? *Animal Welfare* 5, 105–127.

Sheppard, K.C. and Duncan, I.J.H. (2011) Feeding motivation on the incidence of floor eggs and extraneously calcified eggs laid by broiler breeder hens. *British Poultry Science* 52, 20–29.

Sherlock, L., Demmers, T.G.M., Goodship, A.E., Mccarthy, I.D. and Wathes, C.M. (2010) The relationship between physical activity and leg health in the broiler chicken. *British Poultry Science* 51, 22–30.

Shields, S.J., Garner, J.P. and Mench, J.A. (2004) Dustbathing by broiler chickens: a comparison of preference for four different substrates. *Applied Animal Behaviour Science* 87, 69–82.

Shields, S.J., Garner, J.P. and Mench, J.A. (2005) Effect of sand and wood-shavings on the behaviour of broiler chickens. *Poultry Science* 84, 1816–1824.

Simsek, U.G., Dalkilic, B., Ciftci, M., Cerci, I.H. and Bahsi, M. (2009) Effects of enriched housing design on broiler performance, welfare, chicken meat composition and serum cholesterol. *Acta Veterinaria Brno* 78, 67–74.

Sorensen, P. and Kestin, S.C. (1999) The effect of photoperiod:scotoperiod on leg weakness in broiler chickens. *Poultry Science* 78, 336–342.

Sosnowka-Czajka, E., Skomorucha, I., Herbut, E. and Muchacka, R. (2007) Effect of management system and flock size on the behaviour of broiler chickens. *Annals of Animal Science* 7, 329–335.

Spinu, M., Benveneste, S. and Degen, A.A. (2003) Effect of density and season on stress and behaviour in broiler breeder hens. *British Poultry Science* 44, 170–174.

Su, G., Sorensen, P. and Kestin, S.C. (1999) Meal feeding is more effective than early feed restriction at reducing the prevalence of leg weakness in broiler chickens. *Poultry Science* 78, 949–955.

Tolkamp, B.J., Sandilands, V. and Kyriazakis, I. (2005) Effects of qualitative feed restriction during rearing on the performance of broiler breeders during rearing and lay. *Poultry Science* 84, 1286–1293.

Tuyttens, F., Vanhonacker, F. and Verbeke, W. (2014) Broiler production in Flanders, Belgium: current situation and producers' opinions about animal welfare. *World's Poultry Science Journal* 70, 343–354.

Utnik-Banas, K., Zmija, J. and Sowula-Skrzynska, E. (2014) Economic aspects of reducing stocking density in broiler chicken production using the example of farms in southern Poland. *Annals of Animal Science* 14, 663–671.

van Emous, R.A., Kwakkel, R., van Krimpen, M. and Hendriks, W. (2014) Effects of growth pattern and dietary protein level during rearing on feed intake, eating time, eating rate, behaviour, plasma corticosterone concentration and feather cover in broiler breeder females during the rearing and laying period. *Applied Animal Behaviour Science* 150, 44–54.

van Krimpen, M.M. and de Jong, I.C. (2014) Impact of nutrition on welfare aspects of broiler breeder flocks. *World's Poultry Science Journal* 70, 139–150.

Ventura, B.A., Siewerdt, F. and Estevez, I. (2010) Effects of barrier perches and density on broiler leg health, fear, and performance. *Poultry Science* 89, 1574–1583.

Verspecht, A., Vanhonacker, F., Verbeke, W., Zoons, J. and Van Huylenbroeck, G. (2011) Economic impact of decreasing stocking densities in broiler production in Belgium. *Poultry Science* 90, 1844–1851.

Weeks, C.A., Nicol, C.J., Sherwin, C.M. and Kestin, S.C. (1994) Comparison of the behaviour of broiler chickens in indoor and free-range environments. *Animal Welfare* 3, 179–192.

Zhao, J.P., Jiao, H.C., Jiang, Y.B., Song, Z.G., Wang, X.J. and Lin, H. (2012) Cool perch availability improves the performance and welfare status of broiler chickens in hot weather. *Poultry Science* 91, 1775–1784.

Zuidhof, M.J., Robinson, F.E., Feddes, J.J.R., Hardin, R.T., Wilson, J.L. and Newcombe, M. (1995) The effects of nutrient dilution on the well-being and performance of female broiler breeders. *Poultry Science* 74, 441–456.

Zulkifli, I. and Azah, A.S.N. (2004) Fear and stress reactions, and the performance of commercial broiler chickens subjected to regular pleasant and unpleasant contacts with human beings. *Applied Animal Behaviour Science* 88, 77–87.

Zulkifli, I., Rasedee, A., Svaadah, O.N. and Norma, M.T.C. (1998) Daylength effects on stress and fear responses in broiler chickens. *Asian–Australian Journal of Animal Sciences* 11, 751–754.

Zulkifli, I., Gilbert, J., Liew, P.K. and Ginsos, J. (2002) The effects of regular visual contact with human beings on fear, stress, antibody and growth responses in broiler chickens. *Applied Animal Behaviour Science* 79, 103–112.

Zupan, M., Berk, J., Wolf-Reuter, M. and Stuhec, I. (2005) Broiler behaviour in three different housing systems. *Landbauforshung Volkenrode* 55, 91–97.

9

Applied Ethology of Laying Hens

In the UK, there are around 35 million laying hens; nearly half are housed in enriched colony cages, most of the rest in free-range or organic systems, and a small proportion is in indoor barn systems. Young hens (pullets) are reared on separate units until they reach the point of lay at around 16 or 17 weeks of age. At this time, they are transferred to their laying house where they live for a further year. The laying hen sector has not yet reached the level of integration seen in the broiler industry, but there are trends in a similar direction, with colony cage houses holding flocks of tens of thousands of birds, and free-range houses holding flocks of up to 16,000 (usually subdivided into smaller subgroups of 4000 birds). Increased specialization means that larger companies can invest in knowledge, innovation and new facilities, and a large flock size will not necessarily compromise welfare (Chapter 6). However, it is challenging to maintain high levels of hen welfare, and it is a mistake to cut back on stockperson care and inspections on large farms.

In the next sections, we will explore the reasons behind the current diversity in commercial hen housing systems before considering the challenges to hen welfare that arise from the inadvertent consequences of continued selection for high egg production, the behavioural constraints imposed by colony cages and the difficulties of managing non-cage systems to avoid problems of feather pecking, smothering, poor health and high mortality.

Cage Systems

Conventional (battery) cages

Battery cages became the system of choice in the second half of the 20th century, with dramatic improvements in bird health due to the separation of birds from their faeces. In addition, farmers appreciated the sloping wire floors that rolled eggs to the front of the cage for easy collection. Originally, just one or two hens were kept in each cage, but production was reduced only marginally when cages were stocked at very high densities, and before long four to six birds were placed in each cage. Even though the production of individual hens declined, the overall production of the facility was maximized. After transfer from the rearing unit, birds were literally pushed into cages on top of other birds and, being so squashed, the simplest movements became difficult or impossible. Despite the hygienic advantages, the battery cage system soon became an iconic symbol of the problems of 'factory farming'.

In 1988, a minimum legal space allowance of 450 cm^2 per bird was prescribed within the European Community, but this did little to stem the tide of opinion and public outcry over the welfare of battery chickens. The space allowance of each bird was described as 'less than an A4 sheet of paper', which allowed easy visualization of the hens' situation (Dawkins and Nicol, 1989). Dawkins and Hardie (1989) showed that a normal

brown hybrid hen occupied more than 450 cm²
simply by standing. Even at allowances greater
than the legal minimum (e.g. 570 cm² per bird),
simple comfort behaviours were rare or non-existent
(Nicol, 1987a) and, as reviewed in Chapter 4,
hens did not acclimatize to spatial restriction
(Nicol, 1987b). Rather, the hens' motivation to
stretch or flap their wings increased over time.
The consequences of severe spatial restriction also
became better known, as Knowles and Broom
(1990) and Gregory et al. (1991) revealed the effect of limb disuse on bone weakness and subsequent fractures during depopulation. Sherwin
et al. (2010), studying some of the last flocks to
be housed in conventional cages in the UK, found
that nearly a quarter of birds from conventional
cages suffered a bone fracture at depopulation,
far higher than in any other system. A 1996 Scientific Veterinary Committee review concluded
that conventional cages resulted in 'inherent severe disadvantages for the welfare of hens'. This
conclusion was supported by the European Food
Safety Authority (Blokhuis et al., 2005) and the pan-
European project LayWel (http://www.laywel.eu;
Blokhuis et al., 2007). In short, the spatial confinement and barren environment of the conventional cage was viewed as a serious welfare issue
because of its severity (hens could not make simple movements or perform basic behaviours such
as nesting, foraging and perching), duration (the
whole laying year) and scope (all birds in the system
were affected).

In 1999, an EU Directive (1999/74/EC) required an increase in space to 550 cm² per bird
from 2003. Fears that this might increase bird aggression proved unfounded, and this was a relatively straightforward transition for the industry.
However, it was insufficient to allow birds to perform full comfort movements or to counter the
problems of bone weakness and fracture at depopulation. The far more significant outcome of
the Directive was the European ban on conventional cages, which took place on 1 January 2012.
In other countries, a shift to alternative systems
is gaining momentum, with a de facto ban on conventional cages in the states of California and
Michigan, and a phasing out of conventional cages
in Tasmania by 2016 and in New Zealand by
2022. Federal legislation is also proposed in the
USA to phase out conventional cages entirely by
2029 (although at the time of writing, this is currently on hold) and requiring their replacement

by systems that allow more space per bird and
the possibility of exhibiting behaviours such as
nesting, perching and scratching (Jendral, 2013).
However, the ban on conventional cages could
not have taken place without the availability of
alternative systems that offered the relative advantages of conventional cages, such as reduced
parasitism, good hygiene and simplicity of management, but additionally enabled a fuller expression of bird behaviour. One such alternative system
is the furnished cage.

Furnished (enriched or modified) small-group cages

The idea that a cage could be altered to provide for
some of the behavioural needs of the laying hens
came originally from animal welfare scientists
who experimented with various designs that incorporated perches and nesting areas and kept
birds in small, socially stable groups. In the UK,
various prototypes were piloted (Appleby, 1990,
1998; Sherwin and Nicol, 1993a,b, 1994) examining features such as nest shape (moulded plastic
hollows versus flat but sloping floors), degree of
enclosure, nest linings and the feasibility of including a dustbath. Furnished cages were also designed and tested in Sweden (Abrahamsson et al.,
1995, 1996), although during the 1990s other
European countries invested effort in developing
non-cage systems, or lobbied for the proposed
cage ban to be deferred or abandoned.

The results of the first commercial-scale trials
of furnished cages examined the performance and
welfare of hens housed in groups of four to eight
birds, at spatial allowances of between 470 and
875 cm² per bird (Appleby et al., 2002). The trials
were generally successful, with high utilization
of nests for egg-laying during the day and of
perches for roosting at night. Rates of comfort
behaviour increased, and the birds' plumage
and foot condition was improved relative to
birds in conventional cages (Appleby et al., 2002).
Mortality was generally low even when birds had
intact beaks (range 2.9–6.8%), although beak
trimming further reduced the risk of pecking injuries (Croxall et al., 2005). A comprehensive review of furnished cage trials from across Europe
reached largely positive conclusions (Blokhuis
et al., 2005).

Provision for dustbathing is not a requirement of Directive 1999/74/EC, although early designs of furnished cages tried to cater for this behaviour. However, a small box containing a litter substrate was not recognized as an adequate dustbath by most hens, with only 27% of dustbathing taking place in the facility provided (Lindberg and Nicol, 1997). Dustbathing behaviour was more likely to take place on the cage floor, with the notional 'dustbath' used for unintended and problematic purposes such as refuge or egg-laying (Olsson and Keeling, 2002; Wall *et al.*, 2008). Merrill and Nicol (2005) and Merrill *et al.* (2006) wondered whether it might be possible to accept that hens were going to dustbath on the cage floor but to modify the floor to provide greater comfort and feedback for this behaviour. They found that coating the wire with a material that could be pecked, or covering it with artificial turf, perforated to allow faeces to drop through, both encouraged dustbathing and improved plumage and foot condition. Such modifications are a compromise, providing a substrate that is less preferred than sand or peat but better than bare wire (Alvino *et al.*, 2013) (see Chapter 5). However, there have as yet been no attempts to modify commercial cage floors, perhaps due to the cost and hygiene implications.

These pilot studies demonstrated that furnished cages could provide better welfare for hens alongside relatively easy management and good production. By the turn of the new millennium, commercial manufacturers had begun to produce cage systems compliant with the EU Directive. Up to this point, scientists had trialled cages that housed only four or five birds, aware that even small increases in group size to 10 or 16 individuals could increase feather pecking and skin damage (Hetland *et al.*, 2004; Shimmura *et al.*, 2009). However, economic pressures soon led commercial manufacturers to produce cages for 20, 40, 60 or 80 birds. The reduction in capital cost for farmers wanting to invest in a new system won the day, and furnished cages for larger groups of hens became known as colony cages. These now prevail.

Enriched colony cages and Kleingruppenhaltung systems

Colony cages (produced by manufacturers such as Big Dutchman, Tecno, Valli, Facco, Salmet,

Chore-Time and others) meet the requirements of Directive 1999/74/EC by providing birds with 750 cm^2 per bird, perches, nests, and a foraging and scratching area. No explicit provision is made for dustbathing, and birds are housed in groups of between 60 and 100 birds in those European countries that have not banned all cage systems. In Germany, a higher welfare standard was applied and a Kleingruppenhaltung system developed where hens must be kept in groups of no more than 60, with a space allowance of 800 cm^2 per bird (or more for heavier hens), and perches at two heights.

The increase in group number and cage size of the enriched colony compared with the small-group furnished cage brought drawbacks and benefits. In a group of 80, hens may recognize their cagemates, but it is not clear whether stable social hierarchies are formed. This may be socially more difficult than living in a very large (anonymous) flock. Contact with a large number of birds also increases the damage resulting from feather pecking. Stress measured by heterophil: lymphocyte ratio was higher in birds in groups of 60 compared with birds in groups of 40 within the Kleingruppenhaltung system (Scholtz *et al.*, 2008). On the other hand, a larger cage size permits more locomotion (Weitzenburger *et al.*, 2006; Shimmura *et al.*, 2009) and birds in larger colony systems can have lower stress levels than birds in small-group furnished cages (Scholtz *et al.*, 2008). The increased activity in larger colony cages increases bone strength but paradoxically and simultaneously increases the risk of collision and keel fracture (Scholtz *et al.*, 2009). Overall, no significant effects of increasing group size on mortality or production have been reported for increases from eight to 40 (Wall, 2011) or from 20 to 60 (Huneau-Salaun *et al.*, 2011). The general experience with large group sizes has therefore been more positive than expected, and it has not been necessary to implement innovations such as escape routes between compartments (Wall *et al.*, 2004).

How should pullets destined for a life in a colony cage be reared? They might benefit from the greater freedom of a non-cage system during the rearing period, but they might then become frustrated (see Chapter 4) by the subsequent restrictions of cage life. Findings have been mixed. Mortality was higher in furnished cages for birds that had been reared in aviaries, largely due to injurious pecking, which may have been triggered by frustration (Tahamtani *et al.*, 2014). In contrast, no significant differences were detected between cage-reared and floor-reared birds on later plumage condition (Roll *et al.*, 2009) and, while floor-reared birds housed in furnished cages had higher heterophil:lymphocyte ratios, possibly indicating greater stress levels, they also had an improved immunological response to antigen presentation (Moe *et al.*, 2010). Indeed, Moe *et al.* (2010) highlighted the inherent difficulty of interpreting immune parameters such as white blood cell counts as indicators of stress. Overall, the results suggest that pullets are better off in non-cage systems and that this will not compromise their subsequent adaptation to colony cages.

Welfare and resource use in furnished and enriched colony cages

The distinction between furnished cages and enriched colony cages in the preceding sections relates only to group size. New designs of furnished cage and colony cage are still being developed, and in reviewing their benefits relative to conventional cages, it is difficult to draw clear boundaries. For this reason, they will be considered together in this section. The studies come from a range of countries (many non-European) now investigating the potential of modified cages, which seems a highly encouraging development.

Compared with conventional cages, hens in furnished or colony cages show low levels of aggression (Hetland *et al.*, 2003; Shimmura *et al.*, 2006) and more comfort behaviour (Shimmura *et al.*, 2006; Pohle and Cheng, 2009); they are more resistant to experimental infection with *Salmonella* (Gast *et al.*, 2013), more able to dissipate heat during hot weather (Guesdon and Faure, 2004; Guo *et al.*, 2012), have higher bone strength (Vits *et al.*, 2005; Jendral *et al.*, 2008; Tactacan *et al.*, 2009), faster acclimatization at placement (Shimmura *et al.*, 2006) and good rates of production (Roll *et al.*, 2005; Tauson and Holm, 2005; Vits *et al.*, 2005). Intriguingly, hens housed for a year in furnished cages are also more willing to ford a water bath to get to a dustbath than hens kept for the same time in conventional cages (Orsag *et al.*, 2012), perhaps because they are physically stronger or still retain a sense of control over their environment. Overall, furnished and enriched colony cages are substantially better than conventional cages for nearly all aspects of bird welfare. However, it must be noted that no differences in physiological profile have been detected between hens housed in cages with or without the perch, nest or dustbath resources (Barnett *et al.*, 2009), and also that the risk of incurring a fracture during the laying period is slightly greater in modified than in conventional cages due to greater opportunity for collision (Sherwin *et al.*, 2010). However, this is offset by greater bone strength (Abrahamsson and Tauson, 1993; Wilson *et al.*, 1993; Barnett *et al.*, 2009) and a lower prevalence of fracture at depopulation (Sherwin *et al.*, 2010).

Furnished and colony cages are, of course, not problem free. In small groups, subordinate birds may be excluded from facilities by dominant birds (Shimmura *et al.*, 2007) and end up using the nests as a refuge area (Shimmura *et al.*, 2008). Small details of cage design and resource management are critical. Australian studies of furnished cages report nest usage rates of less than 70% (Cronin *et al.*, 2005; Barnett *et al.*, 2009), whereas in European studies, nest usage of over 85% is the norm (Tauson and Holm, 2005; Huneau-Salaun *et al.*, 2011; Wall, 2011; Guinebretiere *et al.*, 2013). Clearly, studies to optimize nest design for all markets are still required. Nests in furnished cages are essentially small boxes with a solid or wire-mesh cage partition on one side and a wall at the back. Nest entrances covered with curtains are situated on one or two sides, and the nest floor may be made of artificial turf, a flat rubber mat or a rubber mat with raised finger-like projections. Curtain designs also vary, and this will influence nest use because a degree of seclusion during nesting is preferred by hens (see Chapter 5). Nest use is improved when artificial turf is used rather than bare wire (Wall *et al.*, 2002; Struelens *et al.*, 2005; Guinebretiere *et al.*, 2013), but not all types of nest floor have yet been compared, and novel possibilities (e.g. perforated netting seems attractive to the hens, while also being more hygienic; Wall and Tauson, 2013) have not yet been commercially adopted.

Perches are widely used for night-time roosting but have mixed effects on plumage condition, with improvements to feather cover on the back but reduced condition of the breast and tail areas (Hester *et al.*, 2013). A small number of hens lay while perching, which can result in dirty and broken eggs. This problem may be reduced if low (7 versus 24 cm) perches are used (Tuyttens *et al.*, 2013), as the resultant increased disturbance of daytime perching leads to a greater utilization of the nest (Tuyttens *et al.*, 2013). However, this solution is something of a trade-off, and it might be better to try to improve the attractiveness of nests for the perch layers. That said, the optimum height of a perch within a cage environment is not known. Low perches are used for night-time roosting and may even be preferred by birds in cages where the distance from the perch to the cage ceiling is a constraint (Chen *et al.*, 2014).

Providing an area for scratching and foraging activities is also problematic. Most designs of colony cages include a mat placed over an area of wire floor where small amounts of feed are distributed (Fig. 9.1), but these behaviours are rarely expressed in their full form (see Chapter 5). In addition, Guinebretiere *et al.* (2012) found that rubber mats were readily destroyed and that adding litter only increased wear and tear on the mat.

Non-cage Systems

The alternative to housing laying hens in furnished or enriched colony cages is to keep them loose in flocks of several thousands. Non-cage systems include indoor barn, aviary or multi-tier houses where birds are kept indoors, and free-range and organic

Fig. 9.1. A hen in the foraging and scratching area of an enriched colony cage. The rubber fingers have eroded leaving only a minimal substrate. (Photo courtesy of Professor Christine Nicol, University of Bristol, UK.)

systems where outdoor range access is also provided. Non-cage systems for laying hens generally provide a ground-level littered area and raised slats or platforms cover the remainder of the house.

The requirement that adequate perches should be provided in non-cage systems has been interpreted in different ways by various EU member states. Most countries have assumed that a degree of elevation is an essential attribute of a perch, given that hens prefer to roost at height (see Chapter 5). However, others have considered a perch to be adequate if it enables foot grasping and, in England and Wales for example, shaped perches are integrated within and at the same height as the normal slatted or tiered floors. The European Food Safety Authority is due to issue advice on this interpretation in 2015.

In the UK, both indoor and outdoor systems can be included within welfare assurance schemes such as RSPCA Freedom Food if they meet the detailed criteria specified by the RSPCA Welfare Standards for Laying Hens. These standards place an upper limit of 16,000 hens per house, and also specify that hens should be divided into colonies (separated within the house by wire-mesh partitions) of no more than 4000 birds. In Germany, indoor barn systems should house no more than 6000 birds. Organically produced birds are usually kept in smaller flocks of 3000 or less. The reasons for limits of flock size are often not clearly given, and there is no strong scientific evidence that the welfare of hen flocks of a few thousand is better than those kept in flocks of tens of thousands. In commercial non-cage flocks, scientific studies have found no increased risk of injurious pecking in adult flock sizes ranging from 250 to 4000 (Oden *et al.*, 2002), 225 to 9950 (Gunnarsson *et al.*, 1999), 800 to 23,000 (Green *et al.*, 2000), 540 to 19,500 (Lambton *et al.*, 2010) or 276 to 15,848 (Gilani *et al.*, 2013). Furthermore, during the rearing period, the risk of gentle feather pecking was actually reduced in larger commercial flocks (ranging from 282 to 38,822) (Gilani *et al.*, 2013). However, there may be some disadvantages of very large flock sizes in free-range systems, as a smaller proportion of the flock tends to access the outdoor range.

Aggressive behaviour is not generally a significant problem in non-cage systems, with infrequent occurrences of actual fights, and no greater risk of aggressive pecking than in caged systems

(Freire and Cowling, 2013). Aggressive threats and pecks have been reported to occur at rates of less than one interaction per bird per hour in many large or commercial flocks (Hughes *et al.*, 1997; Carmichael *et al.*, 1999; Nicol *et al.*, 1999; Oden *et al.*, 2000) but can be higher in smaller flocks of less than 100 or so birds, where birds may attempt to form social hierarchies (Nicol *et al.*, 1999). Aggression can also be localized at points of resource competition (e.g. just outside nest boxes; Lentfer *et al.*, 2013) and so house designs that minimize local crowding will generally be beneficial. In many other respects, it is a significant challenge to manage commercial flocks of laying hens in non-cage systems. The increased freedom for hens to move about and access different areas and tiers (Fig. 9.2) has benefits but also can also result in problems.

Fig. 9.2. Hens have considerable freedom to move within multi-tier systems, selecting the litter floor area for foraging and dustbathing, the curtained areas on middle tiers for nesting, and the elevated positions for daytime vigilance and night-time roosting. However, good system design and high standards of management are needed to avoid injuries and bone fractures due to collisions and poor landings. (Photo courtesy of Isabelle Pettersson, University of Bristol, UK.)

The emergent behaviour of large groups is not always predictable from our knowledge of the behaviour of individual birds. If problems do arise, they can be difficult to contain, as birds interact with each other in unpredictable ways. Some factors for brief consideration next include uneven bird distribution, smothering, the increased risk of bone fractures due to birds colliding with objects within the house, the production consequences of hens laying eggs outside the nest boxes and poor use of the outdoor area in free-range systems. Feather pecking and cannibalism will also be reviewed in this section. Although feather pecking also occurs in cages (with an incident rate similar to non-cage systems), damage is usually greater in non-cage systems. This is because pecking birds can reach many more victims leading to a faster and more far-reaching spread of the problem and because it is difficult to identify and remove the feather pecking birds in large loose-housed flocks (Rodenburg *et al.*, 2013).

Spatial distribution

Birds do not distribute themselves evenly throughout non-cage housing systems (e.g. Lentfer *et al.*, 2013). Some areas are focal points for competition, and individual birds may make use of some areas more than others. A point of debate for some years has been whether individual birds in large flocks organize themselves into subgroups, either for social reasons or because of shared preferences for particular environmental features. Hughes *et al.* (1974) marked birds in flocks of four to 600 with coloured tags and took repeated photographs of their locations. Similarly, Appleby *et al.* (1989) fitted tags to 100 birds in flocks of up 400 and took regular scans of their location. In both studies, there was evidence of location preferences by some birds but no evidence that particular groups of birds stayed together in subgroups. Oden *et al.* (2000) studied flocks of 500 or 600. These researchers caught and marked hens that were in close proximity, dyed the members of each roosting group a particular colour and then observed whether these groups stayed together subsequently. Generally, birds did not stay together during the daytime. Over subsequent nights, birds from the middle sections of pens chose variable roosting positions, but birds from the ends of pens were consistently found roosting in similar locations. However, the work reviewed in Chapter 6 suggests that this consistency is likely to reflect shared resource preferences rather than any form of social bonding.

Non-cage systems provide more options for escape or refuge from other birds than cages or small pens. The increased freedom to move away from troublesome conspecifics may be one reason why some hens prefer to be in larger than smaller groups (Lindberg and Nicol, 1996). However, these studies have been conducted on small experimental groups of birds, and the extent to which the results can be generalized to larger commercial flocks is unclear. As we saw in Chapter 6, the welfare of severely persecuted birds in non-cage systems can be very poor. To avoid being attacked, these victim birds take shelter in unsuitable areas of the house. This was demonstrated in a study by Freire et al. (2003) who fitted radio frequency transponders to birds within a 1000-bird experimental perchery house and recorded individual bird location. They found that the most persecuted birds spent the greatest time in a slatted area beneath tiers of perches, rarely venturing out for food or water.

Smothering

Smothering, whereby birds aggregate tightly together, often piling on top of each other and causing suffocation, is a significant cause of mortality in non-cage laying hens, which can account for up to 40% of all mortality in some flocks (Bright and Johnson, 2011). Smothering can occur in a variety of different situations. For example, large numbers of birds can die following sudden occurrences of panic. Richards et al. (2012) took repeated 24 h video recordings of flocks of 6000 hens and noted approximately three panic episodes per flock per day, many involving just a proportion of the flock, but 15% of panic episodes involved all of the birds fleeing together to one part of the house. Episodes of panic appeared to decrease with flock age and could potentially be reduced by acclimatizing young birds to a wide range of sight and sounds. However, panic is not the only cause of smothering. Bright and Johnson (2011) found that farmers reported some very different scenarios. For example, laying hens can pile on top of each other inside nest boxes, or can engage in a

form of 'creeping' smothering, which can occur in any part of the house or even on the outdoor range. Our own observations show hens circling slowly and pushing underneath each other as if seeking shelter or cover. Although these latter two types of smothering may involve fewer birds than when a whole flock panics, the frequent and unexplained occurrence of small smothering events can lead to high overall mortality and presents a major headache for farmers.

Collisions and bone fractures

Laying hens are heavier than junglefowl, although they do not have increased wing strength to compensate. They are therefore prone to make somewhat clumsy landings, particularly when flying downwards (see Chapter 5) (Moinard et al., 2004). Although the opportunity for exercise afforded in a non-cage system results in stronger bones, the increased risk of collisions and flying accidents negates this benefit, and a high prevalence of keel bone fractures is generally found in non-cage systems (50–86% of birds; Freire et al., 2003; Wilkins et al., 2004; Nicol et al., 2006; Wilkins et al., 2011). The prominent anatomical position of the keel bone makes it most vulnerable to accidental crash landings. Keel bone fractures appear to be painful as they are associated with an increased reluctance of hens to jump or fly downwards (Nasr et al., 2012a), a state that is partially revoked by the experimental use of analgesic drugs (Nasr et al., 2012b, 2013). The positioning of perches and other structures within a non-cage house is therefore crucial to protect hen welfare (Sandilands et al., 2009; Wilkins et al., 2011). New and innovative research at the University of Bern in Switzerland is exploring the potential of 'soft perches' and strategically placed ramps in reducing collision damage.

Floor eggs

In furnished or colony cages, any eggs not laid within the nest area will be laid on the wire floor and will roll to the front of the cage for automated collection. In non-cage systems, however, if birds do not use the nests provided, then the resulting 'floor eggs' are difficult to collect and are more likely to be dirty or broken. Many producers employ staff to walk around the house and outdoor range areas to pick up mislaid eggs, a practice that is expensive for the producer and one that is time-consuming and physically demanding for workers (Scott and Lambe, 1996). It is important that laying hens or broiler breeders recognize and accept nest boxes as soon as they enter the laying house. This can be encouraged by providing perches or raised tiers during the rearing period to allow birds to practice the stepping and hopping movements that may be needed to access nest boxes (Appleby et al., 1988; Gunnarsson et al., 1999) and by making nest access as easy as possible in the commercial house, for example by placing platforms rather than perches just outside the nests (Stampfli et al., 2013). Allowing birds to inspect nests before the onset of sexual maturity is also beneficial (Sherwin and Nicol, 1993a). When the behaviour of individual birds is examined, it can be seen that floor layers are a consistent minority group of birds that appear to find commercially provided nests unsuitable for some reason (Cooper and Appleby, 1996). Providing nests with the features that are most preferred by hens (reviewed in Chapter 5) should therefore reduce the incidence of floor eggs without causing additional problems. In practice, many of the measures taken by producers to try to encourage birds to use the nests during the first few weeks in the laying house (such as confining birds to the slatted areas near the nest boxes, not allowing access to litter substrates, not allowing access to the outdoor range, placing electrified wires on parts of the litter floor where eggs might be laid) can cause other unintended problems, particularly increasing the risk of injurious pecking. When floor-egg management strategies are modified (e.g. to allow newly housed birds access to litter during afternoon periods when most egg-laying has finished), levels of injurious pecking are decreased (Lambton et al., 2013).

Range use

It is disconcerting to find that on many free-range farms, at many times of day, rather few laying hens are observed on the outdoor range (Fig. 9.3a), despite large outdoor areas being potentially available in compliance with the requirements of EU law or assurance certification bodies. Consumers who have paid for free-range eggs would certainly like to see the range better utilized. Producers, however, may hold mixed views. They will have invested in the provision of the outdoor area but may be aware that use of the range can sometimes be associated with problems such as water entering the house via the pop-holes or on the birds' feet, hens laying eggs on the range rather than in the nest boxes, and increased risk of predation or disease transmission. Producers may be less aware of the benefits that good range use brings, including savings on feed costs due to bird foraging (Horsted and Hermansen, 2007), reductions in internal house stocking density and significantly reduced risks of injurious pecking (Green et al., 2000; Nicol et al., 2003; Lambton et al., 2010).

The percentage of a flock that ventures outside has been estimated in many studies, summarized in Table 9.1. The percentage of birds seen out at any one point in time is not the only measure of range use. A low number of birds on the range at any one time could arise either if a small proportion of the flock use the range for extended periods or if a larger proportion of the flock use the range but any individual bird stays out for a short period. Richards et al. (2011) fitted 600 birds (from four flocks of 1500 birds) with radiofrequency identification tags and showed that 50% of the tagged birds were logged at the pop-holes on 50% of the study days. Only 16% of birds were rarely or never logged at the pop-holes. In a recent paper, Gebhardt-Henrich et al. (2014) found that 47–90% of their tagged hens in 12 flocks of 2000–18,000 birds used the range at least once in a period of approximately 21 days while approximately 20% used it daily. Keeling et al. (1988) observed a 600-bird flock and found that only 6% of birds never left the house and only 19% used it rarely. Icken et al. (2011) reported that over a 6 h period from 10:00 to 16:00, tagged hens spent an average of 1 h 29 min (range 2 min to 3 h 16 min) in an outdoor (sheltered) winter garden. These studies show that, even when the proportion of the flock on the range at any one time is relatively low, the majority of birds are obtaining some access to the outdoors for part of the day.

Good range use is also indicated by the extent of the available range area that is used. Flocks vary in the distribution of birds on the range, and many authors have commented that most birds are observed close to the house. Furmetz et al. (2005) found that 25% of ranging birds were more than 20 m from the house. Gilani et al. (2014) found

Fig. 9.3. (a) Poor use of the outdoor range area. Very few birds venture out to a range that has little natural or artificial cover. (b) Good range use encouraged by pasture management. Birds are also attracted to the cover provided by mature trees. (Photos courtesy of Jon Walton, University of Bristol, UK.)

that 37% of ranging birds were found more than 5 m from the house. Good range management facilitates a more even distribution of birds on the range (Fig. 9.3b).

Much of the variation in hen range use can be ascribed to factors that have little to do with management. Ranging is greatly influenced by time of day, with morning and early evening peaks (Bubier and Bradshaw, 1998; Mahboub *et al.*, 2004; Hegelund *et al.*, 2005; Richards *et al.*, 2011). Weather has a large influence on range use (Keeling

et al., 1988; Nicol *et al.*, 2003; Hegelund *et al.*, 2005; Richards *et al.*, 2011), as shown in Table 9.1 where the results of Nicol *et al.* (2003) indicated a fourfold increase in range use under optimal weather conditions compared with poor weather conditions. Ranging can change with bird age, but no consistent pattern has been established. Keeling *et al.* (1988) and Zeltner and Hirt (2003) found more range use in older birds, but Hegelund *et al.* (2005) found reduced range use in older birds. These influences must be considered as they are

Table 9.1. Percentage of hens observed out on range during time point sampling.

Reference	Study scope	Flock size	Percentage out (range, if given)
Gilani et al. (2014)	33 flocks on 28 farms	92–15,848	13% (1–58%)
Sherwin et al. (2013)	19 flocks	1000–16,000	26% of farms < 11% out; 32% of farms 11–25% out; 37% of farms >25% out
Harlander-Matauschek et al. (2006)	Eight flocks in experimental unit	256	30–40%
Hegelund (2005)	37 flocks (organic) on five farms	513–6,000	9% (2–24%)
Nicol et al. (2003)	50 FP flocks on 36 farms (researcher observations)	Average 5000	13.9%
Nicol et al. (2003)	50 flocks where no FP observed on 34 farms (researcher observations)	Average 5000	22.1%
Nicol et al. (2003)	50 FP flocks (farmer estimates)	Average 5000	8.1% (wet); 34% (dry, still)
Nicol et al. (2003)	50 flocks where no FP observed (farmer estimates)	Average 5000 birds	13.7% (wet); 46% (dry, still)
Bestman and Wagenaar (2003)	63 flocks (organic) on 26 farms	Farmers' estimate under best possible conditions	20% of farms <25% out; 38% of farms 26–50% out; 7% of farms 51–75% out; 38% of farms >75% out
Zeltner and Hirt (2003)	Eight flocks	420–511	22%
Zeltner and Hirt (2008)	Eight flocks	20	57%
Keeling et al. (1988)	One flock	600	14–22%
Grigor (1993)		1–7000	40%
Bubier and Bradshaw (1998)	Four flocks	490–2450	5–42%

FP, feather pecking.

important. However, good management and farming practices can moderate the influence of inclement weather (by providing shelters on range, good drainage around pop-holes, shaded areas and windbreaks). In addition, many other farming and management practices influence range use in laying hens.

One of these factors is the age at which birds first experience the outdoor area. Birds are initially fearful of the outdoor range but become less frightened with increasing familiarity (Grigor et al., 1995b). Age at first range access is particularly important in attenuating fear responses. Early range experience has been shown experimentally to increase birds' readiness to emerge into an outdoor environment (Grigor et al., 1995a). Groups of 56 birds aged between 12 and 20 weeks were given experience of handling and/or early outdoor

experience. While handling did not affect ranging tendency, early experience resulted in hens that emerged earlier into outdoor areas as adults, roamed further and were seen to be less fearful. Commercial organic flocks in the Netherlands are sometimes placed on the laying farm at a much younger age than in usual practice, allowing an evaluation of the effect of age of outdoor access on later range use. Bestman and Wagenaar (2003) found that the earlier birds were placed on the laying farm and allowed outdoor access, the more they subsequently ranged. Despite this, Gilani et al. (2014) found no effect of range access during rear, on later tendency to range in commercial UK free-range hens. Restricting birds to the slats when they first arrive in the laying house is a practice adopted by some farmers in an attempt to discourage floor eggs. However, this restrictive practice has many disadvantages. Although there is little or no scientific evidence on the direct consequences of restriction to slats for future ranging behaviour, it seems likely that restricting birds to slats for some weeks will decrease their motivation and ability to explore the floor area when it is made available. There is an increasing trend to allow birds immediate or early access to litter when they arrive at the laying house.

There is good evidence that the proportion of birds ranging tends to decrease with increasing flock size. This general relationship applies over a wide range of actual flock sizes. In a comprehensive study of commercial free-range flocks, including flocks of up to 16,000 birds, Gilani *et al.* (2014) produced a quantitative model that showed the declining percentage of birds on the range expected as flock size increased. Gebhardt-Henrich *et al.* (2014) also found that flock size was negatively associated with the percentage of days spent on the range and the duration of time spent on the range in tagged birds from commercial flocks of flock sizes of 2000–18,000 (Gebhardt-Henrich *et al.*, 2014). Even in smaller flocks, the same type of negative relationship appears to hold (Grigor, 1993; Bubier and Bradshaw, 1998; Bestman and Wagenaar, 2003; Hegelund *et al.*, 2005).

One of the principal reasons for the effect of flock size is that, at equal stocking density, larger flocks are kept in larger houses. Thus, the average distance a bird must travel to reach a pop-hole will be greater, particularly (but not only) in houses where pop-holes are provided on one side only. As a simple example, if ten birds are housed in an indoor pen of 1 m² (1 × 1 m) with one pop-hole provided on one side, the furthest a bird can be from a pop-hole will be approximately 1.1 m. In comparison, consider 150 birds housed in an indoor pen of 15 m² (3 × 5 m) with 15 pop-holes provided on the longer side. Their stocking density is the same (ten birds m⁻²) and there are still ten birds per pop-hole, but now (assuming an even distribution of birds) many individual birds will be approximately 3 m away from the nearest pop-hole. This problem will scale up with increasing house size. Stocking density also has a negative effect on range use (Gilani *et al.*, 2014), possibly because it is more difficult for birds to reach pop-holes if they have to pass many birds.

Birds are far more likely to venture out if the range area is more than just a bare patch of earth or short grass. The positive effect of cover (such as shelters, bushes or other vegetation) on the percentage of the flock using the range is well documented experimentally and in commercial flocks. Range cover can reduce bird fearfulness by increasing their perception of safety from predators, and features of the range that stimulate foraging, feeding or dustbathing behaviours are also attractive to the birds. In a study of 100 commercial free-range flocks, better range use reduced the risk of feather pecking and use of the range was positively associated with the presence of trees, bushes or hedges on the range (Nicol *et al.*, 2003). Gilani *et al.* (2014) found a strongly significant effect of cover on the range in increasing the percentage of outdoor birds that ranged far from the house, i.e. in improving bird distribution on the range, while Sherwin *et al.* (2013) reported a significant positive relationship between the percentage hens using the range and the total area of natural and artificial shelter available on the range. Other studies of smaller flocks have also revealed beneficial effects of range cover in increasing the average number of birds outside (Bestman and Wagenaar, 2003; Hegelund *et al.*, 2005; Harlander-Matauschek *et al.*, 2006). Bubier and Bradshaw (1998) noted that birds housed with a range that provided tall grasses ranged more than birds on three farms where vegetation was short, trampled and barren.

Encouraging hens to use the furthest reaches of the range, rather than clustering in the areas immediately outside the house, is more difficult. Zeltner and Hirt (2003) found that adding roofed sandboxes and avoiding narrow corridors both improved bird distribution on the range.

Taylor *et al.* (2004) reported that small groups of hens were more likely to enter outdoor areas where opaque vertical screens were provided than to enter areas with transparent screens or no screens.

There has been little scientific research on the effect of range area or shape on the propensity of hens to use the outdoor range. This may be because regulations about range area were set early on before major growth in the free-range sector and thus there is little variation in European systems that could be studied. The primary reason for setting regulations on outdoor stocking density was to protect bird health by reducing the risk of pasture contamination with parasites and other infectious diseases and to maintain pasture quality. The European requirement is that there should be no more than 2500 birds ha^{-1}, i.e. one hen per 4 m^2 or 0.25 hens m^{-2} (unless paddock rotation is practised). Levels of worm and protozoan infections are often high in free-range production (Maurer et al., 2013; Sherwin et al., 2013). Mortality due to these conditions is much higher in free-range hens than in hens in other systems (Fossum et al., 2009). There is evidence that good pasture management, including rotation, can reduce the risks (Maurer et al., 2013). Sherwin et al. (2013) found a reduction of approximately 50% in worm egg counts when outdoor stocking density was <1000 hens ha^{-1} compared with >1000 hens ha^{-1}. Perhaps counterintuitively, a higher *outdoor* stocking density can increase range use (Sherwin et al., 2013). This may because the presence of other birds on the range provides a sense of safety and security. The other birds are to some extent a form of range enrichment that can encourage other birds from the house.

Birds access the range via pop-holes or shutters that are open during the day and closed at night. In the UK, a total opening of at least 2 m per 2000 birds should be provided. Sherwin et al., (2013) found more ranging in commercial flocks where a greater number of pop-holes was provided and, similarly, Gilani et al. (2014) found increased range use with more pop-hole space per bird. The total opening width in this study averaged just 0.5 m per 1000 birds for adult hens, possibly because farmers shut some of the potentially available pop-holes during inclement weather. In smaller flocks, the number and size of pop-holes appears, within reason, to have no strong effect on ranging behaviour (Keeling et al., 1988; Harlander-Matauschek et al., 2006).

Injurious pecking

The factors influencing feather and tissue pecking have been studied extensively and recently reviewed in full by Nicol et al. (2013) and Rodenburg et al. (2013). The genetic aspects of this behaviour were also considered in Chapter 1. As mentioned earlier in the current chapter, feather pecking does occur in cage systems, but it is more of a challenge in non-cage systems as it can spread so rapidly. Studies of commercial systems have found that injurious pecking occurs within a majority of laying (Green et al., 2000; Lambton et al., 2010) and rearing (Gilani et al., 2013) flocks.

Feather pecking can be gentle (GFP), where birds peck at the tips of their conspecifics' feathers, or severe (SFP), with feathers pulled out and sometimes eaten. The different types of pecking can be seen at http://www.featherwel.org/injuriouspecking.

GFP is commonly observed during rear (reviewed by Nicol et al., 2013), and it does not necessarily progress or mutate into SFP (Rodenburg et al., 2013). On its own, GFP results in little overt damage, although if feathers are frayed they can become more attractive targets for more severe forms of pecking. SFP has by far the bigger impact on bird welfare and flock productivity, as it is painful for the recipient and results in substantial plumage loss (Gunnarsson et al., 1999; Bestman and Wagenaar, 2003; Drake et al., 2010), skin damage, increased susceptibility to infection (Green et al., 2000), loss of production, increased demand for food and higher mortality (Nicol et al., 2013; Rodenburg et al., 2013). The exposure of bare patches of skin can trigger subsequent tissue pecking, which can result in rapid mortality and cannibalism of living birds or of carcasses. There is a relationship between feather pecking and cannibalism (Cloutier et al., 2000; McAdie and Keeling, 2000; Pötzsch et al., 2001), and specifically between SFP and vent pecking (Lambton et al., 2014). However, severe injurious pecking directed to the vent area or near the preen gland can also arise spontaneously in the absence of prior feather pecking, so the risk factors for each type of injurious pecking should be considered both separately and together. The risk of all of these more serious types of injurious pecking is greatly increased as birds come into lay (Nicol et al., 2013).

All types of injurious pecking appear to be a form of normal pecking redirected inappropriately to another bird. For example, if foraging, dustbathing or exploratory substrates are less attractive than the feathers of a neighbour, then pecks, normally directed towards litter particles, may be directed at feathers instead. This hypothesis is supported by studies showing that absent or poor-quality litter is a major risk factor for injurious pecking (Vestergaard and Lisborg, 1993; Huber-Eicher and Wechsler, 1997; Nicol et al., 2001; de Jong et al., 2013a,b). Ideally, good-quality litter should be present during rear as chicks may develop preferences for pecking substrates at an early age (Blokhuis and van der Haar, 1992; Johnsen et al., 1998). Poor-quality substrates for chicks increase early feather pecking (Huber-Eicher and Sebö, 2001; Chow and Hogan, 2005), as well as later adult pecking tendencies (Nicol *et al.*, 2001; de Jong *et al.*, 2013a). However, the early pecking preferences of young birds are not fixed or permanent. A high-quality pecking substrate for adult birds is protective, regardless of how they have been reared (Nicol *et al.*, 2001; de Jong *et al.*, 2013a,b). Another common feature identified by many studies is a positive association between feather pecking (GFP and SFP) and fearfulness (reviewed by Rodenburg *et al.*, 2013). Of course, birds that have been pecked may become more fearful, but there is also convincing evidence that more fearful birds are more likely to develop feather pecking. This may be because their normal foraging, dustbathing or exploratory behaviour is truncated and disturbed and so their pecking motivation remains high.

In other respects, the different forms of injurious pecking vary. GFP has been related to social exploration (Riedstra and Groothuis, 2002) and dustbathing, and it may develop into a highly repetitive, stereotyped form (Rodenburg *et al.*, 2013). SFP is strongly related to feeding and foraging motivation. Dietary deficiencies, especially relating to amino acid and protein levels, are linked with SFP, as is an inadequate fibre content (van Krimpen *et al.*, 2005, 2009; Rodenburg *et al.*, 2013). Feather eating may be one way that birds can obtain additional fibre in their diets (Harlander-Matauschek and Hausler, 2009). In some cases, birds that initiate severe cannibalistic tissue pecking appear to chase and hunt their companions as if they were prey.

During rear, it is important to give chicks and pullets early access to good-quality pecking substrates and to ensure that stocking densities are not too high (reviewed by Nicol *et al.*, 2013). In addition, providing perches during rear is protective (Gunnarsson *et al.*, 1999; Huber-Eicher and Audigé, 1999; Knierim *et al.*, 2008), possibly because birds learn how to avoid trouble-makers by moving in three dimensions. Other risk factors during rear include the use of bell drinkers (which can make the litter wet) (Drake *et al.*, 2010) and multiple diet changes (Gilani *et al.*, 2013). Indeed, the effect of diet change was staggering, as a sophisticated statistical analysis revealed that reducing the number of diet changes by just one, resulted in a 64-fold reduction in the risk of SFP (Gilani *et al.*, 2013). Every time a diet is changed, there is a risk that birds will find the new formulation relatively unpalatable and turn instead to each other to satisfy their foraging motivation (Dixon and Nicol, 2008). The importance of good range use as a protective factor against injurious pecking during the laying period was reviewed in the previous section. It has also been convincingly demonstrated that feeding pellets rather than mash is a strong and significant risk for SFP (Green *et al.*, 2000; Lambton *et al.*, 2010) and for vent pecking (Lambton *et al.*, 2014), as birds spend far more time foraging with diets presented in mash form. Other environmental and management risk factors identified by quantitative analysis of commercial farms have been reviewed by Nicol *et al.* (2013).

In most countries and most housing systems, chicks will have their beaks trimmed using either a hot blade or an infrared beam, so that plumage damage is reduced if they peck each other as adults. This practice is highly controversial as, while partially effective (Lambton *et al.*, 2013; Nicol *et al.*, 2013), it is considered a mutilation. Hot-blade trimming causes pain and changes in beak sensitivity and function (reviewed by Nicol *et al.*, 2013). The newer infrared method is thought to cause less pain (and therefore is the only approved method in the UK and some Scandinavian countries), but this conclusion is tentative (Dennis *et al.*, 2009; Nicol *et al.*, 2013) and awaits further research. Meanwhile, a number of countries have banned the practice (Sweden, Norway and Finland) or plan to do so (the Netherlands and some states in Germany). In England, a trial examining the welfare of free-range flocks with intact beaks will inform a decision in 2015.

Whether or not beak trimming is banned, translation of the considerable body of scientific knowledge about injurious pecking into workable management strategies that can be employed on commercial farms to reduce the risks is an important goal. Lambton *et al.* (2013) analysed all the available literature on injurious pecking and used this to devise 40 management strategies that could realistically be applied on farms. These authors then evaluated the effectiveness of the strategies on 100 commercial flocks. The most striking finding was that the more management strategies that were employed, the lower the rates of GFP and SFP and the better the birds' plumage condition. The strategies in question were subsequently made freely available on the Featherwel website (http://www.featherwel.org) and include such things as allowing pullets earlier access to litter and range areas on arrival in the laying house, additional fibre in the diet, the provision of harmless pecking blocks and providing shelters on the range.

Which System is Best for Laying Hens?

If health and production were the sole criteria by which to judge a housing system, then the furnished or enriched colony cage would be top of the list. Sherwin *et al.* (2010) applied the same welfare assessment methods across four different housing systems in the UK and concluded that, whatever method they used to combine welfare indicators, the physical health and stress levels of the hens from the furnished cages was better than that from other systems. Birds from furnished cages had the lowest levels of corticosterone, vent damage and keel fractures, for example. Wilkins *et al.* (2011) also noted a fracture rate of 36% in birds from furnished cages compared with over 80% for birds from some non-cage systems. The growing consideration of this system as an alternative to the conventional cage outside Europe is encouraging. For example, the president of the USA United Egg Producers was quoted in early 2014 as saying 'enriched colony housing represents the future'.

However, as we have seen throughout this book, laying hens have complex and sophisticated sensory abilities and behavioural needs, and Sherwin *et al.* (2010) acknowledged that these aspects should also be taken into consideration. Cage systems do not permit birds to fly or flap their wings with ease, to perch at height, to forage for live prey, to perform unconstrained dustbathing on a deep litter substrate or to feel the sun on their backs. They do not allow birds to make use of their learning or navigation abilities (with a consequent reduction in hippocampal cell size in caged hens; Patzke *et al.*, 2009) and they subject birds to an indoor environment where noise levels may be constantly or recurrently high and where natural light is not provided. In short, modified cage systems are by no means perfect.

The problem for consumers and campaigners interested in animal welfare is that reliably better alternatives are not always available. Each system has its own advantages and drawbacks (Lay *et al.*, 2011; Freire and Cowling, 2013). For

example, Rodenburg *et al.* (2008) found that hens in non-cage systems made more use of foraging and perching facilities, and had stronger bones and lower fear levels, but hens in furnished cages (group sizes of 20–50) had lower mortality and lower bone fracture rates. This may come as a surprise to those who assume that free-range systems would clearly win out on the welfare front. In theory, the free-range system, more than any current alternative, has the potential to meet both the physical and behavioural needs of the birds. The problem is that this potential is not consistently realized. An analysis of multiple data sets collected at the University of Bristol found reported mortality rates from over 1000 free-range UK flocks were between 8.1 and 11.2%, with wide variation between farms and rates of 20 or 30% not uncommon. Some degree of severe feather pecking was reported by 40 weeks in 75% of flocks, and fracture rates varied from 45 to over 80%. In summary, in a typical free-range flock, 10% of hens will die before the end of lay, 42% of the survivors will sustain both a fracture and a significant number of severe pecks, 22% will suffer a fracture only and 20% severe pecks only, leaving just 12% of birds unharmed (Nicol, 2013). In comparison, mortality rates of less than 4% are routinely found in colony cage systems. Free-range hens are therefore dying routinely at rates two or three times higher than they should be, and on some farms at rates that are eightfold greater. These significant problems cannot be overlooked just because free-range systems have other advantages for some birds. Good welfare is achieved only when physical and behavioural needs are *both* met. Injured or sick birds cannot make use of or enjoy the facilities provided by a free-range system.

This situation has arisen partly because housing systems have not been designed from fundamental principles, taking current knowledge of the physical and behavioural needs of the hens into account. Where this approach has been tried, for example, in the Dutch Rondeel system (http://www.rondeel.org/), then the welfare of the birds really can be very high in all respects. There has been too much focus on the benefits of an outdoor range area at the expense of a consideration of what hens need. Non-cage housing that provides birds with naturally lit but semi-protected 'winter garden' or 'veranda' areas for foraging, perching, exploration and dustbathing should be developed further and could provide a good solution for farms in countries where the outdoors is often cold and wet. Systems with sheltered areas offering natural light are easier to manage than an outdoor range, providing protection from disease and predation. Another issue is the great variation in performance among different farms. Flocks vary (sometimes unpredictably), but so do farms (often predictably) and assurance schemes need to set targets for improvement if poor animal welfare outcomes are detected on farms.

Ultimately, progress towards better chicken housing systems will be governed by economic and political imperatives, sustainability issues (Leinonen *et al.*, 2012, 2014), food safety, and the attitude and views of all stakeholders (Swanson *et al.*, 2011; Thompson *et al.*, 2011). Clarity and equivalence across international welfare assurance schemes will also stimulate improvements (Main *et al.*, 2014). To finish on a smaller scale, it is worth noting the surprising number of hens that are now kept in tiny backyard or garden flocks in the UK. The British Hen Welfare Trust (http://www.bhwt.org.uk) distributes approximately 60,000 'end-of-lay' hens each year to such homes where many birds live for years beyond their commercial lifespan. The resultant photo gallery (http://www.bhwt.org.uk/spoilt-hens.html) demonstrates that many people find these hens intriguing and entertaining companions.

References

Abrahamsson, P. and Tauson, R. (1993) Effect of perches at different positions in conventional cages for laying hens of two different strains. *Acta Agriculturae Scandinavica* 43, 228–235.

Abrahamsson, P., Tauson, R. and Appleby, M.C. (1995) Performance of four hybrids of laying hens in modified and conventional cages. *Acta Agriculturae Scandinavica* 45, 286–296.

Abrahamsson, P., Tauson, R. and Appleby, M.C. (1996) Behaviour, health and integument of four hybrids of laying hens in modified and conventional cages. *British Poultry Science* 37, 521–540.

Alvino, G.M., Tucker, C.B., Archer, G.S. and Mench, J.A. (2013) Astroturf as a dustbathing substrate for laying hens. *Applied Animal Behaviour Science* 146, 88–95.

Appleby, M.C. (1998) The Edinburgh Modified Cage: effects of group size and space allowance on brown laying hens. *Journal of Applied Poultry Research* 7, 152–161.

Appleby, M.C. (1990) Behaviour of laying hens in cages with nest sites. *British Poultry Science* 31, 71–80.

Appleby, M.C., Duncan, I.J.H. and McCrae, H.E. (1988) Perching and floor laying by domestic hens – experimental results and their commercial application. *British Poultry Science* 29, 351–357.

Appleby, M.C., Hughes, B.O. and Hogarth, G.S. (1989) Behaviour of laying hens in a deep litter house. *British Poultry Science* 30, 545–553.

Appleby, M.C., Walker, A.W., Nicol, C.J., Lindberg, A.C., Freire, R., Hughes, B.O. and Elson, H.A. (2002) Development of furnished cages for laying hens. *British Poultry Science* 43, 489–500.

Barnett, J.L., Tauson, R., Downing, J.A., Janardhana, V., Lowenthal, J.W., Butler, K.L. and Cronin, G.M. (2009) The effects of perch, dust bath and nest box, either alone or in combination as used in furnished cages, on the welfare of laying hens. *Poultry Science* 88, 459–470.

Bestman, M.W.P. and Wagenaar, J.P. (2003) Farm level factors associated with feather pecking in organic laying hens. *Livestock Production Science* 80, 133–140.

Blokhuis, H.J. and van der Haar, J.W. (1992) Effects of pecking incentives during rearing on feather pecking of laying hens. *British Poultry Science* 33, 17–24.

Blokhuis, H., Cepero, R., Colin, A., Elson, T., Fiks van Niekerk, F., Keeling, L., Michel, V., Nicol, C.J., Oester, H. and Tauson, R. (2005) Welfare aspects of various systems of keeping laying hens. *EFSA Journal* 197, 1–23.

Blokhuis, H.J., van Niekerk, T.F., Bessei, W., Elson, A., Guemene, D., Kjaer, J.B., Levrino, G.A.M., Nicol, C.J., Tauson, R., Weeks, C.A. and de Weerd, H.A.V. (2007) The Laywel project: welfare implications of changes in production systems for laying hens. *World's Poultry Science Journal* 63, 101–114.

Bright, A. and Johnson, E.A. (2011) Smothering in commercial free-range laying hens: a preliminary investigation. *Veterinary Record* 168, 512.

Bubier, N.E. and Bradshaw, R.H. (1998) Movement of flocks of laying hens in and out of the hen house in four free range systems. *British Poultry Science* 39, S5–S6.

Carmichael, N.L., Walker, A.W. and Hughes, B.O. (1999) Laying hens in large flocks in a perchery system: influence of stocking density on location, use of resources and behaviour. *British Poultry Science* 40, 165–176.

Chen, D.H., Bao, J., Meng, F.Y. and Wei, C.B. (2014) Choice of perch characteristics by laying hens in cages with different group size and perching behaviours. *Applied Animal Behaviour Science* 150, 37–43.

Chow, A. and Hogan, J.A. (2005) The development of feather pecking in Burmese red junglefowl: the influence of early experience with exploratory-rich environments. *Applied Animal Behaviour Science* 93, 283–294.

Cloutier, S., Newberry, R.C., Forster, C.T. and Girsberger, K.M. (2000) Does pecking at inanimate stimuli predict cannibalistic behaviour in domestic fowl? *Applied Animal Behaviour Science* 66, 119–133.

Cooper, J.J. and Appleby, M.C. (1996) Individual variation in prelaying behaviour and the incidence of floor eggs. *British Poultry Science* 37, 245–253.

Cronin, G.M., Butler, K.L., Desnoyers, M.A. and Barnett, J.L. (2005) The use of nest boxes by hens in cages: what does it mean for welfare? *Animal Science Papers and Reports* 23, 121–128.

Croxall, R., Elson, A. and Walker, A. (2005) Effects of beak trimming on laying hens in small group furnished cages. *Animal Science Papers and Reports* 23, 71–76.

Dawkins, M.S. and Hardie, S (1989) Space needs of laying hens. *British Poultry Science* 30, 413–416.

Dawkins, M.S. and Nicol, C.J. (1989) No room for manoeuvre. *New Scientist* 1682, 44–46.

de Jong, I.C., Reuvekamp, B.F.J. and Gunnink, H. (2013a) Can substrate in early rearing prevent feather pecking in adult laying hens? *Animal Welfare* 22, 305–314.

de Jong, I.C., Gunnink, H., Rommers, J.M. and Bracke, M.B.M. (2013b) Effect of substrate during early rearing on floor and feather pecking behaviour in young and adult laying hens. *Archiv fur Geflugelkunde* 77, 15–22.

Dennis, R.L., Fahey, A.G. and Cheng, H.W. (2009) Infrared beak treatment method compared with conventional hot-blade trimming in laying hens. *Poultry Science* 88, 38–43.

Dixon, G. and Nicol, C.J. (2008) The effect of diet change on the behaviour of layer pullets. *Animal Welfare* 17, 101–109.

Drake, K.A., Donnelly, C.A. and Dawkins, M.S. (2010) Influence of rearing and lay risk factors on propensity for feather damage in laying hens. *British Poultry Science* 51, 725–733.

Fossum, O., Jansson, D.S., Etterlinm, P.E. and Vagsholm (2009) Cause of mortality in laying hens in different housing systems in 2001 to 2004. *Acta Veterinaria Scandinavica* 51, 9–18.

Freire, R. and Cowling, A. (2013) The welfare of laying hens in conventional cages and alternative systems: first steps towards a quantitative comparison. *Animal Welfare* 22, 57–65.

Freire, R., Wilkins, L.J., Short, F. and Nicol, C.J. (2003) Behaviour and welfare of individual laying hens in a non-cage system. *British Poultry Science* 44, 22–29.

Furmetz, A., Keppler, C., Knierim, U., Deerberg, F. and Hess, J. (2005) Laying hens in a mobile housing system – use and condition of the free-range area. In: Hess, J. and Rahmann, G. (eds) *Ende der Nische, Beiträge zur 8. Wissenschaftstagung Ökologischer Landbau*, Kassel University Press, Kassel, p. 313.

Gast, R.K., Gurava, R., Jones, D.R. and Anderson, K.E. (2013) Colonization of internal organs by *Salmonella enteritidis* in experimentally infected laying hens housed in conventional or enriched cages. *Poultry Science* 92, 468–473.

Gebhardt-Henrich, S.G., Toscano, M.J. and Fröhlich, E. (2014) Use of outdoor ranges by laying hens in different sized flocks. *Applied Animal Behaviour Science* 155, 74–81.

Gilani, A.-M., Knowles, T.G. and Nicol, C.J. (2013) The effect of rearing environment on feather pecking in young and adult laying hens. *Applied Animal Behaviour Science* 148, 54–63.

Gilani A.-M., Knowles, T.G. and Nicol, C.J. (2014) Factors affecting ranging behaviour in young and adult laying hens. *British Poultry Science* 55, 127–135.

Green, L.E., Lewis, K., Kimpton, A. and Nicol, C.J. (2000) Cross-sectional study of the prevalence of feather pecking in laying hens in alternative systems and its associations with management and disease. *Veterinary Record* 147, 233–238.

Gregory, N.G., Wilkins, L.J., Kestin, S.C., Belyavin, C.G. and Alvey, D.M. (1991) Effect of husbandry system on broken bones and bone strength in hens. *Veterinary Record* 128, 397–399.

Grigor, P.N. (1993) Use of space by laying hens: social and environmental implications for free-range systems. PhD thesis, University of Edinburgh, UK.

Grigor, P.N., Hughes, B.O. and Appleby, M.C. (1995a) Effects of regular handling and exposure to an outside area on subsequent fearfulness and dispersal in domestic hens. *Applied Animal Behaviour Science* 44, 47–55.

Grigor, P.N., Hughes, B.O. and Appleby, M.C. (1995b) Emergence and dispersal behaviour in domestic hens: effects of social rank and novelty of an outdoor area. *Applied Animal Behaviour Science* 45, 97–108.

Guesdon, V. and Faure, J.M. (2004) Laying performance and egg quality in hens kept in standard or furnished cages. *Animal Research* 53, 45–57.

Guinebretiere, M., Huneau-Salaun, A., Huonnic, D. and Michel, V. (2012) Cage hygiene, laying location, and egg quality: the effects of linings and litter provision in furnished cages for laying hens. *Poultry Science* 91, 808–816.

Guinebretiere, M., Huneau-Salaun, A., Huonnic, D. and Michel, V. (2013) Plumage condition, body weight, mortality and zootechnical performances: the effects of linings and litter provision in furnished cages for laying hens. *Poultry Science* 92, 51–59.

Gunnarsson, S., Keeling, L.J. and Svedberg, J. (1999) Effect of rearing factors on the prevalence of floor eggs, clo-acal cannibalism and feather pecking in commercial flocks of loose housed laying hens. *British Poultry Science* 40, 12–18.

Guo, Y.Y., Song, Z.G., Jiao, H.C., Song, Q.Q. and Lin, H. (2012) The effect of group size and stocking density on the welfare and performance of hens housed in furnished cages during summer. *Animal Welfare* 21, 41–49.

Harlander-Matauschek, A. and Hausler, K. (2009) Understanding feather eating behavior in laying hens. *Applied Animal Behaviour Science* 117, 35–41.

Harlander-Matauschek, A., Felsenstein, K., Niebuhr, K. and Troxler, J. (2006) Influence of pop hole dimensions on the number of laying hens outside on the range. *British Poultry Science* 47, 131–134.

Hegelund, L., Sørensen, J.T. Kjaer, J.B. and Kristensen, I.S. (2005) Use of the range area in organic egg production systems: effect of climatic factors, flock size, age and artificial cover. *British Poultry Science* 46, 1–8.

Hester, P.Y., Enneking, S.A., Jefferson-Moore, K.Y., Einstein, M.E., Cheng, H.W. and Rubin, D.A. (2013) The effect of perches in cages during pullet rearing and egg laying on hen performance, foot health and plumage. *Poultry Science* 92, 310–320.

Hetland, H., Svihus, B., Lervik, S. and Moe, R. (2003) Effect of feed structure on performance and welfare in laying hens housed in conventional and furnished cages. *Acta Agriculturae Scandinavica Section A – Animal Science* 53, 92–100.

Hetland, H., Moe, R.O., Tauson, R., Lervik, S. and Svishus,B. (2004) Effect of including whole oats into pellets on performance and plumage condition in laying hens housed in conventionaland furnished cages. *Acta Agriculturae Scandinavica Section A – Animal Science* 54, 206–212.

Horsted, K. and Hermansen, J.E. (2007) Whole wheat versus mixed layer diet as supplementary feed to layers foraging a sequence of different forage crops. *Animal* 1, 575–585.

Huber-Eicher, B. and Audigé, L. (1999) Analysis of risk factors for the occurrence of feather pecking in laying hen growers. *British Poultry Science* 40, 599–604.

Huber-Eicher, B. and Sebö, F. (2001) Reducing feather pecking when raising laying hen chicks in aviary systems. *Applied Animal Behaviour Science* 73, 59–68.

Huber-Eicher, B. and Wechsler, B. (1997) Feather pecking in domestic chicks: its relation to dustbathing and foraging. *Animal Behaviour* 54, 757–768.

Hughes, B.O., Wood-Gush, D.G.M. and Morley Jones, R. (1974) Spatial organization in flocks of domestic fowls. *Animal Behaviour* 22, 438–444.

Hughes, B.O., Carmichael, N.L., Walker, A.W. and Grigor, P.N. (1997) Low incidence of aggression in large flocks of laying hens. *Applied Animal Behaviour Science* 54, 215–234.

Huneau-Salaun, A., Guinebretiere, M., Taktak, A., Huonnic, D. and Michel, V. (2011) Furnished cages for laying hens: study of the effects of group size and litter provision on laying location, zootechnical performance and egg quality. *Animal* 5, 911–917.

Icken, W., Cavero, D., Thurner, S., Schmultz, M., Wendl, G. and Preisinger, R. (2011) Relationship between time spent in the winter garden and shell colour in brown egg stock. *Archiv fur Geflugelkunde* 75, 145–150.

Jendral, M.J. (2013) Developments in welfare-related laying hen housing practices in Canada and the United States. In: *Proceedings of the 9th European Poultry Welfare Symposium*, Uppsala, Sweden, 17–20 June. World's Poultry Science Association, pp. 27–36.

Jendral, M.J., Korver, D.R., Church, J.S. and Feddes, J.J. (2008) Bone mineral density and breaking strength of White Leghorns housed in conventional, modified, and commercially available colony battery cages. *Poultry Science* 87, 828–837.

Johnsen, P., Vestergaard, K.S. and Norgaard-Nielsen, G. (1998) Influence of early rearing conditions on the development of feather pecking and cannibalism in domestic fowl. *Applied Animal Behaviour Science* 60, 25–41.

Keeling, L.J., Hughes, B.O. and Dun, P. (1988) Performance of free-range laying hens in a polythene house and their behaviour on range. *Farm Building Progress* 94, 21–28.

Knierim, U., Staack, M., Gruber, B., Keppler, C., Zaludik, K. and Niebuhr, K. (2008) Risk factors for feather pecking in organic laying hens – starting points for prevention in the housing environment. In: *Cultivating the Future Based on Science: 2nd Conference of the International Society of Organic Agriculture Research ISOFAR*, Modena, Italy, 18–20 June (poster).

Knowles, T.G. and Broom, D.M. (1990) Limb bone strength and movement in laying hens from different housing systems. *Veterinary Record* 126, 354–356.

Lambton, S.L., Knowles, T.G., Yorke, C. and Nicol, C.J. (2010) The risk factors affecting the development of gentle and severe feather pecking in loose housed laying hens. *Applied Animal Behaviour Science* 123, 32–42.

Lambton, S.L., Nicol, C.J., Friel, M., Main, D.C.J., McKinstry, J.L., Sherwin, C.M., Walton, J. and Weeks, C.A. (2013) A bespoke management package can reduce the levels of injurious pecking in loose housed laying hen flocks. *Veterinary Record* 172, 423–430.

Lambton, S.L., Knowles, T.G., Yorke, C. and Nicol, C.J. (2014) The risk factors affecting the development of vent pecking and cannibalism in free-range and organic laying hens. *Animal Welfare* 24 101–111.

Lay, D.C., Fulton, R.M., Hester, P.Y., Karcher, D.M., Kjaer, J.B., Mench, J.A., Mullens, B.A., Newberry, R.C., Nicol, C.J., O'Sullivan, N.P.O. and Porter, R.E. (2011) Hen welfare in different housing systems. *Poultry Science* 90, 278–294.

Leinonen, I., Williams, A.G., Wiseman, J., Guy, J. and Kyriazakis, I. (2012) Predicting the environmental impacts of chicken systems in the United Kingdom through a life cycle assessment: egg production systems. *Poultry Science* 91, 26–40.

Leinonen, I., Williams, A.G. and Kyriazakis, I. (2014) The effects of welfare-enhancing system changes on the environmental impacts of broiler and egg production. *Poultry Science* 93, 256–266.

Lentfer, T.L., Gebhardt-Henrich, S.G., Frohlich, E.K.F. and von Borrell, E. (2013) Nest use is influenced by the positions of nests and drinkers in aviaries. *Poultry Science* 92, 1433–1442.

Lindberg, A.C. and Nicol, C.J. (1996) Space and density effects on group size preferences in laying hens. *British Poultry Science* 37, 709–721.

Lindberg, A.C. and Nicol, C.J. (1997) Dustbathing in modified battery cages: is sham dustbathing an adequate substitute? *Applied Animal Behaviour Science* 55, 113–128.

Mahboub, H.D.H., Muller, J. and Von Borell, E. (2004) Outdoor use, tonic immobility, heterophil/lymphocyte ratio and feather condition in free range laying hens of different genotype. *British Poultry Science*, 45, 738–744.

Main, D.C.J., Mullan, S., Atkinson, C., Cooper, M., Wrathall, J.H.M. and Blokhuis, H.J. (2014) Best practice framework for animal welfare certification schemes. *Trends in Food Science and Technology* 37, 127–136.

Maurer, V., Hertzberg, H., Heckendorn, F., Hordegen, P. and Koller, M. (2013) Effects of paddock management on vegetation, nutrient accumulation and internal parasites in laying hens. *Journal of Applied Poultry Research* 22, 334–343.

McAdie, T.M. and Keeling, L.J. (2000) Effect of manipulating feathers of laying hens on the incidence of feather pecking and cannibalism. *Applied Animal Behaviour Science* 68, 215–229.

Merrill, R.J.N. and Nicol, C.J. (2005) The effects of novel floorings on dustbathing, pecking and scratching behaviour of caged hens. *Animal Welfare* 14, 179–186.

Merrill, R.J.N., Cooper, J.J., Albentosa, M.J. and Nicol, C.J. (2006) The preferences of laying hens for perforated Astroturf over conventional wire as a dustbathing substrate in furnished cages. *Animal Welfare* 15, 173–178.

Moe, R.O., Guemene, D., Bakken, M., Larsen, H.J.S., Shini, S., Lervik, S., Skjerve, E., Michel, V. and Tauson, R. (2010) Effects of housing conditions during the rearing and laying period on adrenal reactivity, immune response and heterophil to lymphocyte (H/L) ratios in laying hens. *Animal* 4, 1709–1715.

Moinard, C., Statham, P. and Green, P.R. (2004) Control of landing flight by laying hens: implications for the design of extensive housing systems. *British Poultry Science* 45, 578–584.

Nasr, M.A.F., Murrell, J., Wilkins, L.J. and Nicol, C.J. (2012a) The effect of keel fractures on egg-production parameters, mobility and behaviour in individual laying hens. *Animal Welfare* 21, 127–135.

Nasr, M.A.F., Nicol, C.J. and Murrell, J.C. (2012b) Do laying hens with keel bone fracture experience pain? *PLoS One* 7, e42420.

Nasr, M.A.F., Browne, W.J., Caplen, G., Hothersall, B., Murrell, J.C. and Nicol, C.J. (2013) Positive affective state induced by opioid analgesia in laying hens with bone fractures. *Applied Animal Behaviour Science* 147, 127–131.

Nicol, C.J. (1987a) Effect of cage height and area on the behaviour of hens housed in battery cages. *British Poultry Science* 28, 327–335.

Nicol, C.J. (1987b) Behavioural responses of laying hens following a period of spatial restriction. *Animal Behaviour* 35, 1709–1719.

Nicol, C.J. (2013) Free-range v cage: science behind the headlines. *Poultry World* 168, 32–33.

Nicol, C.J., Gregory, N.G., Knowles, T.G., Parkman, I.D. and Wilkins, L.J. (1999) Differential effects of increased stocking density, mediated by increased flock size, on feather pecking and aggression in laying hens. *Applied Animal Behaviour Science* 65, 137–152.

Nicol, C.J., Lindberg, A.C., Phillips, A.J., Pope, S.J., Wilkins, L.J. and Green, L.E. (2001) Influences of prior exposure to wood shavings on feather pecking, dustbathing and foraging in adult laying hens. *Applied Animal Behaviour Science* 73, 141–155.

Nicol, C.J., Pötszch, C., Lewis, K. and Green, L.E. (2003) Matched concurrent case–control study of risk factors for feather pecking in hens on free-range commercial farms in the UK. *British Poultry Science* 44, 515–523.

Nicol, C.J., Brown, S.N., Glen, E., Pope, S.J., Short, F.J., Warriss, P.D., Zimmerman, P.H. and Wilkins, L.J. (2006) Effects of stocking density, flock size and management on the welfare of laying hens in single-tier aviaries. *British Poultry Science* 47, 135–146.

Nicol, C.J., Bestman, M., Gilani, A.M., de Haas, E., de Jong, I.C., Lambton, S., Wagenaar, J.P., Weeks, C.A. and Rodenburg, T.B. (2013) The prevention and control of feather pecking: application to commercial systems. *World's Poultry Science Journal* 69, 775–788.

Oden, K., Vestergaard, K.S. and Algers, B. (2000) Space use and agonistic behaviour in relation to sex composition in large flocks of laying hens. *Applied Animal Behaviour Science* 67, 307–320.

Oden, K., Keeling, L.J. and Algers, B. (2002) Behaviour of laying hens in two types of aviary systems on 25 commercial farms in Sweden. *British Poultry Science* 43, 169–181.

Olsson, I.A.S. and Keeling, L.J. (2002) No effect of social competition on sham dustbathing in furnished cages for laying hens. *Acta Agriculturae Scandinavica Section A – Animal Science* 52, 253–256.

Orsag, J., Broucek, J., Sauter, M., Tancin, V. and Flak, P (2012) Effect of access to dusting substrate on behaviour in layers from different types of cages. *Veterinarija ir Zootechnika* 57, 56–61.

Patzke, N., Ocklenburg, S., van der Staay, F.J., Gunturkun, O. and Manns, M. (2009) Consequences of different housing conditions on brain morphology in laying hens. *Journal of Chemical Neuroanatomy* 37, 141–148.

Pohle, K. and Cheng, H.W. (2009) Furnished cage system and hen well-being: comparative effects of furnished cages and battery cages on behavioural exhibitions in White Leghorn chickens. *Poultry Science* 88, 2042–2051.

Pötzsch, C.J., Lewis, K., Nicol, C.J. and Green, L.E. (2001) A cross-sectional study of the prevalence of vent pecking in laying hens in alternative systems and its associations with feather pecking, management and disease. *Applied Animal Behaviour Science* 74, 259–272.

Richards, G.J., Wilkins, L.J., Knowles, T.G., Booth, F., Toscano, M.J., Nicol, C.J. and Brown, S.N. (2011) Continuous monitoring of pop hole usage by commercially housed free-range hens throughout the production cycle. *Veterinary Record* 169, 338–342.

Richards, G.J., Brown, S.N., Booth, F., Toscano, M.J. and Wilkins, L.J. (2012) Panic in free-range laying hens. *Veterinary Record* 170, 519.

Riedstra, B. and Groothuis, T.G.G. (2002) Early feather pecking as a form of social exploration: the effect of group stability on feather pecking and tonic immobility in domestic chicks. *Applied Animal Behaviour Science* 77, 127–138.

Rodenburg, T.B., Tuyttens, F.A.M., de Reu, K., Herman, L. Zoons, J. and Sonck, B. (2008) Welfare assessment of laying hens in furnished cages and non-cage systems: an on-farm comparison. *Animal Welfare* 17, 363–373.

Rodenburg, T.B., van Krimpen, M.M., de Jong, I.C., de Haas, E.N., Kops, M.S., Ridestra, B.J., Nordquist, R.E., Wagenaar, J.P., Bestman, M. and Nicol, C.J. (2013) The prevention and control of feather pecking in laying hens: identifying the underlying principles. *World's Poultry Science Journal* 69, 361–373.

Roll, V.F.B., Maria, G.A. and Cepero, R. (2005) Ethological parameters and performance of Hy Line White and ISA Brown hens when housed in furnished cages. *Animal Science Papers and Reports* 23 (Suppl. 1), 77–84.

Roll, V.F.B., Briz, R.C. and Levrino, G.A.M. (2009) Floor versus cage rearing: effects on production, egg quality and physical condition of laying hens housed in furnished cages. *Ciencia Rural* 39, 1527–1532.

Sandilands, V., Moinard, C. and Sparks, N.H.C. (2009) Providing laying hens with perches: fulfilling behavioural needs but causing injury? *British Poultry Science* 50, 395–406.

Scholtz, B., Roenchen, S., Hamann, H., Pendl, H. and Distl, O. (2008) Effect of housing system, group size and perch position on H/L ratio in laying hens. *Archiv fur Geflugelkunde* 72, 174–180.

Scholtz, B., Roenchen, S., Hamann, H. and Distl, O. (2009) Bone strength and keel bone status of two layer strains kept in small group housing systems with different perch configurations and group sizes. *Berliner und Munchener Tierartzliche Wochenschrift* 122, 249–256.

Scott, G.B. and Lambe, N.R. (1996) Working practices in a perchery system using the OVAKO working posture analysing system (OWAS) *Applied Ergonomics* 27, 281–284.

Sherwin, C.M. and Nicol, C.J. (1993a) Factors influencing floor laying by hens in modified cages. *Applied Animal Behaviour Science* 36, 211–222.

Sherwin, C.M. and Nicol, C.J. (1993b) A descriptive analysis of the pre-laying behaviour of hens in modified battery cages. *Applied Animal Behaviour Science* 38, 49–60.

Sherwin, C.M. and Nicol, C.J. (1994) Dichotomy in choice of nest characteristics by caged laying hens. *Animal Welfare* 3, 313–320.

Sherwin, C.M., Richards, G.J. and Nicol, C.J. (2010) Comparison of the welfare of layer hens in 4 housing systems in the UK. *British Poultry Science* 51, 488–499.

Sherwin, C.M., Nasr, M.A.F., Gale, E., Petek, M., Stafford, K., Turp, M. and Coles, G.C. (2013) Prevalence of nematode infection and faecal egg counts in free-range laying hens: relations to housing and husbandry. *British Poultry Science* 54, 12–23.

Shimmura, T., Eguchi, Y., Uetake, K. and Tanaka, T. (2006) Behavioral changes in laying hens after introduction to battery cages, furnished cages and an aviary. *Animal Science Journal* 77, 242–249.

Shimmura, T., Eguchi, Y., Uetake, K. and Tanaka, T. (2007) Differences of behaviour, use of resources and physical conditions between dominant and subordinate hens in furnished cages. *Animal Science Journal* 78, 307–313.

Shimmura, T., Eguchi, Y., Uetake, K. and Tanaka, T. (2008) Effects of separation of resources on behaviour of high-, medium- and low-ranked hens in furnished cages. *Applied Animal Behaviour Science* 113, 74–86.

Shimmura, T., Hirahara, S., Azuma, T., Suzuki, T., Eguchi, Y., Uetake, K. and Tanaka,T. (2009) Multi-factorial investigation of various housing systems for laying hens. *British Poultry Science* 51, 31–42.

Stampfli, K., Buchwalder, T., Frohlich, E.K.F. and Roth, B.A. (2013) Design of nest access grids and perches in front of the nests: influence on the behaviour of laying hens. *Poultry Science* 92, 890–899.

Struelens, E., Tuyttens, F.A.M., Janssen, A., Leroy, T., Audoom, L., Vranken, E., De Baere, K., Odberg, F., Berckmans, D., Zoons, J. and Sonck, B. (2005) Design of laying nests in furnished cages: influence of nesting material, nest box position and seclusion. *British Poultry Science* 46, 9–15.

Swanson, J.C., Lee, Y., Thompson, P.B., Bawden, R. and Mench, J.A. (2011) Integration: valuing stakeholder input in setting priorities for socially sustainable egg production. *Poultry Science* 90, 2110–2121.

Tactacan, G.B., Guenter, W., Lewis, N.J., Rodriguez-Lecompte, J.C. and House, J.D. (2009) Performance and welfare of laying hens in conventional and enriched cages. *Poultry Science* 88, 698–707.

Tahamtani, F., Hansen, T.B., Orritt, R., Nicol, C.J., Moe, R.O. and Janczak, A.M. (2014) Does rearing laying hens in aviaries adversely affect long-term welfare following transfer to furnished cages? *PLoS One* 9, e107357.

Tauson, R. and Holm. K.-E. (2005) Mortality, production and use of facilities in furnished cages for layers in commercial egg production in Sweden from 1998–2003. *Animal Science Papers and Reports* 23, 95–102.

Taylor, P.E., Scott, G.B. and Rose, S.P.R. (2004) Effects of opaque and transparent vertical cover on the distribution of hens in an outdoor arena. In: Perry, G.C. (ed.) *Welfare of the Laying Hen*. CABI, Wallingford, UK, p. 414.

Thompson, P.B., Appleby, M.C., Busch, L., Kalof, L., Miele, M., Norwood, B.F. and Pajor, E. (2011) Values and public acceptability dimensions of sustainable egg production. *Poultry Science* 90, 2097–2109.

Tuyttens, F.A.M., Struelens, E. and Ampe, B. (2013) Remedies for a high incidence of broken eggs in furnished cages: effectiveness of increasing nest attractiveness and lowering perch height. *Poultry Science* 92, 19–25.

van Krimpen, M.M., Kwakkel, R.P., Reuvekamp, B.F.J., van der Peet-Schwering, C.M.C., den Hartog, L.A. and Verstegen, M.W.A. (2005) Impact of feeding management on feather pecking in laying hens. *World's Poultry Science Journal* 61, 663–685.

van Krimpen, M.M., Kwakkel, R.P., van der Peet-Schwering, C.M.C., den Hartog, L.A. and Verstegen, M.W.A. (2009) Effects of nutrient dilution and nonstarch polysaccharide concentration in rearing and laying diets on eating behaviour and feather damage of rearing and laying hens. *Poultry Science* 88, 759–773.

Vestergaard, K.S. and Lisborg, L. (1993) A model of feather pecking development which relates to dustbathing in the fowl. *Behaviour* 126, 291–308.

Vits, A., Weitzenbürger, D., Hamann, H. and Distl, O. (2005) Production, egg quality, bone strength, claw length, and keel bone deformities of laying hens housed in furnished cages with different group sizes. *Poultry Science* 84, 1511–1519.

Wall, H. (2011) Production performance and proportion of nest eggs in layer hybrids housed in different designs of furnished cages. *Poultry Science* 90, 2153–2161.

Wall, H. and Tauson, R. (2013) Nest lining in small-group furnished cages for laying hens. *Journal of Applied Poultry Research* 22, 474–484.

Wall, H., Tauson, R. and Elwinger, K. (2002) Effect of nest design, passages and hybrid on use of nest and production performance of layers in furnished cages. *Poultry Science* 81, 333–339.

Wall, H., Tauson, R. and Elwinger, K. (2004) Pop hole passages and welfare in furnished cages for laying hens. *British Poultry Science* 45, 20–27.

Wall, H., Tauson, R. and Elwinger, K. (2008) Effects of litter substrate and genotype on layer's use of litter, exterior appearance and heterophil:lymphocyte ratios in furnished cages. *Poultry Science* 87, 2458–2465.

Weitzenburger, D., Vits, A., Hamann, H. and Distl, O. (2006) Evaluation of small group housing systems and furnished cages as regards particular behaviour patterns in the layer strain Lohmann Selected Leghorn. *Archiv fur Geflugelkunde* 70, 250–260.

Wilkins, L.J., Brown, S.N., Zimmerman, P.H., Leeb, C. and Nicol, C.J. (2004) Investigation of palpation as a method for determining the prevalence of keel and furculum damage in laying hens. *Veterinary Record* 155, 547–550.

Wilkins, L.J., McKinstry, J.L., Avery, N.C., Knowles, T.G., Brown, S.N. and Tarlton, J. (2011) Influence of housing system and design on bone strength and keel bone fractures in laying hens. *Veterinary Record* 169, 414–417.

Wilson, S., Hughes, B.O., Appleby, M.C. and Smith, S.F. (1993) Effect of perches on trabecular bone volume in laying hens. *Research in Veterinary Science* 54, 207–211.

Zeltner, E. and Hirt, H. (2003) Effect of artificial structuring on the use of laying hen runs in a free-range system. *British Poultry Science* 44, 533–537.

Zeltner, E. and Hirt, H. (2008) Factors involved in the improvement of the use of hen runs. *Applied Animal Behaviour Science* 114, 395–408.

Conclusions

It might be thought that the basic behavioural biology of the most common domestic bird on the planet would have been investigated in full many years ago. After all, molecular evidence suggests that humans have been living alongside this species for over 50,000 years. Yet chicken behaviour is a current and burgeoning field of research, with new findings revealing sensory and cognitive abilities that, in some cases, exceed our own. Despite intensive selection for increased productive capacity, the behaviour of domestic chickens differs only in degree from that of its wild junglefowl ancestors. For example, domestic birds are a little less active and exploratory than junglefowl. However, the potential now exists for a far more systematic approach to altering chicken behaviour using quantitative methods that do not depend on a knowledge of specific gene locations. The challenge is to develop an approach that measures behaviour reliably, selects birds in commercially relevant environments, takes social effects into account and acts upon any unintended consequences. Producing a docile bird that is less likely to peck its companions is a justified goal, but problems will arise if, for example, docility is correlated with an increased tendency to succumb to smothering.

In parallel with advances in behavioural genetics, an increasing awareness of the role of epigenetic and early life influences has opened up entirely new areas of research. It seems that stressed parent flocks may produce more fearful offspring, but there is still much to do to ascertain how the hormone levels of parent birds affect egg quality, embryonic development and the future behaviour of their unseen chicks. Commercially hatched chicks also lack the behavioural guidance provided in nature by the broody hen. Recent research has highlighted the strength and longevity of the broody hen's influence, and this means that new ways are now being sought to simulate aspects of maternal care in commercial systems. Dark brooders are just one example.

The chicken brain continues to surprise. Naïve day-old chicks have both an innate sense of how the world should be (with well-developed expectations about the physical and structural properties of objects) and an amazing capacity to absorb new information and make new associations relevant to their environment. Their lateralized brains mean that they are excellent multi-taskers, able to attend to detailed foraging information while remaining vigilant and aware of predatory dangers. Their sensory capacities are also more sophisticated than we once thought. Far from having an undeveloped olfactory sense, it seems chickens can discriminate the faeces of herbivores and carnivores. And, until recently, who knew that chickens could navigate using a magnetic field? And yet, despite this progress in understanding chicken perception and cognition, we still know very little about the emotional lives of chickens or even the extent to which they share emotional systems with other bird and mammal species.

Behavioural science has an important role in the assessment of chicken welfare. By using preference tests and methods of measuring motivational strength, we can discover what matters most to these birds. This approach is perhaps the gold standard of welfare assessment, as it provides the most direct insight into the birds' integration of its own state. Resources that matter greatly (food, nest sites) must be provided to ensure the most basic level of welfare. At the same time, the use of additional resources can be used to assess higher levels of bird welfare. If we see a chicken spending a long time dustbathing, it tells us something important. If a more pressing need must be met (escape from a predator, avoidance of an aggressive companion, drinking in a very hot environment), the chicken would not be spending its time on this low-resilience (luxury) behaviour. Thus, when chickens do perform low-resilience activities, this is generally a good sign that they are healthy and that their most basic needs have already been met. An interesting challenge for animal welfare scientists will be to continue to validate easy-to-measure 'on-farm' indicators against a more fundamental knowledge of bird priorities.

In practice, there is considerable scope to improve the welfare of all types of commercially raised chickens. Although most broiler chickens are slaughtered before all of their behavioural patterns (e.g. night-time roosting) have become fully established, the behaviour of these giant, immature chicks is increasingly constrained by their weight and bulk. This makes normal activities difficult and tiring to perform, and lameness also remains a stubbornly resistant welfare problem. The food restriction of the parent birds is also an area of great concern. The impact of lameness and hunger on the mood of individual birds is now being evaluated using conditioned place preference and judgement bias tasks. The effect of analgesic drugs can also be evaluated to see whether bird mobility is restored and whether birds feel better once inflammation is reduced. Medication of this kind could not, of course, be used on a commercial scale, but such studies may reveal the conditions that have the most

detrimental impact on chicken quality of life, and thereby set priorities for improvement.

A ban of conventional cages for laying hens has been achieved in many countries because of evidence about the effects of extreme confinement on bird health and welfare. Scientific knowledge about the hens' behavioural priorities was an important contribution to this evidence. But the debate is not over and done with. The systems that have replaced conventional cages – furnished cages and non-cage systems – allow greater freedom, but they are also more complex and difficult to manage. There is great variation in the outcomes from these systems, both among flocks and among farms. The behaviour of chickens kept in groups of many thousands is not easy to predict, and behavioural scientists need to work together with the farming communities, epidemiologists, engineers and social scientists to devise solutions that ensure both good production and high welfare. I am convinced that the best systems have yet to be devised. Anyone embarking on a career in chicken behaviour research will find no shortage of important topics.

The capacity of chickens to learn and process complex information may encourage a reconsideration of their moral status. Once we appreciate that these birds do not simply respond to their environment, to each other or to us with a set of simple, fixed or 'unthinking' responses, we may decide that they merit a different position within an ethical framework. Chickens are able to infer their own social position or acquire new information by observing others, they are highly aware of the identity, nature and possibly the intentions of other birds in the vicinity, and they can even apply simple rules of arithmetic and logic. We may admire and appreciate their complexity just as we might admire great paintings or diverse and complex landscapes, and perhaps accord them greater respect or protection on these grounds alone. Alternatively, or in addition, it may be that some aspects of learning and cognition are linked to a capacity to suffer (or indeed a capacity to experience pleasure) that is greater than we previously imagined.

Index

Page numbers in **bold** refer to illustrations and tables.